Spell of the Urubamba

Daniel W. Gade

Spell of the Urubamba

Anthropogeographical Essays on an Andean Valley in Space and Time

Springer

Daniel W. Gade
Department of Geography
University of Vermont
Burlington, VT, USA

ISBN 978-3-319-20848-0 ISBN 978-3-319-20849-7 (eBook)
DOI 10.1007/978-3-319-20849-7

Library of Congress Control Number: 2015952211

Springer Cham Heidelberg New York Dordrecht London

Printed on acid-free paper

Springer International Publishing AG Switzerland is part of Springer Science+Business Media (www.springer.com)

In Memory of Carl O. Sauer
(1889–1975)

Preface

The Urubamba Valley of Peru is one of the most important valleys of Western South America. Formed by a river coursing from the Andean highlands to the Amazon lowlands, the splendor of its environmental diversity, the intricacy of its human–nature relationships and the complex connections of its past with the present captivate me. The Upper Urubamba Valley has long been an important zone of settlement and a corridor of movement. More recently, this deep crease, among the multitude of Andean creases, has become a major destination for travelers seeking an encounter with the remains of a pre-Columbian past bathed in monumentality and mystery.

An author's experience and long-term commitment to a place enhances his ability to gather data and to interweave information and ideas. Since my arrival there in 1963 to conduct dissertation fieldwork in geography, the Urubamba has had me under its spell. The project that I initiated at age 26 involved studying the plant use and folk agriculture along an altitudinal gradient encompassing entirely different environments. The Urubamba charmed me with its snowcaps looming high above, the river sharply changing its water volume between the dry season and rainy period, the spectacular thunderstorms early in the year, and the extraordinary brilliance of the stars in the night sky. Likewise, the human dimension of the valley sparked my great curiosity. The strong peasant culture of country people in the highlands was so different from the semi-nomadic forest Indians in the Urubamba jungle. The Urubamba as a region defined by the river and its valley holds within it a host of individual locales. In turn, these locales fit into a series of zones that make up the whole. Originally part of the attraction of the Urubamba was that it seemed to be lost in time. That is somewhat less true today than it was in 1963. Over the last half century, on each of seven separate research stints, I brought to the valley a fresh set of eyes and a larger accumulation of background knowledge.

This work is organized thematically, rather than spatially or chronologically, since my aim has been to highlight how intellectual curiosity opens an anthropogeographical mode of thinking about place. The first two chapters provide an introduction to the Urubamba and some of the people historically associated with knowledge about it. Chapter 3 on verticality covers the valley's most salient aspect when viewed in its totality from head to toe. Several chapters represent probings of particular parts of the Urubamba: the so-called Sacred Valley as emblematic zone of power and prestige; Machu Picchu, the best known locale to most people; the Vilcabamba as a hidden zone north of Cusco and the two dimensions of the Urubamba tropical zone. Two biotic elements in Andean culture history, one a tree called vilca and the other the spectacled bear, underscore how my association with the valley sparked interests in phenomena whose stories have a wider Andean component.

My approach to geography is one poised on the epistemological borderlands that fuse space, time, ecology and culture, but embrace no indispensable methodologies or grand theories. For their moral and intellectual support and as exemplars of this kind of thinking, I mention Carl O. Sauer (1989–1975), William M. Denevan, Frederick J. Simoons, James J. Parsons (1915–1997) and Philip L. Wagner (1921–2014). These scholars took their research cues from creative imagination, the primacy of intellectual curiosity, and belief in intellectual freedom. In turn, these prompt a sense of enchantment about the world in which we live unencumbered by political ideology, do-gooder impulses or career calculations. Scholars and scientists of the Andes have also inspired me, especially Antonio Raimondi (1826–1890), the remarkable polymathic explorer who filled in so many blanks of knowledge about Peru. At the US Embassy when I first went to Peru, Albert A. Giesecke (1878–1968) provided me with useful information. I learned later that he was a driving force in introducing students and colleagues in Cusco to the possibilities of generating local knowledge and in understanding place. In 1922, Giesecke coined the word *cuscología* to refer to the study of the Cusco region; I would consider this book to be a contribution to that knowledge realm.

The focus on place that anchors this work seamlessly subsumes within it space and time, but it also melts the distinction between nature and culture. Land use, settlement and agriculture encompass culture, economy and society as well as the biophysical frameworks and biotic resources that often receive short shrift in discussions of peasant livelihood. Place also benefits from autobiographical remembering and making sense of a lived experience. Memory sharpens the configuration of place. If this book contains an element of eighteenth-century *Naturphilosophie*, it is because I have incorporated two seemingly contradictory elements: the Urubamba as a reality outside the self and—as happened with Alexander von Humboldt and his South American experiences—one's own retrospection and introspection about the valley. The very fact of writing is a way of reflecting on field experiences and of drawing larger meanings from those reflections.

Information for this book comes from fieldwork and extensive reading. One of the best ways to learn how to write is to read. The following repositories of human knowledge most salient to this project have been Bailey-Howe Library of the University of Vermont; Library of Congress in Washington; McGill University Libraries; Cornell University Library; Bibliothèque des Lettres et Sciences Humaines, Université de Montréal; Sterling Library of Yale University; the Biblioteca Nacional del Perú; and the Biblioteca General of the Universidad San Antonio Abad del Cusco.

Among the many people who helped me were peasant folk who answered my questions or, with equanimity, allowed me to participate in or watch their routines. By observing daily activities, I learned about the close and direct ties that people have had with their environment at several scales—their farm, but also the community, the zone or valley stretch, and the whole Urubamba region. Knowledgeable interlocutors shared their special perspectives. Conversations at different times over the past half century with John H. Rowe (1918–2004) made me appreciate how much a scholarly devotion to the Andean past can uncover so much about this part of the world. A kindred interest in Andean natural history provided the basis for a friendship with César Vargas Calderón (1903–2002) of Cusco, Peru. Mario Escobar Moscoso, *simpatico geógrafo cusqueño*, helped enormously in my efforts to learn in depth Andean culture and landscape. Margarita Giesecke Sara Lafosse de Rubio (1948–2004) provided access to her grandfather's papers, which led me to the Machu Picchu story. Stuart White of Cuenca, Ecuador provided an account of his amazing experience in the Vilcabamba Valley. Courtney Thompson of Freeport, Maine and a person of many talents, provided technical support on the manuscript. I wish to acknowledge my deep gratitude to her for getting this work into its final form. My biggest debt is to Mary Killgore Gade, keeper of the hearth and love of my life. She has been a superb traveling companion with boundless curiosity, a critical copy editor and formidable cartographer. She has contributed enormously to this book. I take responsibility for any errors that may remain in this book.

Burlington, VT, USA Daniel W. Gade

Contents

Chapter 1
The Urubamba in Panoptic Perspective

Abstract A far western tributary of the Amazon system, the Urubamba River has formed a main valley that anchors this discussion of Andean land and life. The valley floor at its southern point exceeds 4300 m asl; the valley ends below 300 m above sea level (asl) where the Urubamba joins the Tambo River to form the Ucayali. This 862 km-long depression manifests notable diversity of climate, vegetation and land use. In the temperate environment above 1800 m asl, human settlement and agriculture were important both before and after the Spanish Conquest (1532 AD). In contrast, the tropical environment in the Urubamba below 1800 m asl, except for a small periphery, had minimal Inca occupation. In the late sixteenth century, the Spaniards began to establish coca and sugar cane haciendas. Observational data and documents from the past introduce a diachronic understanding of the Urubamba as a valley region. Fieldwork beginning in 1963 validated that the place is wondrous in its ecologically diversity and traditional in its way of life, but also that it was on the cusp of change as land tenure became the major issue.

1.1 The Valley in Context

Midway in the vast extent of the Andes stretching 7000 km from north to south stand the high peaks of the Cordillera Oriental in the Department of Cusco. This super chain, comprised of Paleozoic sediments and Cenozoic magmatic rocks, owes its geological origin to the repeated subductions of the Nazca Plate, which triggered crustal upheaval. Runoff from these mountains feeds rivers flowing through depressions made by the action of running water and, at an earlier time, by the gouging of moving ice. One of those creases in the land, the Urubamba Valley, is the focus of this book (Fig. 1.1). The Urubamba is not only a storied place in the culture history of the Andes; it also is of major contemporary significance for agriculture, settlement, transportation and tourism. More than any other place in the Central Andes, this valley has impressed innumerable visitors for its exceptional beauty and rich evidence of a remarkable prehistoric past. The fundamental unity of the Urubamba is defined by the hydrographic system of the main river and its tributaries. The substantial diversity of climate and vegetation characterizing the Urubamba contributed to a diverse land use.

D.W. Gade, *Spell of the Urubamba*, DOI 10.1007/978-3-319-20849-7_1

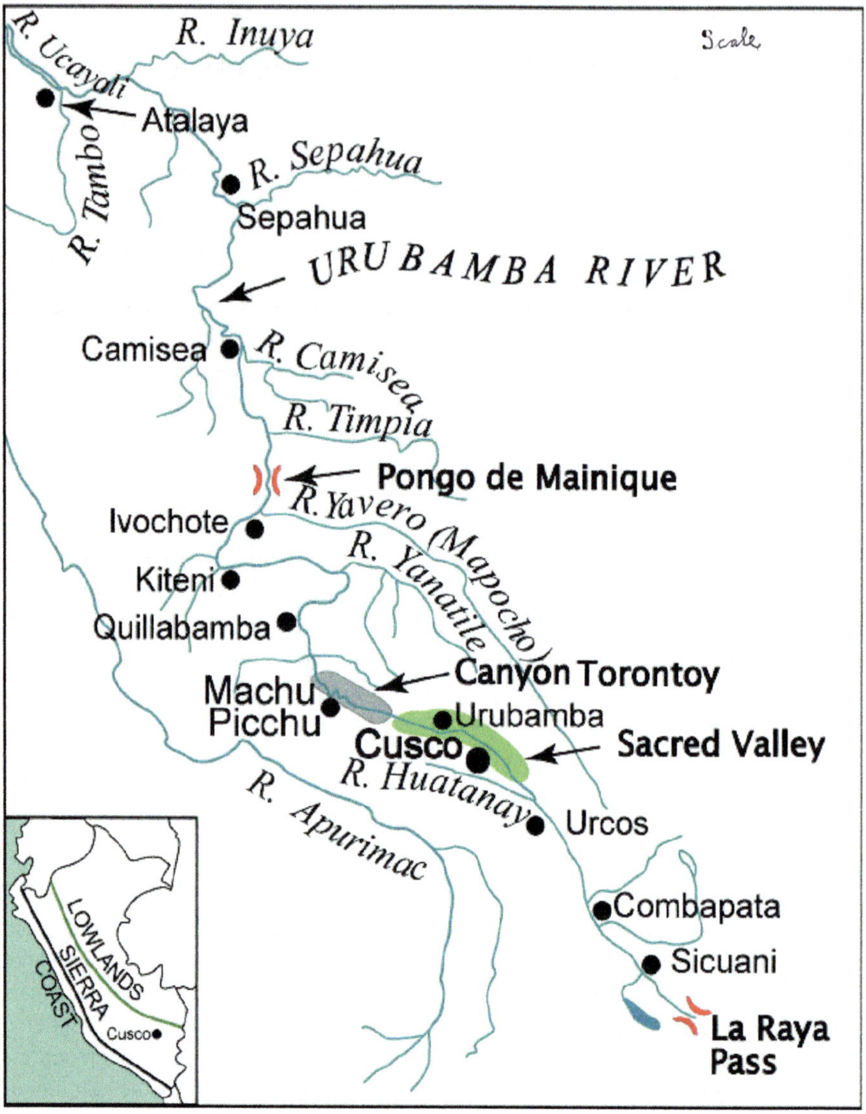

Fig. 1.1 The Urubamba starts in the south in the glacial meltwaters above La Raya Pass and ends in the north at its confluence with the Tambo River. (Map by M. K. Gade)

This fetching, kaleidoscopic valley has been the focus of my attention as a scholar, researcher in the field, and traveler for half a century. The host of indelible memories of experiences, people and specific locales which I accumulated in this sinuous space, defined my professional life activity as a field-oriented cultural-historical geographer. Its landscapes and folk have a significance that the casual visitor cannot appreciate. My investment in time, personal relationships, financial resources, and intellectual effort connect me deeply to the Urubamba. Each of my trips unveiled novel facets and fomented new insights into knowledge I had already acquired. That progression gave me sound reasons to return repeatedly to the valley, but not so frequently that changes were overlooked. At the time of my initial project in the early 1960s, the Urubamba, still largely preindustrial and paleotechnic, was a zone of contestation and upheaval over the land question. To many people in the valley today, that tumultuous period is treated as ancient history. In this book I have attempted to unite a diverse array of valley knowledge with a personal perspective where it seems appropriate. The Urubamba Valley is comprised of several parts: the river, which largely accounts for valley geomorphology; the floodplain, which may be wide or narrow; the slopes, step or gentle, on both sides; and the mountains or plateaus that rise above.

1.2 Focus on the River

1.2.1 Course of the Urubamba

Not until after the 1806 expedition of the Franciscan cleric Ramón Busquets and his companions did some geographically-minded individuals become aware that the Urubamba River is part of the Amazon system. In a flotilla of 10 canoes, Busquets went from Echarati to the Ucayali, on a trip that he paid for with his life. That expedition made evident that the Urubamba River moves in a convoluted course, first west and then north until it joins the Tambo to form the Ucayali. The sources of the Urubamba River are the glacial meltwaters above 5200 m asl, and from there it flows 650 km before it joins the Tambo and forms the Ucayali. In the context of the Amazon drainage basin, the Urubamba is but one hydrological contributor in the far western reaches of that basin. In the total Amazon drainage of over 5,316,000 km^2, the Urubamba drainage covers 76,200 km^2 (Fig. 1.2).

From La Raya Pass (4340 m asl) to its confluence (217 m asl) with the Tambo River, the vertical drop exceeds 4000 m. Two tributaries lying 30–40 km more distant than La Raya Pass have claims to being the source. One is the Hercca, which flows out of Lake Langui Layo (58 km^2 at 3950 m asl) and enters the Urubamba on its west side. The other powerful tributary is the Salcca, flowing out of Lake Sibinacocha (30 km^2 at 4900 m asl) and into the Urubamba on its east side. These two large lakes and other smaller ones feed the river in the dry season. In the rainy season from December to May, precipitation runoff is the source of most river water. At Pisac, the Urubamba discharges more than ten times the amount of water in February than it does in August. Thanks to lakes, springs (ca. 400), glaciers and somewhat greater precipitation, the Urubamba carries seven times more water than does the Apurimac River (ONERN 1986:208).

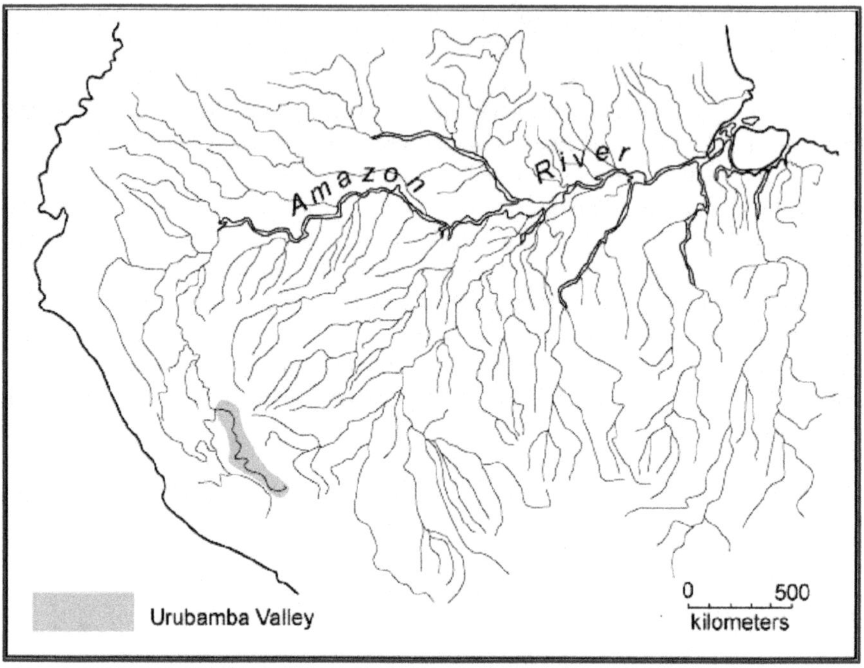

Fig. 1.2 Location of the Urubamba drainage in the context of the total Amazon system. (Map by M. K. Gade)

1.2.2 Toponymy of the River

The name Urubamba derives from the Quechua prefix *uru* meaning worm and *bamba* referring to a flat expanse. Use of that name for the river postdates the colonial period when it was customary to apply local names to different stretches of the river, corresponding to the villages along it. For example, one map had five names for this watercourse: Rio de Vilcanota (called that near its headwaters), Rio de Sicuani, Rio de Calca, Rio de Urubamba and Rio Santa Ana, named for the most important hacienda in that zone near present-day Quillabamba (Colpaert 1865). In the lower valley, which had no villages during the colonial period, the name Amaybamba sometimes referred to a tributary (now called Lucumayo or Huayapata) and sometimes to the Urubamba.

In the sixteenth century, that parochial nomenclature coexisted with the name Vilcamayo or its variants Huilcamayu or Wilkamayu (Glave and Remy 1983:54). In the seventeenth century, Mollinedo referred to the "Villcamaio" flowing by Andahuaylillas (Villanueva et al. 1986:159). The eighteenth century geographers Antonio de Alcedo (1966) and Cosme Bueno (1767) also called the river Vilcamayo. In the nineteenth century, Clements Markham used both Vilcamayo and Vilcanota when referring to the stream. Antonio Raimondi in 1865, E.W. Middendorf in 1888 and Hiram Bingham in 1909 each called the river the Vilcanota (Raimondi 1874;

Middendorf 1973; Bingham 1911). Vilcanota, a word meaning "house of the sun" in Aymara, reflects the fact that Aymara was the main language of Canchis until the seventeenth century (Bertonio 1984). Yet the site at La Raya Pass that gave its name to the river was an Inca pilgrimage destination organized by Quechua speakers. The French explorer Francis Castelnau expressed amazement at the "large number of different names for this stream: Vilcanota, Yucay, Vilcamayo, Urubamba" (de Castelnau 1850–1859: IV:273).

Only in the twentieth century did Urubamba, as applied to the main river, become a synonym for Vilcanota. In 1899 Luis M. Robledo, an entrepreneur in the rubber business, used the name Urubamba in his lectures at the Centro Científico del Cuzco and the Sociedad Geográfica de Lima in which he laid out the unity of the valley from La Raya to the Tambo by describing the hydrography of the entire river system (Robledo 1900).[1] Robledo's aim was to capture a vision of Cusco's geography that integrated its highlands and tropical lowlands. His business interest was to bring rubber out of the lower Urubamba to Cusco so it could be exported. Robledo broke the river valley into three sections: the Alto Urubamba extending from the La Raya headwaters to the entrance of the Torontoy canyon; the Medio Urubamba, from Torontoy to the Pongo de Mainique; and the Bajo Urubamba from the Pongo to the confluence with the Tambo. However, these divisions and labels never were widely adopted. One anthropologist defined the Lower Urubamba ("Bajo Urubamba") as the stretch between the Camisea in the south and the town of Atalaya at its confluence with the Rio Tambo in the north (Gow 1991:27). Others quite commonly used the expression Alto (Upper) Urubamba to refer to the river south of the Pongo up to its confluence with the Yanatile. Inconsistent and contradictory use of sectional names for the river and its valley makes it preferable to abandon those toponymic divisions.

Names used for the river are still inconsistent. One Peruvian government agency calls it "Vilcanota" from its source at La Raya to the town of Urubamba, and "Urubamba" in the stretch from that town down to the river's end at the Tambo (ONERN 1986 I:211). Some authors followed that usage (Dourojeanni 1987:84), but such a division reflects no consistent historical pattern or major changes in the size of the river to prompt the change in names. By applying the term Urubamba to the whole 862 km stretch of the river and its valley from its hydrographic source at La Raya to its end at the confluence with the Tambo, the confusion of multiple names is avoided and the integrity of the drainage basin is recognized. Luis Robledo's suggestion was a good one. The toponym Vilcanota, though still much used, should be decommissioned just as other early names for parts of the river have fallen into oblivion. The one useful sectional division is between the "upper valley" of the Urubamba above 2200 m asl and the "lower valley" below that elevation. That distinction provides a useful shorthand regionalization between the highland part of the valley with its largely temperate environment to the south and its semi-tropical and tropical habitat to the north.

[1]An English-language summary of Robledo's paper delivered to the Sociedad Geográfica de Lima in January 1899 made foreigners who were interested in Peru aware of its importance. Luis Maria Robledo Ocampo, born on Hacienda Masapata in La Convención, became known in Cusco as an explorer and as an entrepreneur in Peru's eastern region. He died in the revolution of 1910.

1.2.3 Human Use of the River

In common with other mountain streams, nowhere is the main river in the upper valley used for navigation, a pattern found everywhere in the realm of Andean mountain streams. Only kayakers seeking thrills choose to contend with the strong currents. From Cocabambilla downstream, navigation becomes possible, though risky, especially in the rainy season when the waters surge, until below the Pongo de Mainique.

Water in the riverbed of the upper Urubamba is surprisingly little used. People have gotten their water from tributaries high above that has been directed by gravity flow into canals and from there to crop fields. An exception is in Quispicanchi where the water from the left side of the Urubamba moves into a 6-km long canal to irrigate ca. 300 ha. Irrigation water is most critical in September and October after sowing has taken place, but rain is still scarce. Also, the now heavily polluted stretch of the river in Calca and Urubamba Provinces is little used for washing clothes or bodies. Fishing in the upper valley has disappeared as a commercial endeavor. Some fish survive in the lower Urubamba where large volumes of water dilute the pollution brought down from the Highlands. Overfishing through use of dynamite or barbasco and episodic toxic concentrations have caused fish die-offs and have greatly reduced the traditional fishing activities of July and August.

Lateral tributaries provide drinking and irrigation water for communities on the valley floor. These lateral valleys are referred to in Spanish as *microcuencas* (small basins) to distinguish them from the main valley. Historically, small side streams also powered gristmills. For example, once horizontal water mills on the Hercca River ground grain for Sicuani. Beginning in the 1930s, turbines on that same fast-moving stream generated hydroelectricity. Similar installations were in half a dozen towns up and down the valley, but always on lateral streams where currents were strong. In 1963, to provide Cusco with power the first hydroelectric installation on the main Urubamba was built below Aguas Calientes where the river drops 354 m in a 2-km stretch.[2] In the 1980s, further growth of metropolitan Cusco prompted the enlargement of that facility. With that increased capacity, many towns and villages as far south as Urcos and as far north as Echarati received electricity for the first time. Plans to construct more dams and reservoirs on the main Urubamba will, if implemented, affect the whole region in many ways.

1.2.4 The River Speaks

The Urubamba can be heard when not seen. Its sound changes with its location and the seasons. Near its source the Urubamba is a babbling brook whose water bed—narrow enough to jump across—will course 5000 km into the Atlantic Ocean. José

[2] Cusco first received electricity in 1914 through the entrepreneurship of Cesar de Luchi Lomellini, an Italian immigrant, who financed a water-powered turbine and transmission line from it.

de Acosta, the Spanish chronicler in the sixteenth century, believed that the Urubamba started at a spring of grayish-colored lye water (*agua de lejia*) described by him as having a burnt odor (Acosta 1940:121–122). In fact, what he observed was meltwater carrying powdered stone flowing from the glacier high above. Several decades later, another Spaniard, Reginaldo de Lizárraga (1987:190) also understood the water coming down near La Raya Pass was the source of the Urubamba, a river he called the Rio de Quiquijana. In the early seventeenth century, the idea prevailed that the main river in the eastern lowlands, the one into which the Urubamba flowed, was the Marañon, not the Amazon (de Lizárraga 1987:170).

Below its juncture with the Rio Hercca near Sicuani, the Urubamba flows stealthily and silently through the flats of northern Canchis. At Tinta, the stream is 19 m wide; only 15 km downstream at Combapata, the addition of the cold milky waters of the Salcca widens the Urubamba to 30 m wide and doubles its depth. All through Quispicanchi, the main river moves mutely through a generally open valley, but that changes when the Sacred Valley section is reached. Near San Salvador, the river begins to murmur as it flows through a steep-sided quebrada from 1 to 4 km wide. Coursing past one pre-Hispanic monument after the other, that riverine sonority is a reminder that the Incas anthropomorphized important Andean streams as "speakers." A song in Quechua with the line, "nanaj mayupas rakhu takin wan añayñispanñan wiraquchata," translates as "the brisk torrent, with its hoarse chant, is singing the praises of Viracocha."

Riverine decibels reach their highest in the Torontoy canyon where the river falls 1200 m in 32 km. Constricted by the narrow passage, the thrashing turbulent rapids emit a deep roar akin to distant thunder. On the floor of the gorge at Aguas Calientes, the deafening sound bounces off the cliff faces on both sides. In that aqueous pandemonium, one must speak loudly to be heard. Some 400 m above, however, in the aerie that is Machu Picchu ruins, the cacophony heard below fades to silence. I imagine that the Inca Pachacutec, who ordered the town built, loved that serenity high above the river din. As the waters move on their northward course, at Santa Teresa the Urubamba is finally released from its batholithic prison. There, where two major affluents, the Santa Teresa and the Sacsarayoc, enter the main stream on the same side, the streams carved out a huge ampitheatre, and the Urubamba loses its timbre. But downstream the river enters another constriction between the steep hills covered with coffee and coca, a murmuring tonality picks up. It does not last, for some 30 km onward, the Urubamba receives, on the right, the waters of the Lucumayo (formerly the Amaybamba) and, about 1 km downstream from there on the left, the Vilcabamba enters and the river sound is reduced to a slight whisper. It is all about reverberation: the narrower the valley, the louder the noise. Between Maranura and Quillabamba the valley again constricts to form an echo chamber.

The Urubamba remains generally topographically constricted until its confluence with the Yanatile. From that point onward, the valley widens and the river deepens as increasingly large tributaries carve spaces and vast quantities of water add to the main channel. The last big clamor of the Urubamba begins about 250 km downstream from the Yanatile where the river rushes through the ramparts of the Pongo de Mainique. Between its cliffs reaching up to 300 m asl, the swirling water during the rainy season rushes fast and furiously, drowning out the calls of the cock of the rock (*Rupicola peruviana*) nesting on its cliffs. Once the Pongo is traversed, calm returns to this jungle river which flows in majestic silence to its terminus near Atalaya.[3] Where the Urubamba joins the Rio Tambo, the Ucayali begins, a river 300 m wide and twice as voluminous and deep as the Urubamba. The sound of the Urubamba also varies with the season. During January and February, the river is louder as the water volume increases enormously after sometimes diluvial rainfall. The sound geography of the Urubamba enables one to imagine with eyes closed one's geographical location and season. Since the sound of the river is a function also of the arrangement of mountains that frame it, those two features form a dyad, for the waters of the river come mostly from the surrounding mountains. Making cosmic associations was an element of Inca religion. In the Imperial period, the Urubamba Valley was a pilgrimage path in a rite of regeneration (Urton 1981:64).

1.3 Physical Geography of the Valley

1.3.1 Land Forms

Within the larger Cordillera Oriental, four ice-covered mountain chains loom above the Urubamba Valley.[4] Snowlines have varied through time; for the coldest part of the Pleistocene, glaciologists estimate that the lower limit of ice was at 3650 m asl (Mercer and Palacios 1977). Glaciers atop the cordilleras—La Raya, 88 km^2; Vilcanota, 539 km^2; Urubamba, 23 km^2; and Vilcabamba, 175 km^2—are major sources of water for the floor of the Urubamba Valley (Mercer and Palacios 1977). The Cordillera de La Raya is 60 km long and its highest peak is Chinchina (5489 m asl). The Cordillera de Vilcanota makes a 120 km-long arc in Canchis and Quispicanchi; of its ten peaks reaching over 6000 m, the highest is Ausangate (6354 m asl), the destination of an annual syncretic pilgrimage into its snows. For scientists, however, the most famous mountain in that chain is Quelccaya (5470 m asl), lying 60 km from the Indian village of Succa Pulca near the pass east of Sicuani (Bowen et al. 2005). This peak now holds Peru's largest glacier, covering 44 km^2 and glaciologists have studied Quelccaya intensively since Lonnie Thompson first reached it in 1974. In 1983, Thompson drilled a core into the icecap down to bedrock from which he reconstructed much about the climate history of the Central Andes. Melting has fragmented the ice patches in this cordillera into 400 different glaciers.

Though lower in elevation, and much shorter in length, the Cordillera de Urubamba is the best known chain. Its peaks frame 23 km on the east side of the Sacred Valley and though not all are visible from the valley floor, as the sources of lifegiving water they are a vivid part of the valley imagination. In that section

[3] Atalaya is actually on the Rio Tambo about 1 km from its confluence with the Urubamba.

[4] The Cordillera de Vilcanota and Cordillera de Urubamba, which begins north of Urcos, are sometimes considered as one chain.

Fig. 1.3 Much of the Urubamba valley mid-section has an alluvial floodplain excellent for agriculture. But the steep slopes are also cultivated in many places. (Photo by D.W. Gade)

landforms on the west side include a high plateau with low hills, lakes, and some areas of large flat expanses. The most northerly of the four chains, the Cordillera de Vilcabamba, lies about 100 km north of the city of Cusco. This range, aligned east–west, has a dozen impressive peaks, the most iconic of which is Salcantay (6271 m asl). That peak, together with Pumasillo (6070 m asl), and Soray (5950 m asl), have glaciers whose melt waters add volume to the Urubamba River and to a smaller extent the Apurimac to the west. Many lakes above 3900 m asl serve as reservoirs for irrigation below.

The floor of the main Urubamba Valley stretching 181 km from La Raya Pass to Ollantaytambo has a general declivity of less than 1 %, smoothed by running water and glacial action during the Pleistocene (Fig. 1.3). The width of the valley floor ranges from 2 to 8 km and its flatness also varies. Between Sicuani and Tinta, an ancient lake bed left a floodplain surface so flat that the marshy conditions made certain areas unsuitable for agriculture. Between San Salvador and Chilca, 40 km below Ollantaytambo, the valley assumes an east and west orientation with a flat bottom and steep slopes. This stretch is now called the Sacred Valley. Between 2350 and 1500 m asl, the Urubamba River cut a spectacular gorge ("Canyon of Torontoy") through a Permian batholith consisting of granite and granodiorite. Despite its narrow floor and high rainfall, this gorge bottom accommodates the most intensive tourist activity in all of Peru, thanks to the ruins of Machu Picchu being perched above and overlooking it. From 1500 m asl to the junction with the Yanatile at 730 m asl slates and clayey schists in places have eroded into a series of high river terraces with 30° slopes. Unlike the upper valley, the floodplain below 2200 m asl is, with a few exceptions, too filled with gravel to support much agriculture. In the local folk classification, the segment from 1500 to 730 m asl is called the "valle" to distinguish it from the "quebrada," an expression often used for the Sacred Valley. Below 730 m asl elevation, a second, much longer stretch of westerly orientation runs for about 70 km between Chahuares and Kiteni. Through that valley stretch, fluvial action carved out a landscape of low hills to the Pongo de Mainique. At the Pongo, the river has cut a gorge 2 km long and less than 30 m wide through another granitic batholith. Beyond the Pongo, the Urubamba meanders about 250 km over interminable plains until, at Atalaya (287 m asl), the Tambo and Urubamba, each ca. 300 m wide, join to form the Ucayali.

Thanks to the flat floor of its upper section, the Urubamba Valley became a transportation funnel and a zone of agricultural settlement. By comparison, the Apurimac Valley and Colca Valley (Department of Arequipa), have narrow bottoms with little alluvial soil and few possibilities for irrigation. Whereas conditions for agriculture in those two valley bottoms are unfavorable, the upper slopes offer usable land. The Urubamba Valley has always held large numbers of people and natural events periodically work havoc. Floods occur when the timing of heavy rainfall and glacial melt coincide.

Earthquakes trigger landslides with devastating effects. The 1650 tremor which decimated Cusco also caused major landslides in the valley. Heavy rainfall causes landslides, of which the most devastating one in recent memory was that of January 1998. Santa Teresa, a town of 450 dwellings, was totally destroyed, buried and pushed into the river by a wall of mud and rock from the mountain above. Sufficient warning reduced the number of people killed to 17. A few weeks later, in February 1998, debris washing down the Rio Aobamba just below Machu Picchu destroyed the railroad track to Santa Teresa. The same excessive rainfall buried the Machu Picchu hydro plant under 28 million m^3 of rock, mud and water. Nevertheless, in 2001, the government rebuilt the plant on the same site. In February 2010, flooding of the rail line to Machu Picchu trapped 4200 tourists and residents. Helicopters flew 993 sorties to airlift them to Cusco.

1.3.2 *Climate*

From south to north, temperatures increase predictably with decreasing elevation; that is the most salient fact of the Urubamba climate.[5] (Fig. 1.4). Above 3900 m asl, temperatures rarely exceed 14 °C in any month of the year and at night they fall below 0 °C in the dry season. The much more extensive mesothermal section from

[5] Climatic understanding of the Urubamba suffers from incomplete and sometimes unreliable records. The only long-term and reliable station is outside the valley at Cusco (3400 m asl) where the average temperature is 12.8 °C and the coolest average monthly low temperature of −1 °C occurs in July. Cusco receives 707 mm, more rain than any place in the Upper Urubamba Valley.

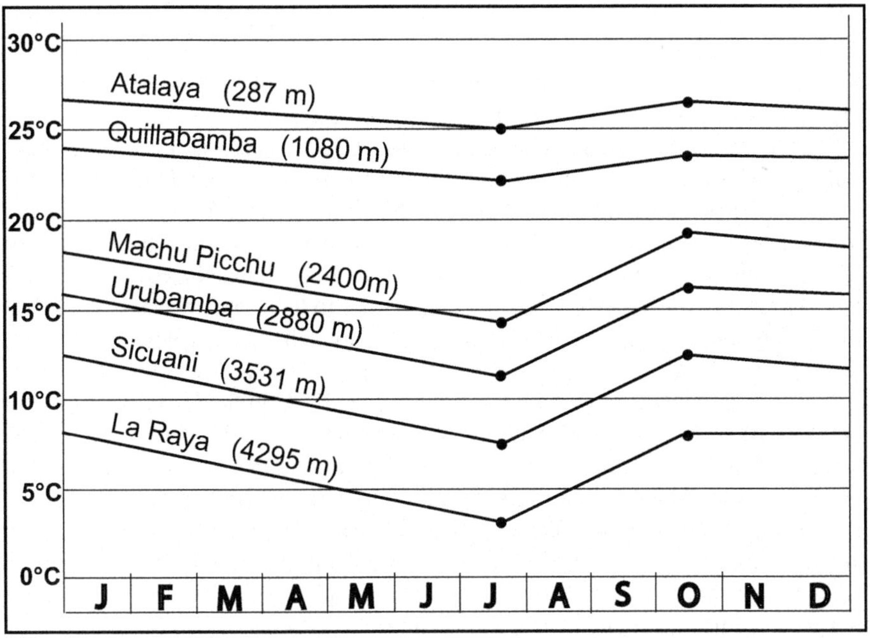

Fig. 1.4 Mean monthly temperature variation during the course of a year at six meteorological stations from La Raya (4295 m asl) to Atalaya (287 m asl). Throughout the valley the warmest month is October. (Graph by M. K. Gade)

3200 to 2400 m asl has warm days especially in October and November when the temperature rises to 22 °C. Annual average temperatures correlate with elevation: Sicuani (3574 m asl): 11.8 °C; Combapata (3464 m asl): 11.9 °C; Checacupe (3445 m asl): 12.0 °C; Urcos (3149 m asl): 13.5 °C; Calca (2926 m asl): 14.8 °C; and Urubamba (2863 m asl): 14.5 °C (Avalos et al. 2011).[6] A temperature/elevation anomaly occurs between Calca and Urubamba; the former also normally receives more than 100 mm more rainfall than the latter. Freezing weather at night more than the warmth of the day influences the kinds of crops grown. At Urcos (3149 m asl) frosts occur from May to August and often in September. At Combapata (3464 m asl) 80 strong freezes occur during the 5 months from May to early October. At Maranganí (3700 m asl), 100 freezing nights can be expected between late April and early November.

Precipitation in the upper valley is highly seasonal everywhere. Amounts vary and there is no neat progression with altitude. Sicuani receives 594 mm; Pisac, 586 mm;

[6]All weather statistics in this chapter are from SENAMBI, the acronym for the Servicio Nacional de Meteorología e Hidrología del Perú, the entity responsible for generating weather and climate data for the country. Reliability of meteorological data in the valley remains problematical, due to the sparse network of stations in a zone of rapid temperature variations and questionable local records.

Calca, 540 mm; and Urubamba 435 mm. Furthermore, amounts vary considerably from one year to the next. In Quispicanchi, total receipts may be from 650 to 850 mm; in Calca and Urubamba, precipitation falls in the 360 to 600 mm range. Orographic exposure accounts for local differences.

The macrothermal lower valley starts at about 2200 m asl, but it is the high rainfall there (2009 mm) that most defines the *ceja de la montaña*. Evapotranspiration is only 1200 mm, making water abundant, the air humid, and vegetation luxuriant. Yearly average rainfall increases from 1283 mm and 40 % humidity at Quillabamba to 1600 mm a year with 80 % humidity in Atalaya. In Quillabamba, average January temperatures of 26 °C and 25 °C are quite similar to those at Atalaya, but absolute maximum temperatures in the latter reach to 37 °C. Greater cloudiness at Atalaya compensated for the 800 m difference in elevation to explain what otherwise seems counterintuitive. Maximum daily temperatures of 35 °C in the lower valley below 1000 m asl occur when no clouds obscure the sun's rays. Thermal values aside, sharp seasonal differences in precipitation characterize every place in the valley. When the sun migrates south of the equator, the rainy season begins, regardless of the elevation. Rains coming between November and April are often torrential and are accompanied by lightning, thunder, and sometimes hail in the highlands. Blue skies mostly prevail from May through September; clouds, if they form, rarely release moisture except in the tropical north.

1.3.3 Vegetation

The spontaneous plant cover of the Urubamba varies in response to climate, but even more to human impact (Fig. 1.5). Above 3900 m asl puna grassland covers the floor and slopes and comprise three main species: *Stipa ichu*, *Calamagrostis vicunanum* and *Festuca dolichophylla* (Fig. 1.6). Although the puna has patches of arborous growth, trees may never have dominated this microthermal section of the valley. From 3800 m asl to 2300 m asl, the native forest that once covered the slopes had been removed well before the Spanish Conquest. A few patches of highland forest have survived in lateral valleys but 99 % of the area is devoid of wild-growing trees. In uncultivated land, shrubs and herbs cover the landscape. By contrast, a rain-soaked elfin forest still covers most slopes between 2300 m asl and 1680 m asl. High precipitation, acidic soils, and little flat land discourage cultivation in this zone. This cloud forest changes abruptly at 1600 m asl to a dry tropical forest which extends down to 700 m asl. Many slopes are covered with herbaceous cover as a result of burning and cutting. Approximately 80 % of the dry tropical forest has undergone anthropogenic impact. Below 700 m asl, the selva alta rainforest dominated the lower valley to the Pongo de Mainique until the 1960s. A half century of colonization has now cleared much of this forest within 3 km of the river except in protected preserves and upper parts of tributary valleys beyond the reach of woodcutters. The selva baja variant of the rainforest, more diverse than the selva alta,

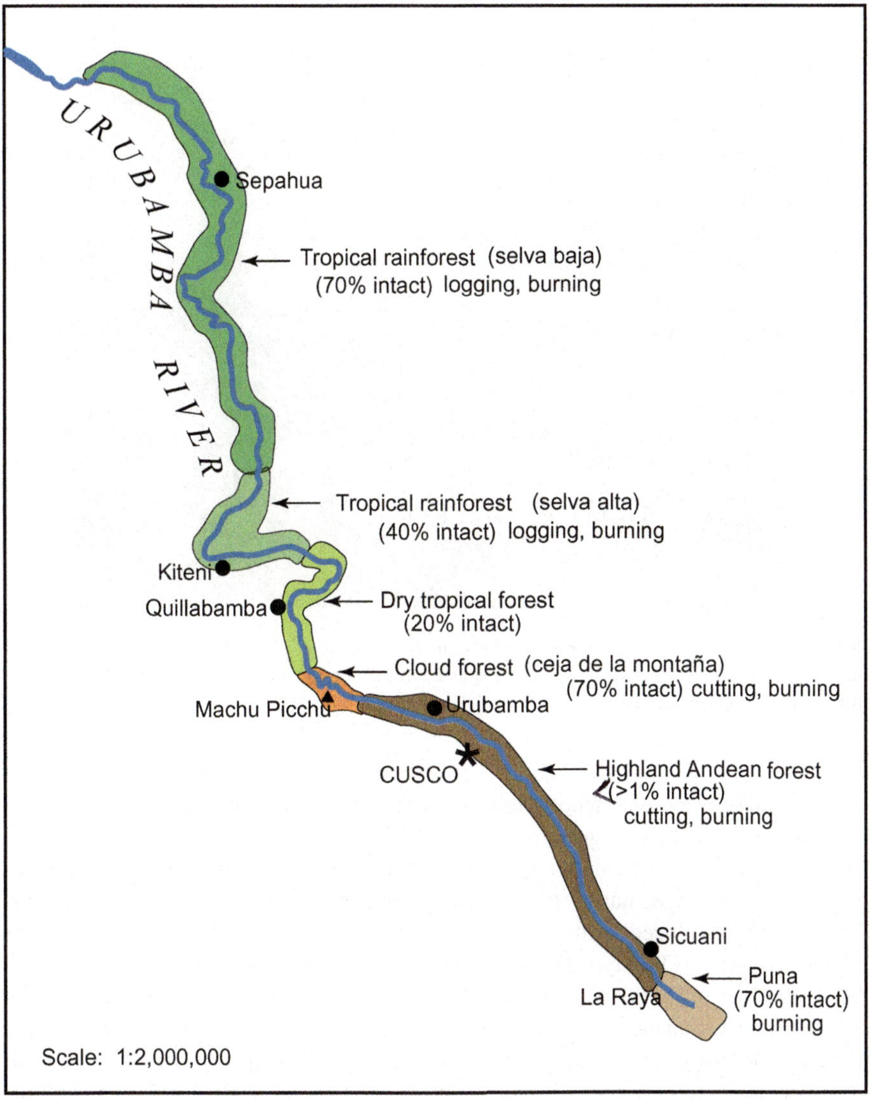

Fig. 1.5 Major vegetation zones around the main river of the Urubamba and estimated degree of anthropogenic impact. (Map by M. K. Gade)

covers much of the Urubamba below 400 m asl. Tree species there are more tolerant of standing water. At a local scale, considerable diversity exists corresponding to pedologic, geologic or microclimatic factors. For example, from Huayllabamba to Pachar on the left side of the river, cretaceous sediments yielding limestone and salt favor the establishment of calceophytes and halophytes.

Fig. 1.6 The puna above 4300 m asl in the Urubamba where grazing of sheep and alpacas is the main activity. Agriculture in this zone is sparse; in the foreground, a field of bitter potatoes in fallow. (Photo by D. W. Gade)

The valley also holds what might be called "cultural vegetation," mainly in the form of tree plantations. Agroforestry, an old Andean tradition, predates the Conquest. However, rather than the native species planted by the Incas, one introduced species, *Eucalyptus globulus*, native to Australia, has come to dominate tree plantations. Seeds were introduced to the valley between 1875–1880 and planted on haciendas under irrigation (Tamayo Herrera 1978:172). Within the valley, Quispicanchi between 3100 and 3300 m asl proved to have the ideal environment for growing that species. Below that elevation, the higher evapotranspiration and temperatures in the Province of Urubamba somewhat impede its growth. The uppermost altitudinal tolerance extends to between 3700 and 3800 m asl. Commercialization of eucalyptus mostly began after the 1950 earthquake which created a heavy demand for wood in the city of Cusco. In the 1960s, haciendas often maintained profitable eucalyptus woodlots (Dickinson 1969:305–306). Land planted in this species was still limited when I first went to the valley. In 1963, Urubamba and Quispicanchi Provinces had 39 ha of eucalyptus; in 1980, that area had increased more than 30 times. After agrarian reform, peasant communities and small landowners used of credit made available to plant this tree. Eucalyptus production has constantly increased in the valley, reflecting high demand for wood. The tree is now an important land use in the Urubamba Valley between 3600 m asl and 2500 m asl, though it grows both above the former and below the latter.

1.4 Human Presence in the Urubamba

1.4.1 Settlement of the Valley

About 4000 years ago, agriculture and livestock husbandry began to replace hunting and gathering in the valley. Alluvial soils gave the valley a substantial resource advantage for the creation of a settled life. Before the Inca hegemony began to reshuffle the deck, the upper valley had six major ethnic groups. From south to north, these were the Canchis and Canas, both of whom are believed to have spoken Aymara until late in the Inca period. The Cavina people dominated present-day Quispicanchi area. In the Sacred Valley, the Cuyos, Pinagua, and Ayamarca had a cultural character that strongly resembled that of the Cusco Valley. The Incas considered those folk to be extensions of themselves. No defined ethnic group characterized the tropical zone between 2000 and 800 m asl. Forest-dwelling tribes dominated the lower Urubamba and later became defined as two groups: the Matsigenka and, farther downstream, the Piro. More pre-Columbian ruins are found in the upper Urubamba than in any other valley of Highland Peru. The lower valley below 1800 m asl has, in contrast, almost none. Since reference to places is so often made to contemporary administrative units, a map is included here to show the boundaries of provinces within the Department of Cusco and of the various departments in Southern Peru (Fig. 1.7).

Most pre-Conquest constructions above 2000 m asl in the valley are certifiably Inca. Some are Wari (400–800 CE) in origin. Except in Ollantaytambo, little about contemporary villages in the valley is pre-Spanish. In 1572, Viceroy Francisco de Toledo imposed his scheme of bringing people, who had up to then been living semi-dispersed in kin-based semi-dispersed allyus, into clustered villages called *reducciones*. With the exception of Lurucachi in Canchis, these clusters have survived and several have grown into much larger towns. Sicuani, now the largest city in the valley and the second largest city in the department outside Cusco itself, began as a reducción. Destined to be more than a village, Sicuani in 1792 had roughly 5000 people, the largest population center on the royal road between Cusco and Puno. Very much of an Indian town, Sicuani was comprised of people from different places and customs (Garrett 2005:160). The whole Urubamba Valley had a heavily indigenous population through the colonial period. In 1689, Indians numbered 38,054 and only 1559 were mestizo or white. Even the city of Cusco, the regional center of Spanish power, had 8372 Indians at that time out of a total population of 10,834 (Garrett 2005:61).

In Canchis and Quispicanchi, the Toledan *reducciones* were subject to the labor obligation called the mita. Every year, a certain number of men from each community in those provinces and others were required to go to Potosí and work in the silver mines. Forced migration of Indians to the "Cerro Rico" of Potosí drained life out of this zone of the valley, for many mitayos never returned (Stavig 1999:183). This policy instigated the biggest protest against Spanish authority in the colonial period. Though most reducciones survived, settlement also dispersed outside those nucleations.

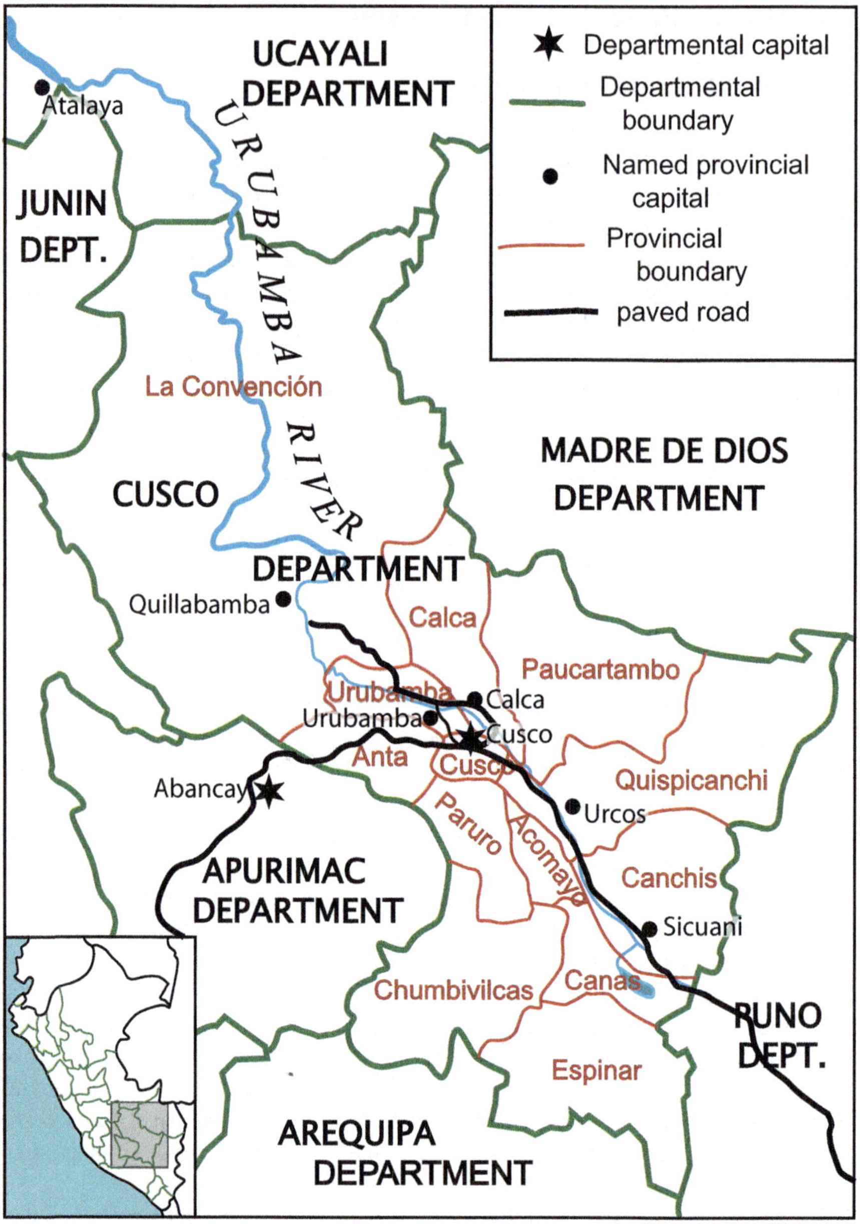

Fig. 1.7 Provincial boundaries in the Urubamba watershed and departmental boundaries for Southern Peru. (Map by M.K. Gade)

Already in the seventeenth century, farmers built isolated dwellings close to their fields. With road construction, dwellings were built along axes of transportation. Land reform after 1970 led to even more dispersion as peasants received parcels that had been in large estates, and they wanted to live near them. In the tropical zone of the valley, the Spaniards founded no reducciones. The only real town, Quillabamba, was largely carved out of Hacienda Santa Ana and most of the flat land on that river terrace is now urbanized. Downriver in that same zone is Echarati, a town formed mostly in the 1970s on land that had belonged to Hacienda Echarati and now rapidly growing into a city.

Since before the Conquest, the most common building material in the valley has been mud bricks (adobes). Stone was for monumental building, not for peasant dwellings, a distinction continued in the colonial period. Mud and straw were available virtually everywhere and adobe bricks require little skill to make or to build with. However, the rainy season takes its toll on the adobe, leading to rapid deterioration. Fired brick and poured concrete, considered to be "noble materials" have become favored in several valley towns. Adobes, once universal in the upper Urubamba, are now used mainly by poor people. The traditional roofing material of thatch has largely disappeared in the valley above 1000 m asl. Whereas small indigenous villages in the 1960s still had roofs thatched with ichu grass and sometimes other plant species, mestizo towns and haciendas typically had roof tiles, betraying Spanish influence. At that time, sheet metal (*calamina*) often covered school buildings. Now calamina caps many dwellings. The rural hut (*choza*) in the valley served to store crops as much as to shelter people. Residents slept and cooked in their choza, but without light, water, plumbing or clean air, people spent most waking hours outside. With the arrival of electricity, peasants wanted more substantial dwellings, since they could comfortably spend more time indoors. In the 1960s in the valley only nine towns (among which were five provincial capitals) and three haciendas had electrical power. In most cases lights came on only at night.

The Urubamba valley floor above 2400 m has had, by Andean standards, a high population density; more than 50 inhabitants per km^2. By contrast, on the plateau of Canas and Chumbilvilcas to the west where extensive livestock raising has dominated the rural economy, population densities have been less than a third of the Urubamba Valley. Much valley land is irrigated; crop fields on the steep valley slopes are not. The "riego"/"secano" contrast is more fundamental than any difference in soils or temperatures. Cultivation of steep slopes reflects population pressure; in the 1960s, it was also that haciendas appropriated the best land, pushing peasants to cultivate unirrigated slopes above the floor. Cultivation of exceedingly steep slope land, was hazardous. Stories circulated of campesinos who lose their footing in those eyrie-like fields and fall (*desbarrancarse*) to their death in the ravine (*barranca*) below. Cattle, less sure-footed than goats, have also died that way.

1.4.2 Evolution Toward the Urubamba Present

Inhabitants of the valley live very much in the present. They recognize the Incas as their ancestors, but they know nothing about the ethnic groups that predated the Inca imposed hegemony. Schools have, in general, done a poor job in awakening an appreciation of and an interest in the Inca and pre-Inca remains around them. Pre-conquest sites were destroyed by peasants carting away stones from ruins to build huts. Dispossession at the hands of colonial masters contributed to their disregard for their own past. In a supreme act of bureaucratic arrogance, Spaniards moved native people from their ancestral lands into grid-pattern villages of Spanish design.

Upheaval of the cultural pattern set the stage for the hacienda system that replaced the encomienda in the late sixteenth century. Europeans either purchased or received these lands as grants. Opportunity for land acquisition increased when epidemics killed native people in large numbers. The colonial policy of land consolidation (*composición de tierras*) allowed unused lands to be taken over and auctioned off to Spaniards. Dispossession continued through deception and legalisms that native people did not understand. In 1720, thousands died in the biggest epidemic ever known in the valley, and its cause remains uncertain. Spaniards claimed much of the land abandoned by Indian mortality. By Peruvian independence in 1825, people of full or partial Spanish descent who depended entirely on indigenous labor owned much of the valley floor. The land use system gave all rights and privileges to the usurpers who were in complicity with the legal and judicial powers.

Further dispossession occurred when the railway penetrated the upper Urubamba. With the arrival of the train to Sicuani in 1893, the sheep wool and alpaca fleece of Canchis Province found its market in Arequipa where entrepreneurs had started textile factories. For the first time, the high country, which had always been of minimal interest to whites (*mistis*) became economically attractive (Jacobsen 1993:198–258). Non-Indians bent on establishing livestock haciendas used casuistry to take puna land away from Indians within Cusco Department. In the high provinces of Chumbivilcas, Canas, Espinar, Paruro and Acomayo, natives resisted these land thefts. Indian herders were less subservient than Indian farmers and frequent violent confrontations occurred when herders resisted the abuses of mestizo hacendados in that isolated zone. Rumors about the "*peligro indígena*" made hacendados in those remote zones fear for their lives (Deusta Carvallo 1981; Poole 1994). By contrast, on the densely populated floor of the Urubamba Valley, native rebellions were rare. The military forces quartered in Cusco easily swept into the nearby agricultural valley and put down organized protests. The Tupac Amaru uprising of the 1780s, which affected many places in the Urubamba drainage, was an exception to effectiveness of centralized control (Fig. 1.8).

Settlement of Spaniards in the Urubamba, as well as the valley's importance as a transportation corridor, fostered acculturation of its Indian population. The drift away from indigenous ways is the major cultural change in the valley over the past half millennium. The first big trauma that caused the acceptance of European ways occurred when Spanish authorities demanded particular kinds of tribute (Cook 1975). From Calca to Sicuani many communities were required to pay the Spanish overlords tributes of maize, potatoes, and wheat. Several communities had a tribute requirement of poultry, forcing Indians to raise these birds, whether they wanted to or not. The authorities demanded the valley *reducción* of Caicay to supply chuño from the puna lands where potatoes were grown and processed above the town. In cold, high Maranganí where freezing temperatures made agricultural yield uncertain, the only tribute item was homespun clothing, reflecting the presence of fleece-bearing animals and a long weaving tradition.

Fig. 1.8 A layer of human bones exposed in 1968 by intense rains at Sangarará recall the violent protests against the Spanish colonial government and its mita policy. Led by the legendary figure José Gabriel Condorcanqui Noguera Tupac Amaru (1738–1781), the rebels assaulted this town west of the Urubamba. When creoles and the Indian elite sought sanctuary inside the church, the rebel forces torched the church on November 19, 1780 and about 100 people died. Their bones are a symbol of that struggle against Spanish rule and the church as a corrupt colonial institution. (Photo by D. W. Gade)

1.4.2.1 Acculturation of Native People

In the late eighteenth century, the great majority of valley inhabitants were still Indian. In 1795, the indigenous population in Pisac was 95 %; Calca, 86 %; and Urubamba, 68 % (Mörner 1975). A century later, Indian demographic dominance declined somewhat. When the American archaeologist Efraim Squier was in Ollantaytambo in 1874, the locals spoke only Quechua; *ollantaytambinos* are now bilingual. In La Convención in 1961, 88 % of the population spoke Quechua and 33 % went barefooted (Perú 1966). Today Spanish is preferred by the younger generations, suggesting the eventual displacement of Quechua. Religion is more diverse than it once was. Protestant sects, which had a small presence in the 1960s, are now flourishing (Fig. 1.9).

In the rugged high country above the valley floor, acculturation has been much less. The upper slopes on the east side of Quispicanchi still maintain a strong indigenous character. To reach the community of Chillihuani, located between 4000 and 5000 m asl, one must climb a 16-km ascent from the valley town of Cusipata. Chillihuani has two ecological zones: on the lower puna, quinoa, cañihua, kiwicha, potato (converted into chuño and moraya), ullucu, añu and oca are grown (Bolin 1998). From there a second 16-km climb takes one to the upper puna where *comuneros* herd llamas and alpacas. The two zones face different threats—hail that destroys

Fig. 1.9 In a scene from 1963 a foreign missionary baptizes a peasant woman in the Urubamba River. Protestants are now much more numerous in the valley of the Sierra than they were in the 1960s. (Photo D. W. Gade)

crops in the lower region and lightning that kills animals in the higher one. Theft and fear of theft are common in both. At least until recently, the 1500 people of Chillihuani nevertheless had a strong sense of their cultural identity and a social solidarity that was not apparent in the valley blow. The "elegant and respectful demeanor" of these folk perched high above the valley floor showed respect, strength and courage in contrast to the materialistic and atomistic mestizo values apparent in the valley bottom (Bolin 1998). As the modern world seeps into the fabric of the rural Andes, few communities will remain isolated. A plan to dam the Salcca River for hydroelectricity would, if implemented, forever change peasant life everywhere in Quispicanchi and Canchis.

1.4.2.2 Changes in Transportation

Valley people, like in most peasant cultures, walk a good deal, for the private automobile is still a luxury reserved for the very few (Fig. 1.10). Everywhere, burros were the main beasts of burden for moving goods within a community. Llama trains when seen came from roadless communities above the valley. Horses and mules for riding were surprisingly few and, when used, were to reach places, especially the lateral valleys, where no roads yet existed (Fig. 1.11). During the 1960s, the stake body

Fig. 1.10 Peasants walk a lot. Here people are on the main path up the Salcca Valley returning from market in Combapata on the Urubamba. (Photo by D.W. Gade)

Fig. 1.11 Travel into lateral valleys beyond the main Urubamba depression has often involved riding horses. (D.W. Gade; photo by M. Escobar)

Fig. 1.12 Trucks were once the main mode for human transport in the Urubamba. Buses have now largely replaced them for carrying passengers. (Photo by D.W. Gade)

truck served for the transportation of most people in the main valley. Many once roadless communities now have at least a rudimentary track accessible to trucks during the dry season. Roads for motorized vehicles in this part of Peru first appeared in the 1920s when President Leguia instituted road conscription as a kind of personal labor tax on peasants. That program made truck transportation possible in the main Urubamba Valley in the 1930s.

In the 1960s, trucks had some competition for passengers from buses, but the former had some advantages. The truck beds with high wooden slat sides accommodated cargo as well as peasants carrying large packages (*bultos*) (Fig. 1.12). Without fixed seats, more people could always be crowded in. Time of departure was an approximation, for *señor chófer* left only when he had, in his own judgment, enough passengers. Along the way drivers picked up or dropped off people in every village. A teenaged *ayudante* collected fares and performed other tasks as directed by the driver. Unlike buses, trucks had no passenger lists to present at the police checkpoints. Almost all truck drivers were cholos of indigenous background. They personalized their vehicles with macho bravura ("el Puma") or religious devotion ("Niño Jesús," "Virgen de Copacabana," etc.). No one questioned the driver's decision to stop for an hour to eat a meal or to visit a local girlfriend for a quickie along the way.

None of the roads in the Urubamba Valley were paved in 1963–1964. In the rainy season, thick mud and deep ruts were part of travel. In the dry season, clouds of choking dust followed in its track. The elevated chassis of trucks enabled them to maneuver over big rocks and through bridgeless streams that would damage the undercarriage of a bus. Since trucks did not undergo regular maintenance checkups, frequent mechanical breakdowns and flat ties made delays common. When landslides stopped movement, the driver expected passengers to help dig a path through the obstruction. In order to reach their destinations, passengers obliged. Poorly maintained roads and imprudent driving practices led to occasional accidents.

The arrival of trucks triggered a series of economic and social changes. As conveyors of large quantities of merchandise, trucks facilitated setting up a shop in a village, carrying a seriously ill person to the hospital in Cusco, and enabling a seller to take produce to distant markets where her goods were in high demand. When trucks brought commercial flour in bulk from Cusco, many water-driven grist mills near a village became obsolete. Potato production on the puna without road access is dependent on pack animals, but those tubers put on the market cannot compete on the market with potatoes from locations served by trucks (Brisseau-Loaiza 1972).

Fifty years later, the truck to carrying passengers had given way to more comfortable buses with fixed schedules and much improved roads. Goods, however, now move in trucks even more than in the past. The two railroads in the valley that in the 1960s carried both goods and people have become tourist trains with no local traffic. In 1998, all train service to La Convención ceased when a gigantic avalanche destroyed a large section of track north of Machu Picchu. In the south, the privatized the Ferrocarril del Sur stopped carrying peasants through Canchis and Quispicanchi and replaced the old trains with semi-luxurious train cars for tourists.

1.5 The Urubamba Valley as a Region

The Urubamba Valley includes those parts of its tributary valleys that are closely connected to the main depression encompassing the valley floor and sides above it. Other tributaries, however, form more distinct units of their own. Thus, the Huatanay River Valley flowing through metropolitan Cusco is not the Urubamba though it flows into it. Likewise, the Salcca, which reaches far upstream from Combapata into puna country and glaciers, is part of the Urubamba only in a hydrographic sense. In the same way, the towns of Chinchero, Maras, Yanaoca, Catcca, and Santa Barbara all lie within the drainage of the Urubamba, yet are not physiographic parts of the valley as I have conceptualized it. Delimited in that way, the geographic Urubamba Valley had a population of 260,000 in 1987 and 371,000 only 20 years later (Perú 2005).

Since so much exchange is based on complementary trade outside the valley itself, the Urubamba depression is not a self-contained economic region. Weekly markets that function largely to exchange products from different ecological zones are most important in Canchis and Quispicanchi. The two provincial capitals of Sicuani and Urcos have daily markets and also much larger Sunday markets. San Pedro, Checucupe, Pitumarca, and Quiquijana hold markets on Sunday, enabling peasants from remote areas to also attend mass. San Pablo, Combapata and Maranganí have theirs on Saturday and Tinta has a Thursday market at which livestock is sold to agents of the slaughterhouse in Cusco and Arequipa. Towns serve the retail needs of only limited sections of the depression and the high country above them. The biggest

Fig. 1.13 A view of Cusco. In the foreground the Plaza de Armas is in front of the cathedral. Between 1963 and 2015, Cusco's population grew five times. Buildings now climb even the steep slopes. (Photo by D.W. Gade)

urban centers are all provincial capitals: Sicuani (33,815); Quillabamba (23,992); Urubamba (10,000); Calca (9106); and Urcos (5995). Although it lies outside the Urubamba depression itself, the city of Cusco functions as the first-order center serving almost all the valley. In 2015, its metro area had close to 400,000 people (Fig. 1.13). As the departmental capital, Cusco is in frequent communication with Lima. Its retail and wholesale roles are important to valley merchants and consumers. Cusco's institutions serve quite a few valley folk. Many students from valley towns attend the university in Cusco. Before the valley had its own secondary schools, young people, living with relatives, attended *colegio* in Cusco. Two exceptions to Cusco's economic dominance of the valley exist. The Urubamba north of the Pongo is now increasingly oriented toward Pucallpa, and southern Canchis has long had its strongest ties with Arequipa.

1.6 Doing Fieldwork in the Urubamba

The Urubamba Valley was the setting for my first independent research effort. One's initial foreign fieldwork automatically becomes the baseline intersecting time and the self against which later field experiences are measured. Irrevocably an experience in the past, a fieldwork stint nevertheless holds a magical aura that makes it almost mythic in one's scholarly imagination (Vitebsky 2012:181). Even though my research scope branched out over the years to include Paraguay, Brazil, Southern Europe, Ethiopia and Madagascar, the Central Andes had a transformative effect on me. Meeting the challenges confronted during my first stay in Peru prepared me for subsequent challenges there and elsewhere.

I did not come to the Andes to prove some grand theory. When that is the objective, the research design is more straightforward but it also risks forcing data into a pre-determined structure. My effort in 1963–1964 essentially recorded tacit information—that which had never been written down—and turned it into formal knowledge. From that I generated a large mass of data from which I crafted a lengthy narrative. Years after my work had been published, the meaning of those findings, as well as the experience itself, held my attention. At some point I realized that this first Peruvian fieldwork period and others that followed had value beyond the knowledge acquired. Close encounters with places and the people who live there honed my discernment of the knowledge frontier and of the culture I had entered. Fieldwork involved personal incidents, sometimes quite trivial, that became part of my memory. The experience of collecting information in a foreign place is also about the self. It funneled me into a hall of mirrors revealing not only the work accomplished but also my own strengths and weaknesses as an investigator.

Without question my counter-enlightenment sensibility explained much about how I viewed the Valley and the Andes in general. A diachronic approach and reference to the past came naturally to me. Pre-Columbian ruins everywhere in the upper valley physically brought the past to life and, at the same time, exposed transience. Andean folk culture and palaeotechnic agriculture bespoke of a people seemingly in a time warp. My temperament favored the idea of the fixed traditional community dominated by a subsistence orientation and survival of old traits whereas others might see a community of people with fluid identities on the path toward development. Those who do fieldwork see what they want to see. As artisans of our own existence, we create a world suited to our being. Understanding an author's perspective and intent seems fundamental (Gade 2011).

1.6.1 Impressions of the Place

My starting point in any locale was to first analyze the landscape, which provided something of its past and its current processes. The Urubamba conveyed an intimacy, for unlike flat treeless plains on which only distant horizons demarcate the view, deep valleys frame space neatly. Land use had a sharp vertical contrast. The humanized valley floor, the grassy side slopes and snowy peaks above represented discontinuities that effortlessly elicited questions about the land and the people on it. Daylight in a valley is nuanced. Well before the sun is overhead, a soft light permeates the valley. Gradually the sun rises high enough to cast its first rays on the valley floor. In the late afternoon, the sun disappears, but the light lingers until darkness descends about six o'clock. The cloudscape adds to a feeling of enclosure, especially at the height of the rainy season when thunderheads form on an almost daily basis and release their moisture in the afternoon or evening. At this time of the year, the early morning mists have a magical quality.

Away from the road, the lack of moving machinery reinforced a tranquility. The prevailing sound was the tinkling of bells as a herd of sheep grazed up and down a slope. Leading the flock was a child singing a wayno to entertain herself. Andean rural life had many such scenes and sounds. The tendency to problematize peasant existence denies or ignores its positive dimension of living at one with the land the many advantages of being part of a community with kith and kin. Outsiders from the developed world have been prone to see drudgery, hardship and deprivation. They convince themselves and others that the peasantry must change. Even those who have never seriously observed the content of rural life come to that conclusion by checking how Andean people figure low on the published indices of modernization.

1.6.2 The Urubamba as a Zone of Many Wonders

Fieldwork requires one to determine what is important and what is not. In that filtering process, only part of what is seen and heard gets recorded. My gaze, however, was often wider than that of most scholars, perhaps because my education stretched into many different corners of thought and practice. In my inclusive eyes-wide-open filtering, I had a compulsion to write things down, as so much of what I was seeing begged to be recorded. Still, I had to prioritize my research boxes. My main effort was to capture the diversity, distribution and uses of plants up and down the valley. Beyond that ethnobotanical focus, other phenomena grabbed my attention out of sheer fascination. I was interested in landscapes as comprised of interlocking elements, each of which has its own history and geography. In all, I indulged myself in satisfying a range of curiosities about the Andean countryside and its inhabitants, and succeeded in developing several side projects. Admittedly, however, some matters, though important, appealed to me less. Land tenure, a major concern in Peru at the time was wrapped up in legalisms and politics. Kinship also did not inspire me even though sorting it out for the valley would have revealed much especially about ownership patterns.

1.6.2.1 Bridges

Bridges became an interest—and preoccupation—that grew out of a recurring need during the course of my fieldwork: the need to cross rivers to observe crops and talk with farmers on the other side. Except near its headwaters, fording the Urubamba and most of its tributaries was not an option even during the dry season. Rafts, perhaps because of the current, were few and undependable. In a few places, a large hanging basket or cage (*oroya*) tied to a rope pulled by a person on the bank allowed one to cross. Traversing the Urubamba required bridges and so were indispensable to

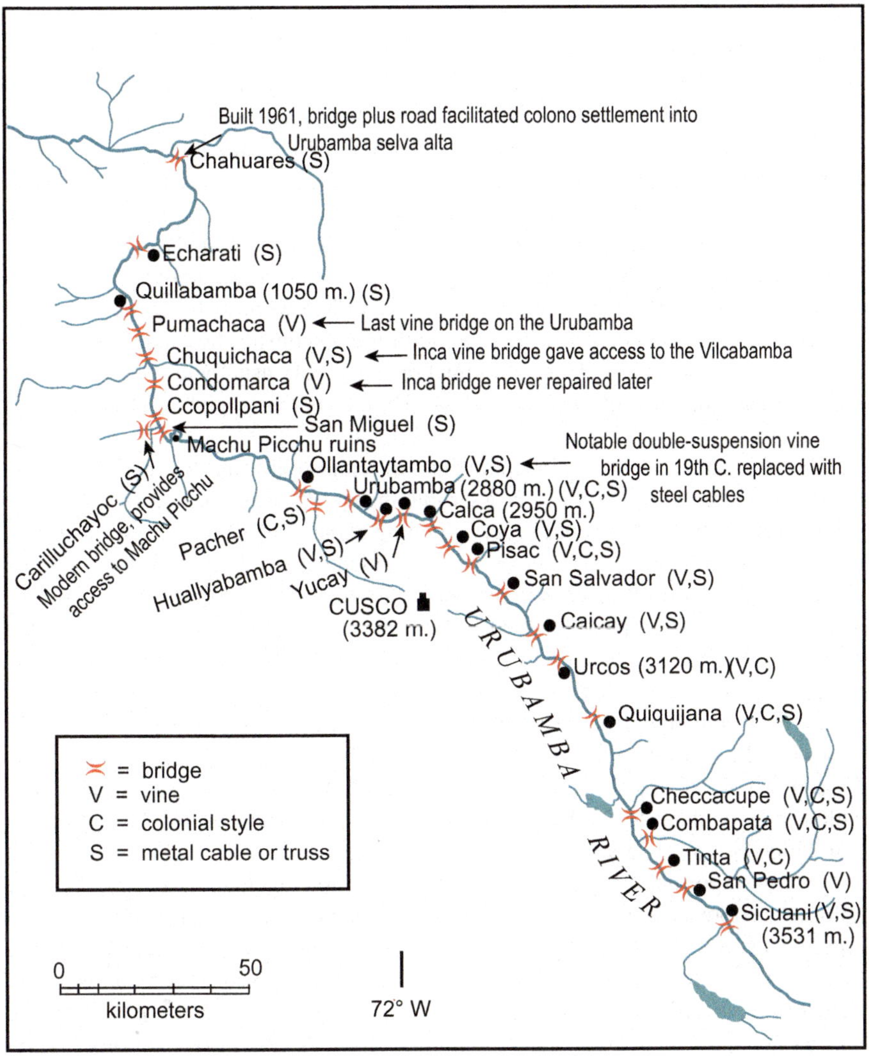

Fig. 1.14 Major bridges, historic and contemporary, in the Urubamba Valley. (Map by M. K. Gade)

my valley surveys. In 1963–1964, no vegetal fiber structures had survived anywhere in the valley. Replacing them were suspension bridges with a wooden plank floor and metal cables. There were also half a dozen steel truss bridges with metal beams.

My concern with bridges grew into a special interest of those in the past (Fig. 1.14). Many of them were part of the historical record; other sites could be reconstructed from the presence of abandoned piers or abutments. Until the late nineteenth century, the Urubamba Valley had 13 hanging bridges of vegetal fiber, usually described in the sources as *puentes de crizneja*. They consisted of strands of plaited vegetal fiber from the pliable branches of several plants. One plant used was willow (*Salix babylonica*), a small tree grown in the valley to stabilize river banks and to provide the raw material for baskets and bridges. The Incas may have brought willow to the Peruvian Sierra from Central Chile where it is ostensibly native. Another source of bridge material was lloque (*Kageneckia lanceolata*), a wild growing native shrub of the region with tough, pliable branches. Whatever the exact raw material, these structures all had a fragility that required frequent replacement. Certain Indigenous communities near bridge sites had the knowledge of twining the strands and hanging them across streams. Depending on the amount of use they received, bridges often had to be remade yearly. The work of bridge replacement occurred during the dry season, when agricultural responsibilities and the danger of drowning were much less. This communal responsibility usually lasted over 2 days: the first one to plait the plant fibers and the second day to suspend the strands across the stream.

Crossing a hanging bridge in a poor state of repair was scary. A strand could suddenly break, throwing the bridge crosser into the river and probable death. In a good state of repair, the bridge could be a frightening experience. When the wind swayed the bridge, people and pack animals were petrified especially if the bridges had no parapets on the two sides. If a person lost her footing, she would be hurtled into the abyss below. Mules much preferred to ford the river, but the strong current in the rainy season made that dangerous for the animal and the cargo. Even without wind, mule drivers had to use ingenuity to induce their equines to cross a bridge. One ploy was to place a mare in front leading the way. The heavy weight of mules and packs, as well as the cutting impact of horseshoes, rapidly wore the plaits of a vine bridge. By contrast, a llama and its packs weighed a third of that of a mule with packs. The callous pads on llama feet did less damage to the surface. Many bridges had guardians who helped in the crossing. Some bridge crossings were free.

Several important bridge crossings over the Urubamba in both the late Inca and colonial periods connected Cusco with the Altiplano to the south. The Inca *capac ñan* became the royal road in the colonial period. The Cabildo de Cusco, not any local entity, were responsible for certain bridge crossings and the tambos where travelers and their animals found fodder and overnight shelter. Between Cusco and La Raya Pass, nine tambos lined the road, six of which were also near bridge crossing sites: Urcos, Quiquijana, Combapata, San Pedro de Cacha, Sicuani and Lurucachi. The bridge at Urcos had heavy traffic, first of llamas then mules carrying coca from the hot valleys east of Paucartambo to Potosi. By tradition, the Indians of Andahuaylillas 15 km distant, not the Indians of Urcos, had the responsibility of repairing it (Villanueva et al. 1986:159). At Quiquijana, the river divided the town. In 1596 the Spaniards built a stone bridge ("cal y canto") at Quiquijana, making it an important point for wine merchants from Arequipa on the trail to the Paucartambo Valleys. In 1626, a flood destroyed the bridge and to ensure the river could be crossed, a vine structure quickly replaced it. Less than 30 km upstream at Combapata, the Cabildo of Cusco in 1561 constructed a classic cal y canto bridge spanning the Salcca River and its powerful current. Made of travertine, the bridge had a central pier and two arches. Spanish authorities ordered Indians at Combapata to build a hanging vine bridge next to it for their use (Yépez Valdez 2005:78). Checacupe also had a cal y canto bridge using Inca foundations. In Tinta, Spaniards constructed a cal y canto bridge in 1603, but already in 1631, a flood had seriously damaged it. In 1675 the corregidor organized the construction of another bridge to take its place.

In the Sacred Valley section, eight bridge sites crossed the Urubamba since at least the late Inca period: Caicay, San Salvador, Pisac, Coya, Calca, Urubamba, Huayllabamba and Ollantaytambo. At Pisac, it was not the village of Pisac, but the hacendados of Paucartambo who collected the toll to cross it and who paid for its repair. At Urubamba colonial authorities replaced the vine bridge with a cal y canto structure. When, however, floodwaters in February 1744 destroyed that bridge, a hastily constructed vine structure served as a substitute crossing (de Esquivel y Navia 1980:II:297). But the fragility of vine bridges made them much more vulnerable to malfeasance. During the Tupac Amaru uprising in the eighteenth century, rebels burned all the vine bridges in the valley and the colonial authorities forced Indians to immediately replace them. Bridge replacement was a normal and expected rural activity until the twentieth century. On his 1865 trip through the valley, Antonio Raimondi commented that Indians of Calca and Coya replaced their vine bridges every 2 years with plaits made from lloque branches (Raimondi 1898a). In 1888 Ernst Middendorf (1973:II:201) wrote of crossing a similar vine bridge at Pisac (Raimondi 1898b). Harriet Adams (1908:618) traveling in 1908, mentioned vine bridges in the Sacred Valley, though by then iron cable bridges had replaced some hanging plant fiber structures. At Huallyabamba, that change in technology occurred in 1900 (Adams 1908:618). Within a decade of her visit, automobiles came into use, requiring bridge crossings with a wider passage space, flat floor boards, and greater weight tolerance. Suspended metal cables placed a certain distance apart satisfied those requirements.

Bridges made places more important than otherwise they would have been. At Pachar, the local authorities had a cal y canto bridge constructed over the Rio Huaroocondo that became vital to movement on the west side of the valley. Farther down from Pachar was the lloque bridge across the Urubamba near Ollantaytambo. There, the Incas had positioned a rectangular monolithic rock in the middle of the river to serve as a central pier (Fig. 1.15). By creating that island, two shorter strands could be suspended rather than one long strand. To reduce erosion on the pier, the Inca placed another boulder 18 m upstream to divert the current to each side. When mules became the main beast of burden in the colonial period, a roadway of wooden poles was laid to make a surface on which animals could safely walk. When the colonial bridge at Urubamba collapsed in the eighteenth century, the Ollantaytambo bridge became the main crossing from Cusco to the tropical valleys until the former was repaired. When in the 1880s the Junta de Alcabala built the *camino de herradura* through the Urubamba canyon, the Ollantaytambo bridge again became a critical crossing site. In 1898, the Junta, whose funds came from taxing coca leaf sent to market, replaced the lloque strands on that bridge with iron cables. That allowed a plank floor (*tablero*) to be suspended, facilitating mule traffic carrying heavy loads of coca, aguardiente, and coffee.

Fig. 1.15 The bridge site across the Urubamba near Ollantaytambo, first used by the Incas who built two hanging structures of woven tree branches on either side of the central pier. Increased mule traffic in the late nineteenth century led to replacing the plant material with iron cables. (Photo by D.W. Gade)

On that canyon trail beneath the Machu Picchu ruins which were not yet described, the French engineer Lacabaret organized the construction of a metal-cable suspension bridge crossing the Urubamba at San Miguel. Only three decades later, that indispensable link from the valley to Cusco was already in poor condition through lack of maintenance. Five kilometers upstream from San Miguel was the temporary bridge across the Urubamba to reach Machu Picchu. Hiram Bingham described it as four logs bound together with vines, stretched across the stream "a few inches above the roaring rapids" (Bingham 1913:712). For several years those who came after Bingham used this flimsy structure to get to the ruins. When, in February 1912, Agustín Lizárraga tried to cross this ad hoc installation, it collapsed and he was swept to his death. Repaired, an unknown individual later dynamited it in order to direct visitors enroute to Machu Picchu to a river crossing he controlled (Flores Ochoa 2011:23).

The tropical Urubamba below Huadquiña had three suspension bridges of vegetable fiber and of Inca origin: Condomarca (ca. 1300 m asl), Chuquichaca (1200 m asl) and Pumachaca (ca. 1150 m asl). The Incas may have destroyed the Condomarca bridge to prevent Spaniards from using it (Kaupp and Fernandez Carrasco 2010:55). Chuquichaca near present-day Chaullay gave access to the Vilcabamba Valley and was one of the most important hanging bridges over the Urubamba. An enormous boulder in the middle of the river a few meters from the mouth of the Vilcabamba provided the elevation needed on which to suspend the structure to each bank. Even in the dry season, fording the river in that location was impossible. Termite damage made vine bridges in that hot and humid zone short-lived. In 1916, a metal cable bridge was placed on the old Inca foundation. Then, in the 1960s, a more modern truss bridge replaced the metal cable structure. In 1998 a mudslide swept away that bridge as well as the boulder in the river. Nearby a modern truss structure now crosses the river. The third bridge, known as Pumachaca, located near Hacienda Santa Rosa, was at some point in the colonial period, no longer repaired (Kaupp and Fernandez Carrasco 2010:167–169). In 1997, a landslide carried away one of the two supports that marked the crossing spot of a 37-m long hanging structure.

Below Chaullay a wooden bridge built at Pavayoc in 1875 provided a crossing to Quillabamba located on a river terrace above the floodplain. In 1876, that bridge, possibly decimated by termites, fell. Two decades passed without a bridge until the Frenchman Lacabaret in 1896 supervised the construction of a suspension structure with metal cables. Bridges of a similar design were built elsewhere in the region financed by the coca taxing authority. Hacienda Ccollpani, upstream from Huadquiña, suddenly became much more accessible. Below Quillabamba, entrepreneurs during the rubber boom built a metal cable bridge between Rosalina and Putucusi over a tributary of the Urubamba. In 1961, the public works of Cusco Department constructed a metal truss bridge over the Urubamba at Chahuares, part of a push to give access for colonization of lands down the left bank of the Urubamba to Cirialo. Bridges of several kinds, some for pedestrians only, span the Urubamba in what had long been Matsigenka territory.

Since they sit low over the river, metal truss bridges are vulnerable to floodwaters, and in 2010, floodwaters destroyed the bridge of that type at Pisac. But more such bridges have been built in the valley. In 2007, the elected authorities of the Province of La Convención had a very controversial metal truss bridge constructed near the town of Santa Teresa. Known as Carilluchayoc, the bridge provides access to a 20-km long road up to a place known as Hidroeléctrica, several kilometers below the town of Aguas Calientes. The bridge and road provides a backdoor to Machu Picchu for travelers on a shoestring budget. Rather ride the expensive tourist train, they can reach the ruins by taking a bus over the Málaga Pass and walking the several kilometers from Hidroeléctrica. Strong opposition to the Carilluchayoc bridge came from the railroad, since it threatened its monopoly on travel to Machu Picchu. The Instituto Nacional de Cultura also fought this bridge as a violation of the buffer zone created to prevent individuals from setting up unauthorized businesses near the Sanctuary.

In the upper valley, six foot-bridges across the Urubamba present in 2011 were not there in the 1960s. When haciendas owned most of the valley floor they did not want *abigeos* crossing the river and absconding with their livestock. An Inca-style suspension bridge ("Yunkapunku") made of plaited grass strands was installed in the District of San Salvador in 2004 at an old Inca bridge abutment. Two communities, Qollotaro (in Caicay District) and Salloq (in Andahuaylillas District) make the strands out of cabuya (*Fourcroya andina*) (Valencia Espinosa 2012). Essentially a project to attract tourists on a bicycling circuit, funding for the bridge came from the Instituto Nacional de Cultura. Local people in the district no longer knew that technology and so asked the bridge builders of Q'eswachaka in Quehue District of Canas Province east of the Urubamba depression to come and teach them to make and lay strands (*sogas*). The Q'eswachaka bridge still functions because the Cusco brewery and the Patronato Cultural Machu Picchu pay community members (*comuneros*) to regularly rebuild that Inca style structure to keep the skills alive and to attract tourists. Unlike the Inca style bridges in the Urubamba, Q'eswachaka is comprised of cables made of grass.[7] Another valley bridge rehabilitated with funds from the INC is the colonial bridge across the Salcca near Combapata. One of the most aesthetically pleasing bridges in the Andes, its double arch was separated by a central pier which had a cross and a round oratory. Two other bridges, one far from and the other on the main road, stand nearby. At Quiquiijana another double arch bridge of the colonial period has not been repaired, but it remains there, 80 m from the metal truss bridge, as a reminder of the historical importance of this crossing site.

1.6.2.2 Flowers and the Lord of the Earthquakes

Another side project during my fieldwork of the early 1960s started with ethnobotanical observation of ornamentals in the valley. Andean people are fond of flowers, an affinity I had once assumed was a cultural universal until later when I went to Africa. African cultures seemed to be indifferent to flowers as objects of beauty or symbols of celebration or grief. Africa contrasted with two major indigenous New World cultures, Central Mexico and the Central Andes, both of which have appreciated flowers before and after the Spanish Conquest (Parsons 1992). Each area domesticated or ennobled some of the familiar garden ornamentals grown in North America. Native to Mexico are marigold, cosmos, zinnia, dahlia and morning glory; from the Andes come begonia, nasturtium, petunia and fuchsia.

In the Peruvian Andes, flowers are cultivated in gardens and also picked from the wild. Most showy plants bloom there during the rainy season, which may explain why some species are considered symbols of fertility. For special festivities, unmarried young women, place blooms on their hats. Flowers are part of the ritual offerings assembled on homemade altars to the mountain gods. One wild-growing Andean flower of spiritual value is *pfallcha* (*Gentiana primuloides*), often blue, but also occurring in several other colors. Another blooming herb, *ñukchu* (*Salvia oppositifolia*), has a specific religious significance the one day of the year on which the crucified Christ figure called Lord-of-the-Earthquakes, or "Taitacha Temblores" by the peasants, is taken from his usual place of honor in the cathedral and is paraded through the streets of Cusco. In anticipation of this procession, children scamper up

[7] In Canas Province, Department of Cusco, six Indian communities, of which the largest is Quehue, have traditionally had the responsibility for remaking this bridge called Q'eswachaca over the Apurimac River. At an elevation of 3654 m above sea level, the natural vegetation is ichu grass (*Stipa ichu*), so the raw material for the bridge was also grass. Each community still contributes twined rope for its renovation even though, in the 1960s, a metal truss bridge was built a few meters from this site.

slopes to pick basketsful of the blood-red ñukchu growing spontaneously on hillsides, and later will throw the flowers as the image passes. Different ideas circulate about precisely when this blossom became attached to this event. Did the tradition of throwing red flowers go back to the Inca period when the Inca himself might have been showered with flowers as he passed in procession as a sign of respect and veneration? None of the chronicles mentioned that, nor is it clear if in the early colonial period ñukchu was associated with the cult that has grown up around the Taitacha figure. The flowers included in several canvases of the Taitacha image painted during the colonial period are not ñukchu. A recent study made of the colonial history of this cult came no closer to establishing an historical linkage between the flower and the event (Stanfield-Mazzi 2007).

The "Taitacha Temblores" name means "little father of the earthquakes." Since the colonial period, Lord of the Earthquakes has been a major religious cult in the city of Cusco. The material focus of this particular veneration is the image of a very distraught crucified Christ. For 364 days of the year, the image has its candle-lit side chapel in the penumbral cathedral. On one day, Holy Monday of Holy Week, the Christ figure is removed for its yearly sortie. Placed on a heavy litter and accompanied by a brass band playing a mournful dirge, Taitacha is brought outside via the main portal for his procession. Small boys positioned in elevated side niches shower the image with its first ñukchu blossoms. In the light of day on the street, the statue depicting Christ in his hour of greatest agony is a rich tableau of contrasting hues. The black color of the image, caused by centuries-old encrustations of candle smoke, is heightened by the gleaming silver of the opulent litter. Piles and wreaths of carmen blossoms of ñukchu, symbolizing the blood flowing from Christ's wounds, cover the image and the litter on which it stands. The procession moves at a snail's pace around Cusco's central plaza where cusqueños stand ready to throw handfuls of ñukchu flowers onto the litter. Huffing, puffing, grunting and sweating, the porters are visibly under strain from the heavy load. As night falls, the dirge ends as Taitacha Temblores is returned for another year to his place within the cathedral.

The annual procession is the most moving public display of spirituality in Cusco and, unlike most other religious occasions, its celebratory dimension has almost no profane elements. Anxiety associated with an unpredictable and much feared earthquakes—of which Cusco has had several—explains the devotional fervor. Cusqueños especially remember in their collective history the 1650 devastating earthquake which hit the city and panic ensued. Dozens of aftershocks created a desperate uncertainty about what was likely to happen next. If ever there was a need for divine protection, this moment of impending death and destruction demanded an act of expiation. The bishop ordered the Christ image to be removed from the cathedral and carried in procession. The tremors ceased and a miracle was declared, an example of unequivocal of cause and effect. In 1707 seismic shocks again stopped after the ecclesiastical authorities removed the image from its side altar and carried it in procession. In the popular mind, that was not a coincidence. In 1950, a major earthquake destroyed part of the city and while the Taitacha procession did not prevent the tragedy, it did give survivors solace in their loss and trust in divine mercy.

In Cusco, Taitacha Temblores has been a cult with wide devotion among the faithful and has a confraternity of its own.

Lord-of-the-Earthquakes is a strong folk tradition for cusqueños, but it is also about the psychology concerning natural hazards. In the Andes seismic movements occur with some frequency and engender an acute sense of insecurity. Most tremors felt are harmless; some are not. A strong shock can flatten buildings, kill people and destroy whole cities. Powerlessness to prevent them has invited supernatural intervention as the only feasible form of protection. To the scientifically minded observer, this association of religion and a natural hazard reflects unwarranted credulity without realizing that the folk Catholic response offers an important psychological safety valve. Some other cultures have, during such catastrophic events, resorted to human sacrifice. In the medieval Christian tradition, clerics interpreted earthquakes as a divine retribution. That view persisted in Peru into the twentieth century. It represented divine payback for human waywardness, without, however, detailing which particular sins had warranted divine punishment. As for ñukchu, its association may, in fact, not be that old. Nothing in the Inca or colonial periods suggests a tie with any kind of procession. Now, ñukchu at Lord of the Earthquakes has grown into something larger than it was. Increasingly artistic wreaths are now made of ñukchu.

My interest in the ethnobotanical association of an Andean flower prompted me to think through the nexus between religion and disaster events, and led me to write the essay that became the first article written in English on the Lord of the Earthquakes cult (Gade 1970c). Perhaps ironically, a journal published by the Benedictines, an order that dominated monasticism until the middle of the twelfth century, provided the forum to ponder a superannuated medieval tradition. Although the Lisbon earthquake of 1755 opened a Pandora's Box of bitter debate, the Church never clearly resolved the role of divine will in natural disasters. Advances in tectonic knowledge forced theologians to rethink the retribution argument in order to avoid cognitive dissonance. In the Andes scientific understanding still had not fully penetrated the culture in the 1960s. Much more than ecclesiastics, the popular mind of peasant folk resisted naturalistic interpretation. The door to an important philosophical and historical reflection was opened to me by a species of salvia dotting the hillsides.

1.6.2.3 Other Side Projects

Several other side projects occupied my peripheral thinking about the valley. I became interested in the temporary watch huts that peasants constructed in their crop fields to protect their maturing crops from thieves (Gade 1970b) (Fig. 1.16). Peasants told me that stealing crops was a significant rural social problem at that time. The forces of law and order rarely intervened in these furtive episodes. Shortly after I published an article on it in Spanish, a Peruvian Marxist scholar sharply retorted that peasant thievery reflected the concentration of land in the hands of haciendas (Sabogal Weisse 1970). He did not explain why, then, poor peasants stole more from one another than from the hacendado.

Fig. 1.16 As crops ripen, boys spend nights in watch huts to protect maturing maize. In the distance, serious gully erosion indicates the long-term consequences of cultivating steep slopes. (Photo by D.W. Gade)

I became interested in grist mills, an ancient technology that then survived in the Central Andes (Gade 1971). In the early 1960s, the Urubamba Valley had 94 mills, 84 of which were constructed of adobe and had a horizontal wheel moved by water. The Province of Canchis had 62 of those 84 mills. The horizontal wheel is a simple technology, requiring little specialized knowledge to operate or repair, since no gears are involved. My main study area for mills was the community of Cachuma in Canchis Province. Water from the Rio Cachuma which tumbles down the mountain side into the Urubamba was directed to millraces to turn the wheels of 12 mills (Fig. 1.17). Upon my return there in 2011, I found them all defunct and most of them in ruins. That abandonment coincided with the great decline in wheat cultivation in the 1990s. Growing barley and selling it to the Cusco brewery had become more profitable. Some farmers continued to grow wheat, but had their grain ground into flour in a commercial mill in the town of San Pedro de Cacha on the Cusco-Sicuani road, and when the electric grid reached that village in the 1990s, electric motors made all gristmills obsolete.

Other side projects grew out of my curiosity about three domesticated plants native to the Andes. However, the Urubamba Valley was not the place to study them, so, for a time, my attention was directed elsewhere. One plant was achira (*Canna edulis*) which took me into the deep chasm of the Apurimac to document its cultivation as a crop of ritual importance (Gade 1966, 2012). A second crop was cañihua (*Chenopodium pallidicaule*), a plant in the same genus as quinoa.

Fig. 1.17 An abandoned gristmill in the Cachuma Valley whose fast-moving stream joins the Urubamba near San Pedro de Cacha (Canchis). In the 1960s more than a dozen mills in this one valley were still functioning. (Photo by D.W. Gade)

Although it had once been cultivated in the Urubamba high country, it was on the Altiplano where I discovered cañihua growing. In the 1960s it was still an important crop around the town of Ayaviri and elsewhere on that high plateau hovering around 4000 m asl (Gade 1970a). A third plant, tarwi (*Lupinus mutabilis*), occurred in the Urubamba Valley, but I also observed it around Paucartambo and near Lake Titicaca (Gade 1969).

Animals were also among my Andean interests. The guinea pig (*Cavia porcellus*) or, in Peru, *cuy*, was common in most peasant households. They scurried around the hearth area of the hut, nibbling, chirping and fighting, until the day they were slaughtered for food. Festive occasions traditionally called for roasted or fried cuy. Over time I became fascinated with these little creatures, their coat variations and habits. They are very active at night, as I learned during my fieldwork when I was trying to sleep in native huts. Later I assembled documentation to understand their place through time in Andean culture (Gade 1967). Given their long prehistory of domestication in the Andes, it was puzzling how poorly documented they were in the archaeological record. The observation that absence of cuy bones among kitchen middens can be blamed on the presence of domestic dogs provided a partial explanation (Valdez and Valdez 1997). Guinea pig bones most often appear in the context of human graves where these animals were buried with the dead (Sandweiss and Wing 1997).

The tapir (*Tapirus terrestris*) is a large mammal with the anatomical anomaly of having three digits on the back feet and four on the front feet. In the 1960s, tapirs were still found in a radius around Koribeni mission where Matsigenka Indians hunted them with guns. One day when hunters came back with a dead tapir and divided its parts among families I got a close-up look at its anatomy. I subsequently learned that, during and after the colonial period, tapir feet had a special value as an anti-epileptic (Gade 1999, 2003). That tiny detail triggered an inquiry into the history of human epilepsy and the assumed remedies for it derived from animals.

1.6.3 Arrangements and Contingencies of Fieldwork

Though the bulk of my fieldwork was in the Urubamba, I had several reasons to spend snatches of time in Cusco. Though outside my study area, this city was where road and rail transportation between the upper and lower section of the Urubamba converged. Cusco also at that time had two daily flights from and to Lima, making it the only place in the southern Peruvian highlands with reasonably dependable mail service. Government offices, archives and a library were there. Several university professors had specialized knowledge applicable to my project. Owners of many haciendas in the valley resided mainly in Cusco. Cusco also had medical facilities and pharmacies. Several other foreign researchers located in the town became part of a support system. My minimalist *pied à terre* on the Calle Fierro had no hot water, but at least I had access to a shower of sorts. Before the 1950 earthquake, the Nueva Alta barrio had been a neighborhood of muleteers and I and three families shared a patio space. The landlord, a shady lawyer of the sort locally known as *tinterillo*, lived downstairs. He brought servants (*muchachas*) from the countryside and frequently beat them. From time to time their screams echoed in the patio.

Cusco in 1963 was a manageable city of about 80,000 quite traditional people. Men who considered themselves members of the "*gente decente*" category wore a coat and tie as an outward manifestation of their status. Such was the social code that when I contacted them, I dressed like that as well. The city core still bore evidence of the devastating 1950 earthquake, which in a few seconds destroyed 3000 building. Ironically the collapse of Santo Domingo church exposed Inca walls and awakened an interest in the pre-Columbian inheritance forgotten for three centuries. Cusco's tourist vocation mandated rebuilding in neo-colonial and neo-Inca styles. Outside the historic center, municipal planners laid out new streets for residential districts and built a university campus on the fringe. The 1963 Cusco increased threefold the pre-1950 perimeter. The city's strong sense of its own importance flowed from its role as the Inca capital and later as a regional center of administration. Though modest by present standards, in 1963 tourism formed a not insignificant part of the economy. Its hotels and restaurants provided employment in a job-scarce city.

Fig. 1.18 Publicity for the beer manufacturer in Cusco now warns of the danger of excess consumption of alcohol. Formerly billboards proclaimed the healthfulness of drinking the beer, since it was made from pure water. (Photo by D.W. Gade)

The brewery that made "Cerveza Cuzqueña," provided 80 % of the local manufacturing jobs with 300 blue and white collar workers. Beer in Cusco profited enormously from the absence of potable water from the tap. Bottled water had not yet appeared and when it did, only tourists could afford to buy it. Bottled beer in Cusco satisfied the need for a clean liquid to satisfy thirst. In 1872 a hacendado of Austrian origin, Gustavo Manglesdorff, started a brewery in Cusco that became a model for five other local breweries by the end of the nineteenth century. In 1941, the government forced the largest brewery owned by Germans to yield its shares to Peruvians who formed the Cervecería del Sur. Since then, this one firm, which has its own water wells, has produced more than 90 % of the beer consumed in Cusco Department.

In the 1960s, the brewery's advertisements proclaimed "drinking beer is drinking health," which indirectly alluded to the dubious cleanliness of chicha, the traditional beverage of Cusco peasantry. Its oldest form of preparation involved chewing maize kernels, a process called in Quechua *mukuy*. An enzyme in human saliva fermented the concoction. Even the Spaniards of the sixteenth century considered muku-made chicha revolting. Clever advertising of bottled beer promoted unbridled consumption. Between 1956 and 1986, its production in Cusco increased at least 20 times, and it became the preferred binge beverage of the rural peasantry. As per capita consumption of beer soared, the negative consequences of drunkenness became apparent. Traffic accidents and domestic violence were said to have increased. Realization that beer was not a harmless beverage led the Cusco brewery to remove its suggestive posters of "*beber cerveza es beber salud*" and replace them with admonitions for moderate consumption (Fig. 1.18).

Nevertheless, by 2012, per capita consumption of bottled beer in Peru reached 40 L/year, more than twice that of the 1960s.

The sharp rise of visitors in the 1970s catapulted Cusco as the center of South American tourism from the Western Hemisphere. Monumental stonework and other associations with the Incas, not the Spanish colonials, were the biggest draws. The remains of the Inca Temple of the Sun snared many more visitors than did the sixteenth-century Dominican church built on top of it. The Inti Raymi pageant ("Festival of the Sun"), a twentieth-century resurrection and totally contrived tourist pastiche, gets more visitors than the more genuine traditional feast of Corpus Christi. By 2013, though Cusco's population had increased fivefold, dependency on tourism in the city was considerably greater than in the 1960s. Despite their prime destination being Machu Picchu, tourists lodged overwhelmingly in Cusco. In 2007, 86 % of the tourists who came to Cusco went to Machu Picchu (Dávila Rojas 2009). Everyone who visits those Inca ruins passes through Cusco. Now almost every inch of the historic core is given over to tourism and a square meter of property in the core sells for US$6000. In the process, Cusco attract tourist-related investment and is increasingly cosmopolitan. Brought by people from the outside the city now incorporates more cultural aspects of the world than locals could ever conceive of themselves (Lawbon and Chion 2012).

The old churches still serve locals, but the money to keep them in repair now comes from the tickets that the municipality requires tourists to buy. In 2012, tourism in Cusco raised 12 times more money than it did in 1963. For the Department as a whole, 27 % of the gross domestic product came from tourism.[8] Cusco serves as the staging area for trips to the Sacred Valley, Inca Trail, and Manu National Park in the jungle of Madre de Dios Department. The Inca sites of Choquequirao and Vilcabamba la Vieja are more recent destinations. For decades, they have been visited by groups on high adventure; when a road is built to Vilcabamba and a funicular to Choquequirao above the floor of the Apurimac canyon—both have been proposed—visitors will arrive in mass.

Tourism has complex consequences for Andean people. Catering to visitors brings increased prosperity and accelerates acculturation to Western ways. Given the low degree of entrepreneurship in the manufacturing sector, tourism provides the means for families to rise out of extreme poverty. Tourists expect and want things and scenes traditional and folkloric: llamas, native dress, Andean flute music and some foods. This interest helps to maintain and even revive certain aspects of Andean culture. Tourism also has given people of full or partial indigenous background a greater sense of pride of who they are. Well into the 1950s, Indians were despised, maltreated and exploited. That attitude is much less in evidence today, partly because prejudices have contracted, but also Indians living as Indians draw tourists to Peru.

[8] Statistics revealed that tourism in 2007 represented 27 % of the regional income and that the average foreign tourist spent US$987 compared to US$298 by Peruvian tourists. Of the 922, 487 tourists in Cusco in that year, 68.5 % were foreign and 31.5 % were domestic. Visitors spent an average of 3 days in Cusco including, for the great majority, a trip to Machu Picchu.

At the same time, as the author José Maria Arguedas believed, tourism destroys cultural integrity. One manifestation of that is the international market for textiles whose quality has fallen sharply. Woven textiles produced for tourists are colorful and folkloric, but no longer well made. Country folk who move into towns and participate in the cash economy buy for themselves factory-made clothing. Tourism also encourages individuals to take on street performer roles. A photogenic child in chullu and poncho leading a llama may earn more in tips during one day than a construction worker. Tourism in Cusco has evolved hard edges designed to maximize profit. Tourists are viewed as economic units rather than individuals.

The modus operandi in the fieldwork for my first valley project involved naturalistic field observation, collecting plants, and interviewing (Gade 2015). Plant use and processing, whether in the chacra, household space or market, dominated my attention. Information collected came from my own effort. In my notion of the doctoral dissertation, hiring local assistants to interview did not seem correct or necessary. Peasants rarely refused to cooperate even when they were not sure of my motives. Physical discomfort—fleas, dust, mud, intestinal problems and fatigue—were manageable, although later I realized how much easier it was to do Andean fieldwork in one's 20s than in one's 60s. Sleeping arrangements were often improvisional. Interviewing skills improved over time. I learn to jot down key words from my encounters with informants in order to later recall what had been discussed. Writing up field notes was time consuming and required an ambiance free of distractions. The next step, analyzing those field notes, came later when I could create enough of a bubble around myself to think clearly about the meaning of the information received.

1.6.4 *The Land Question in the Valley of the 1960s*

My initial plunge into the world of the Andean peasant was an unforgettable experience. I spent the first week in Lima to get a sense of what resources there I could count on for my dissertation. The flight to Cusco from Lima on a propeller DC-6 so poorly pressurized that passengers had to suck on oxygen tubes, delivered me to a peasant world that I had never experienced before. The intimate connection of people to land struck me immediately. Stoicism in the face of adversity characterized peasant life. But I had not been prepared for the degree of social turmoil which overlaid my project area when I arrived. Struggle over land tenure in the Urubamba Valley was the defining issue in Peru during the whole decade of the 1960s. Whereas now the Internet instantly brings news from around the world, events that transpired months before my arrival and reported in the Lima press had not reached me. I learned that 2 years earlier a national debate in Peru over the land question had ended with no change in the status quo. Failure to resolve a course of action left an atmosphere of malaise that hung over much of the Peruvian Sierra.

Most of the land in the Urubamba Valley was in haciendas, but ca. 85 legally-recognized peasant communities also owned land. In addition, individuals owned small parcels here and there. The haciendas were the focus. Organizations of peons and tenant farmers on these estates called for change. Hacendados were defensive, for most estates did not make full productive use of their large land holdings. Moreover, some hacendados were abusive or condescending to their workers and set unreasonable work requirements. On some estates, peons had to work 120 days a year for the hacendado and were paid next to nothing. I had much to learn in 1963 about the land tenure issue. Although I sympathized in general with the need to distribute land to the peasants who actually farmed it, selective rather than wholesale reform seemed wiser. Not all estate owners treated their workers badly nor used their land inefficiently.

One estate, in particular, became the great symbol of the need for land reform in the lower valley. In 1963, Hacienda Huadquiña was the largest and one of the oldest estates in La Convención. Huge in size and long isolated, Huadquiña had always functioned as a law unto itself. Even as late as 1783, the owners had the "abominable costumbre" that all who went to hear mass in the hacienda chapel had the obligation to work free for the rest of that day (Oricain 1906). My first visit there forced me to consider the tie between land use, land tenure and culture. Only 73 ha of a huge expanse totaling 58, 450 ha were cultivated. The owner, Alfredo Romainville Garzón, had inherited Huadquiña from his mother who, in turn, had gotten it from her father Mariano Vargas. A merchant from Arequipa, Vargas bought the property in 1858. As reported by a French traveler, Vargas grew one main commercial crop, sugar cane, which was converted into rum (Grandidier 1861:150). In 1899, he introduced the first metal sugar press to La Convención to extract cane juice. After Don Mariano's death in 1902, his daughter Carmen, who had married Eduardo Romainville Centeno, inherited the property valued at 90,000 soles. Eduardo's father, Pierre aka Pedro, was a French immigrant who himself married a hacienda heiress.

Hiram Bingham, who came to Huadquiña shortly after he explored Machu Picchu in July 1911, wrote that "Señora Carmen Vargas of Hda". Huadquiña inherited from her father about 1000 sq. miles of land lying between the Urubamba and Apurimac Rivers. Some of the land is occupied by sugar plantations, other parts are given over to the raising of sheep and cattle, while a large portion is still tropical jungle (Bingham 1913:530). Bingham had arrived there to survey other Inca ruins on her vast property. In a later book Bingham wrote "Huadquiña is a splendid example of the ancient patriarchic system…those whose home is on the estate regard Señora Carmen with an affectionate reverence which she well deserves" (Bingham 1922). Carmen Vargas de Romainville, who also had a house in Cusco, may well have been the *grande dame* Bingham suggested. Apparently his assertion about the hacienda owner accurately reflected the sentiment of her peons (Anrup 1990:89). In 1913 Doña Carmen had a turbine placed in a stream to generate electricity for her casa hacienda and in 1915 she became the first person in Cusco to own a car (Aguilar Callo 1999; Tamayo Herrera 2000:II:38).

In 1951, Alfredo, one of Carmen's six children, became the lord of the manor. Alfredo acquired a reputation in the region for cruelty and by the 1960s had become known as the "the monster of the Convención Valley (Mayer 2009:255). Peons referred to him by the insulting Quecha name of *huancco* which means a cut-off body part. A rumor, so outrageous it seemed apocryphal, circulated that an Indian peon who refused to kiss Señor Romainville's feet had had her arm cut off. Even if this story is false, Huadquiña functioned as a throwback to a time when landowners in the Sierra claimed ownership of their workers and the right to do with them as they pleased. Alfredo ordered women and children to pick coca leaves from his cocales without pay. To reinforce social boundaries, the owner did not permit his peons to wear shoes nor their children to attend school. The hacendado's sons had their way with the adolescent daughters of peons and reportedly fathered many of their illegitimate children.

By 1963, worker docility had come to its end. When word spread that Romainville had whipped a union leader who had accused the hacendado of stealing neighboring land, a peasant revolt engulfed La Convención. The cauldron of anger and threats to the Romainville family forced them off their property. When I arrived in June 1963 peons had already occupied nine estates in La Convención. Most hacendados had left the valley out of fear for their lives. In 1964 the government land reform agency expropriated Huadquiña; the estate was in a legal limbo and had become derelict. None of its 362 peons remained on the property to harvest the sugar cane. In this period, the Romainville family was at their house in Cusco. Several people in the town of Santa Teresa near the hacienda told me that the landlord would be killed if he dared return there.

Formed in 1961, the federation of peasant unions had taken up the cause against the abuses of the Romainville family and other hacendados in the valley. Their solidarity in 1962 influenced the government to formally abolish free labor service to haciendas, sundering the region's long unwritten contract between landlord and serf. Gone was what Enrique Mayer (2009:254) called the "dutiful paternalism of the former and the respectful obedience of the latter." The system had started in the colonial period when haciendas took advantage of Indian distress. Native people who fled their communities to escape the mita found places on estates as yanaconas, accepting serfdom in exchange for exemption from the dreaded mine labor.

Farther down the valley, in the provincial capital of Quillabamba, peasants joined forces under the direction of Hugo Blanco. A charismatic Trotskyite from Cusco, Blanco had studied agronomy in Argentina and was himself the grandson of landowners. He spoke fluent Quechua and effectively voiced peasant grievances. Once the hacendado abuses were fixed, the main labor issue was to reduce the work ("*condiciones*") demanded by the hacendado who owned the plots that the peasants had use of. The government at first refused that concession and issued a warrant for Blanco's arrest. Bearded like Fidel Castro, he and a group of supporters fled into the hills. Just as the neo-Inca Tupac Amaru had done 400 years earlier, Blanco and his companions ensconced themselves in the isolated Vilcabamba zone where they established a medical post, cooperative and school. Before my arrival, the Peruvian military had captured Blanco and dismantled the peasant federation seen as a threat to the social order. When I arrived in 1963, a contingent of 200 soldiers had been stationed for a year in the tropical hacienda zone of the Urubamba to ensure peace. The national government had designated La Convención as the most explosive zone in the country and feared that a spark ignited there would engulf the whole country in chaos.

Fig. 1.19 Luis González Willys in 1964 at Hacienda Potrero near Quillabamba. In his youth Sr. Gonzalez attended the University of Notre Dame (USA). He was the only alcohol producer in the valley to have escaped the agrarian reform of the 1970s. (Photo by D.W. Gade)

In that atmosphere of social contestation, I sought to carry on my innocent program of recording non-political data, interviewing people, mapping fields and gardens, and collecting plants. My countryside ambulations raised suspicions and prompted an interrogation by the Peruvian Investigative Service (PIP). I was able to show them letters that established my bona fides and attested to my scientific purpose. One document came from the prefect of Cusco Department, the archaeologist Luis Felipe Pardo; the other, signed by Fred Harrington, then chief executive officer of the University of Wisconsin at Madison, came complete with a gold-colored seal.

On one trip to La Convención, Wesley Craig, an amiable rural sociologist from the Brigham Young University, accompanied me. Wesley was preparing a dissertation from Cornell University on land tenure changes in the region. At Hda. Potrero near Quillabamba, we interviewed Sr. Luis González Willys, one of the few hacendados who had not fled the valley (Fig. 1.19). About 85 ha of the hacienda's 4870 ha were under cultivation, worked by 219 men.

On his own accord González had given plots of land to his peons and, as compensation for working for the hacienda, paid them 25 soles a day (then $1.00) instead of the standard daily wage of 15 soles. Even with those inducements, he could not find enough workers for his sugar cane operation and rum distillery. Sr. Gonzalez showed us around his estate and insisted that we sample his *aguardiente de caña*. I drank it with gusto, complimenting Don Luis on how smooth (*liso*) his distillate went down. Wesley, the Utah-based Latter-day Saint, bowed to the social pressure of the occasion and, grimacing, gulped down the rum in his glass. Principles sometimes get shunted aside when a scholar seeks certain information.

As I learned about valley agriculture and tied it to social realities, it became apparent that overturning the old order was, by then, less about sugar cane than coffee. Starting in the 1950s, peons (*arrendires*) started to grow coffee trees on the plots rented to them by the hacendado. The owners agreed to that arrangement, for the hilly land on which coffee was grown left free the flat land they used for sugar cane. High international prices for arabica beans made coffee an attractive cash crop. Increasingly arrendires, heavily occupied with growing this export crop, sub-let portions of their parcels to others who grew subsistence crops. Coffee trees, requiring 6 years between planting and the first harvest, represented a long-term investment. Thus arrendires were perpetually anxious, lest the hacendado abrogate the rental agreement. Insecurity of tenure evolved into a feeling that they, whose efforts increased the value of land of the estate, should not have to work for the hacendado to pay for the use of that space. To press their case for more secure contracts, arrendires hired lawyers and paid them from coffee profits (Fiorivanti 1976). For both social and economic reasons, La Convención played a key role in advancing agrarian reform in Peru. One consequence of the breakup of the hacienda was the demise of sugar cane as a crop. Its cultivation, harvest and processing required a technology, economy of scale, and abundance of labor that became unavailable. Hacienda Potrero was an exception; descendants of Luis Gonzalez Willys, the hacendado I knew, continued to produce aguardiente until 2012 when the property was sold.

Social conflict also stirred violence and demands for reform in the upper valley. Beginning in September 1963, strikes, roadblocks, and land invasions reflected peasant discontent with the status quo. Unions mobilized members to arbitrarily withdraw their labor from haciendas. Skirmishes multiplied in the first 6 months I was in the country. On Christmas Eve 1963, the owners of Hda. Ninabamba near Urcos used an automatic rifle to kill seven peasants and injure 23 others. Then in early 1964 an altercation at Hda. Onocora near Sicuani had the tragic aftermath of 19 dead and 15 wounded. Violence contributed to the rapid unraveling of a land tenure system that had been in place for more than 300 years. In that dicey social situation, I needed both groups of informants: peasants to tell me about their crops, but also hacendados to tell me about their property, its history, and details about the zone of the valley they lived in. Sitting on the fence seemed necessary to the success of my project, but it also reflected my ambivalence.

1.6.4.1 Valley Interlocutors

Some hacendados were reasonable people who had a positive role in valley agriculture. They maintained the genetic purity of maize races and were repositories of considerable crop biodiversity. In 1964 I found among the unharvested crops at Hda. Paltaybamba in the Vilcabamba Valley, then in a state of abandonment after its owner, Sr. Pancorbo, had fled in fear of his life, a greater diversity of tree cotton

(*Gossypisumn barbadense*) than I had seen anywhere else in the region. Some of that selection may have dated from the Inca period when colored cottons were important in making the mnemonic string device called the quipu.

Haciendas were also points of introduction of new crops. An unknown hacendado introduced Granada maize and gramalote, a pasture grass in the genus *Paspalum*, to the lower valley in 1888. Pedro Duque Pérez, who had acquired Hda. Santa Ana in 1893, made trips to Colombia where he had come from and brought back seeds of 17 kinds of plants. Among them was a cultivar of cacao ("blanco"), a papaya cultivar and the tangerine, which subsequently became the most important citrus cultivated in the lower valley. In 1896 another hacendado, Benjamin de la Torre, brought seeds of the mango to Hacienda Maranura and in 1907 tea to Hacienda Huyro. Both became economically successful introductions that benefited the whole valley. As individuals, hacendados varied in their treatment of workers, and not all of them exploited their peons. Across the valley from Huadquiña at Hacienda Ccollpani Grande, Julio Tomás Rivas had none of the grim reputation of the Romainville family. Before the rail line and vehicle road, mule drivers typically spent the night on the Rivas estate on their way to Cusco from Quillabamba.

Interviewing peasants also was important to the success of my project. Their information often required cross-checking, for they had a tendency to state what they thought you wanted to hear. A foreign researcher in the Peruvian Sierra had to be attentive to peasant perceptions of the gringo in their midst. In the Andean folklore of the *pishtaku* or *naqaq*, it was said that a foreigner arrived in a village to steal children, kill them, and boil their bodies down for oil to be used as fuel in jet planes. Anecdotes abounded of communities collectively being convinced that the stranger had evil intent. Going to a hamlet with a plant press I allayed suspicions about why I was there when I showed people how plants could be dried. That demonstration led the locals to assume that, since I was involved with materia medica, I must be a harmless *herbolario*.

1.6.4.2 Peasant Characteristics

An aspect of Andean campesino psychology was the almost Pavlovian reaction to any request: "manan kanchu" or, in Spanish, "no hay," meaning "there isn't any." In fact, however, the phrase was not about lack. Given the history of exploitation that the rural Andean folk have suffered from outsiders, I viewed this retort as an understandable defense honed over decades, maybe centuries. Nevertheless, the negativity of the "no hay" syndrome elicited many reactions. Anthony Dell (1927:203), who came to the valley in the 1920s, wrote "Like every traveler off the beaten track in Peru I began to loathe the words "no hay" and the blithe way in which people said them." Similar experiences may have influenced the sociologist Pierre van den Berghe (1992:426) to characterize Andean culture as dour, hostile and devoid of *joie de vivre*. Most disheartening to me was how those who considered themselves superior exploited non-literate country people. Estate owners and managers were among them, but more unconscionably, teachers, priests, and health care workers (*sanitarios*) abused their positions to extract labor, money or food from peasants (Paponnet-Cantat 1995). The anthropologist Alfred Métraux asserted that "*les Andes sont sinistres*" (van den Berghe 1992). In my view, social acceptance or acquiescence of abuse better merited the adjective "sinister."

Racial or, more properly, ethnic conflict characterized the Urubamba Valley. Two pieces of Peruvian literature lyrically captured the quandary of life in the Southern Sierra. One narrative that captured the Indian/non-Indian divide was *Aves sin nido* (translated into English as "Birds without a Nest") by Clorinda Matto de Turner (1973), first published in 1889. In the novel, civil and ecclesiastical authorities in the village dehumanized native people, treating them as inferior beings. Called Killac in the book, the village in question was modeled on Tinta in Canchis Province where the author had lived with her British husband who worked as stationmaster for the railroad. Even as late as 1964, injustice and impunity were the order of the day in Tinta. Matto de Turner's book, now considered to be a key work of late romanticism in Latin American literature, captured the exploitive human relations of the Urubamba Valley. The other work, *Rios profundos*, by José Maria Arguedas (1958), had even more ethnographic authenticity. Arguedas was not only an accomplished anthropologist in his own right, but he had grown up among native people as Matto had not. *Rios profundos*, a *Bildungsroman* of his own experience, was apocalyptic in tone and went from the sublime to the grotesque.

1.6.4.3 Reservations About Development

Empathy for the lives of peasants never turned me into a social activist or developmentalist. The Marxist rhetoric surrounding that period had filtering through it another kind of potential tyranny. A few years after the collectivist enterprises that included state farms and production cooperatives took over many estates, they failed. Peasants simply wanted their own plots. When viewed historically, the peasants themselves used the Marxist ideology that preached revolution in order to achieve their own objectives. Nor was I taken with the modernization theory that dominated social science thinking at the time. Many scholars had come to Cusco to study peasants from that perspective. Agrarian reform was their mantra, which they saw as a form of economic development. The developmentalist critic Albert Escobar (1988) argued that economic development is a Western invention to dominate poor countries. An irony of development discourse is the gap between rhetoric and implementation. In my experience, development scholars get grants and move funds which meet their own lifestyle expectations more than they trigger the changes the researchers so ardently promote.

My one foray into a developmentalist scheme was not to promote it but to forestall it. One year after I completed my dissertation, I proposed to prepare a manual of Andean peasant agriculture for Peace Corps Volunteers sent to the countryside. My project involved explaining the logic of why peasants do what they do by describing and explaining their traditional crops, livestock, tools and farming methods. With that information, Volunteers would better understand peasant rationale and temper the desire to institute change. For example, hybrid maize was not suitable for the Andes, for it presented the risk of contaminating Andean maize races and force farmers to buy new seed every year from commercial seed companies. I also wanted to defend traditional practices and show them to have a rational basis. One practice was mixing crops in the same field to reduce risk if one of the crops does not yield. As part of the plan for the manual, I intended to interview Volunteers at their living sites and to record their views about peasant life and economy. The Peace Corps central office in Washington approved my proposal to spend 6 months in southern Highland Peru.

Once in Peru and after some weeks interviewing in the campo, I got word that the in-country director of the Peace Corps in Lima was alarmed by my project. Since most PCVs sent to the rural Andes had little knowledge of farming before they arrived, having a U.S. governmental agency sponsor a researcher documenting PCV ignorance seemed like a bad idea. Six weeks into the project, Wesley Jones, the U.S. Ambassador at the time, got into the fray, and demanded that the Peace Corps in Washington cancel my contract, which was done. The doctoral dissertation of Glen Sheffield (1991) detailed the administrative conflict of my project.

Volunteers personally benefited from their 2-year stints in a foreign country. They learned the language well, made useful contacts and, for some, had an experience that transformed their lives. But the Peace Crops promoted change in peasant life without being held responsible for the consequences if things went awry. Andean peasant livelihoods have been conservative precisely because a family cannot afford to take many risks. Following the advice of an outsider "expert," if bad, could lead to disaster by cutting the slender thread of their subsistence security. After the revolutionary government took over in 1968, the Peace Corps organization in Peru received outspoken criticism and the government in Lima cancelled its contract. In 2001, Peru's then-president Alejandro Toledo, who had personal ties to the United States, invited them back.

1.6.4.4 Personal Comments on Fieldwork in a Peasant-Dominated Valley

In retrospect, the risks I took while doing my early fieldwork in Peru astound me now. Walking across high railroad trestles, negotiating narrow trails carved into the sides of mountains above steep abysses, crossing a raging river in a basket hanging on a threadbare rope are all memories of what I did to get my story. Other recollections are of truck drivers who flaunted their bravado by constantly passing other vehicles on blind curves. Numerous encounters with unfriendly dogs prompted me to always keep stones in my pocket. The human element, so rich in indigenous tradition, pulled me into its orbit. Exactly 27 days after my arrival, that sense of enchantment, even rapture, ceased. A convergence of depression and unpredictable rage grabbed hold of my psyche with all the earmarks of culture shock. Keeping body and soul together were admittedly difficult, but more specifically, I was personally dissatisfied with my progress. Once I recognized the roots of my negative emotions, a temporary change in venue enabled me to gain a different perspective.

After that, my outlook became healthier and I renewed my determination to get on with the project.

After almost 8 months in Peru, I became well-adapted to the Andean rhythms of daily life, the sibilant cusqueño Spanish, the starchy food, cold nights, frequent stomach disorders, and the plaintive wayno music that spoke to the soul of rural Andean folk. The adaptation process also involved rationalizations. I accommodated myself to the social expectations of a hierarchical society, a requisite for acquiring the knowledge I wanted and needed. Understanding the logic of the patron-client relationship enabled me to deal with the *do ut des* imperative of the culture. I also rationalized the peasant distrust as a manifestation of the "image of the limited good." That phrase was the title of a famous paper by an anthropologist who explained that peasants perceive that resources are being continuously redistributed but never increased (Foster 1965). Catholic folk beliefs and practices made sense when viewed as concerning divine power rather than ethical behavior. Understanding the rural Peruvian concept of time required thinking in cyclical rather than linear terms. Andean peasants were decidedly not primitives with prelogical minds, yet their close identification with seasonal rhythms and natural forces suggested the kind of *participation mystique* between people and their environment found in paleotechnic cultures. There were times when I felt my own participation mystique that came from an intense closeness with the river, mountains and sky, cobalt blue in the day and a starry firmament at night.

It took 8 months to find the keys that finally unlocked for me Andean rural culture. Certain things passed me by, but being an outsider with a sense of wonder and curiosity about a different part of the world, I took nothing for granted. I was enchanted by that place. Conceptualizing and completing my own project successfully provided me the sense of intellectual autonomy which remained with me. The existential decisions shaped my life course, making me partial to the freedom to study what I will. I am a solitary navigator floating like a cork on the waters of my curiosity. Scholars who burden themselves with prescriptions about what they should study vs. what they want to study follow a mode of thought outside themselves. One anthropologist chastised scholars, he called them romantics, for not participating in local improvements and cultural changes, or not working to relieve the hardships and deprivations of those they study (Bolton 2011:255).

Only years after a field experience does the realization sink in that the research one has conducted captures a particular time. In the history of the Central Andes, paleotechnic methods in agriculture in the early 1960s dominated even hacienda production. Periodic returns to the field enabled me to better appreciate how and why changes have occurred. Experience and aging lend themselves to maturity of judgment. Interpretation of events benefit from reflection long after they occur. The day I arrived in Cusco in 1986, an explosion at the Cusco rail station killed several people and injured many more. That terrorist attack, meant to harm local tourism, cast an economic pall over the city and region for more than a year. Foreigners in Peru became targets of terrorists, but for the first time foreigners also became subjects of suspicion.. The highly publicized case of Lori Berentsen of Cambridge, Massachusetts, who aided and abetted the MRTA (Tupac Amaru Revolutionary Movement), raised the specter of outsiders working to bring down the Peruvian state. The imagined new order that was supposed to emerge following, first, the destruction of the hacienda system and then the terrorism of 1982–1994, did not occur.

References

Acosta J (1940) Historia natural y moral de las Indias. Fondo de Cultura Económica, Mexico City

Adams HC (1908) Some wonderful sights in the Andean highlands. Natl Geogr Mag 18(9):597–618

Aguilar Callo VR (1999) Angostura: una hacienda señorial. Revista Universitaria-UNSAAC 137:209–224

Anrup R (1990) El taita y el toro. Etnografisk Museen, Göteborg

Arguedas JM (1958) Los rios profundos. Editorial Losada, Buenos Aires

Avalos G et al (2011) Atlas climático de precipitación y temperatura del aire en la cuenca del Rio Urubamba. SENAMHI, Lima

Bertonio L (1984) Vocabulario de la lengua aymara (1612). Centro de Estudios de la Realidad Económica y Social, Cochabamba

Bingham H (1911) Across South America. Houghton Mifflin, Boston

Bingham H (1913) In the wonderland of Peru. Natl Geogr Mag 24:387–573

Bingham H (1922) Inca land: explorations in the highlands of Peru. Houghton Mifflin, Boston

Bolin I (1998) Rituals of respect: the secret of survival in the high Peruvian Andes. University of Texas Press, Austin

Bolton R (2011) Chijnaya, the birth and evolution of an Andean community: memoirs and reflections of an applied anthropology in development. AltaMira Press, Lanham, pp 215–263

Bowen M et al (2005) Thin ice: unlocking the secrets of climate in the world's highest mountains. Henry Holt, New York

Brisseau-Loaiza J (1972) Le rôle du camion dans les relations ville-campagne dans la région du Cuzco (Pérou). Cahiers d'Outre-Mer 97:27–56

Bueno C (1767) Jerarquía ecclesiástica peruana: descripción de las provincias pertenecientes al Obispado del Cuzco. Lima

Colpaert E (1865) Mapa del Perú. In: Garreaud PE (ed) El atlas geográfico del Perú. Lima

Cook ND (ed) (1975) Tasa de la visita general de Francisco de Toledo. Universidad Nacional Mayor de San Marcos, Lima

Dávila Rojas CA (2009) Importanica del turismo en la región Cusco. Economia Politica y Gestión (Cusco) 25: enero

de Alcedo A (1966) Diccionario geográfico-histórico de las Indias occidentales o América (1786–1789), 4 vols. Ediciones Atlas, Madrid

de Castelnau F (1850–1859) Expédition dans les parties centrales de l'Amérique du Sud: de Rio de Janeiro à Lima, et de Lima au Para, 6 vols. P. Bertrand, Paris

de Esquivel y Navia D (1980) Noticias cronológicas de la gran ciudad del Cuzco, 2 vol. Fundacion Augusto N. Wiese, Lima

de Lizárraga R (1987) Descripción del Perú, Tucumán, Rio de la Plata y Chile (ca. 1598). In: Ballesteros I (ed) Historia 16. Madrid

Dell A (1927) Llama land, east and west of the Andes in Peru. George H. Doran Co, New York

Deusta Carvallo J (1981) Intelectuales y compesinos en el sur andino. Allpanchis 15(17–18): 41–60

Dickinson JC III (1969) The eucalypt in the Sierra of southern Peru. Ann Assoc Am Geogr 59(2):294–307

Dourojeanni M (1987) Gran geografía del Perú: naturaleza y hombre. Editorial Mantes, Lima

Escobar A (1988) Power and visibility: development and the invention of management in the third world. Cult Anthropol 3:428–443

Fiorivanti E (1976) Latifundio y sindicalismo agario en el Perú. Instituto de Estudios Peruanos, Lima

Flores Ochoa JA (2011) Tradición oral y etnohistoria: el "descubrimiento de Machupiqchu". Revista Vanguardia en los Andes 2:9–24

Foster GM (1965) Peasant society and the image of the limited good. Am Anthropol 67:293–315

Gade DW (1966) Achira, the edible canna, its cultivation and use in the Peruvian Andes. Econ Bot 20(4):407–415

Gade DW (1967) The guinea pig in Andean folk culture. Geogr Rev 57(2):213–224

Gade DW (1969) Vanishing crops of traditional agriculture: the case of Tarwi (*Lupinus mutabilis*) in the Andes. Proc Assoc Am Geogr 1:47–51

Gade DW (1970a) Ethnobotany of Cañihua (*Chenoodium pallidicaule*), rustic seed crop of the Altiplano. Econ Bot 24(1):55–61

Gade DW (1970b) Ecologia del robo agricola en las tierras altas de los Andes centrales. América Indígena 30(1):3–14

Gade DW (1970c) Coping with cosmic terror: the earthquake cult in Cuzco, Peru. Am Benedictine Rev 21(2):218–223

Gade DW (1971) Grist milling with the horizontal waterwheel in the central Andes. Technol Cult 12(1):43–51

Gade DW (1999) Nature and culture in the Andes. University of Wisconsin Press, Madison

Gade DW (2003) Tapir magic in the Andes and its shamanic origins. J Lat Am Lore 21(2):201–220

Gade DW (2011) Tracking gristmills in Portugal retrospection and reflexivity on place, technology and change. J Cult Geogr 28(3):433–458

Gade DW (2012) Cultural geography and the inner dimensions of the quest for knowledge. J Cult Geogr 29(3):337–358

Gade DW (2015) Plants and people in the Andes: doing ethnobiology in the 1960s. J Ethnobiol 35(2)

Garrett DT (2005) Shadows of empire: the Indian nobility of Cusco. Cambridge University Press, New York, pp 1750–1825

Glave LM, Remy MI (1983) Estructura agraria y vida rural de una región andina: Ollantaytambo entre los siglos XVI-XIX. Centro de Estudios Rurales Andinos Bartolomé de las Casas, Cusco

Gow P (1991) Of mixed bloods kinship and history in Peruvian Amazonia. Clarendon, Oxford

Grandidier E (1861) Voyage dans l'Amérique du Sud; Pérou et Bolivie. Michel Lévy Frères, Paris

Jacobsen N (1993) Mirages of transition: the Peruvian Altiplano, 1780–1930. University of California Press, Berkeley

Kaupp R, Fernandez Carrasco O (2010) Vilcabamba desconocida: exploraciones 1997–2003. Editorial Grafica Rivera, Cusco

Lawbon M, Chion M (2012) Rooted cosmopolitanism: spaces of multiplicity in Cusco, Peru. Int J Urban Reg Res 36(3):539–553

Matto de Turner C (1973) Aves sin nido. Ediciones PEISA, Lima

Mayer E (2009) Ugly stories of the Peruvian Agrarian Reform. Duke University Press, Durham

Merier JH, Palacios O (1977) Radiocarbon dating of the last glaciation in Peru. Geology 5:600–604

Middendorf EW (1973) Peru: observaciones y estudios del pais y sus habitantes durante una permanencia de 25 años (1893) 3 vol. Universidad Mayor de San Marcos, Lima

Mörner M (1975) Continuidad y cambio en una provincia del Cuzco: Calca y Lares desde los 1680 hasta los 1790. Historia y Cultura (Lima) 9:79–117

ONERN (1986) Inventario y evaluación de los recursos naturales de la zona alto andina del Perú. Departamento del Cusco. Oficina Nacional de Evaluación de Recursos Naturales, Lima

Oricain P (1906) Compendio breve de discursos varios sobre diferentes materias y noticias geográficas (1790). In: Maurtua V (ed) Juicio de limites entre el Perú y Bolivia. Imprenta de Henrich y Comp, Barcelona pp 319–377

Paponnet-Cantat C (1995) How health care really works: the case of an Andean community in southern Cusco, Peru. Anthropologica 37:123–139

Parsons JJ (1992) Southern blooms: Latin America and the world of flowers. Queen's Q 99(3):542–561

Perú (1966) Sexto censo nacional de población 1961. Dirección Nacional de Estadistica y Censos, Lima

Perú (2005) Censo 2005: X de población y V de vivienda. Lima

Poole D (ed) (1994) Unruly order: violence, power, and cultural identity in the high provinces of Southern Peru. Westview, Boulder

Raimondi A (1874) El Perú 4 vol. Impr. Del Estado, Lima

Raimondi A (1898a) Itinerario de los viajes de Raimondi en el Peru: valle de Lares, Santa Ana y regreso por Mollepata y Limatambo. Bol Soc Geog Lima 8:1–16

Raimondi A (1898b) Itinerario de los viajes de Raimondi en al Peru: Cuzco, Quispicanchis, Lucre, Pisac, etc. hasta Marcapata. Bol Soc Geog Lima 8:241–277

Robledo LM (1899) Conferencia sobre la Hoya del Urubamba. Boletin del Centro Cientifico del Cuzco 3(3):29–60

Robledo LM (1900) La via fluvial del Urubamba. Boletin de la Sociedad Geográfica de Lima 8:417–450

Sabogal Weisse JR (1970) El robo a los Andes. América Indígena 30(4):1019–1037

Sandweiss DH, Wing ES (1997) Ritual rodents: the guinea pigs of Chincha, Peru. J Field Archaeol 24:47–58

Sheffield GF (1991) Peru and the Peace Corps, 1962–1968. Ph.D. dissertation, University of Connecticut

Stanfield-Mazzi M (2007) Shifting ground: elite sponsorship of the cult of Christ of the earthquakes in eighteenth-century Cusco. Hisp Res J 8(5):445–465

Stavig W (1999) The world of Tupac Amaru. University of Nebraska Press, Lincoln

Tamayo Herrera J (1978) Historia social del Cuzco republicano. Lima, Privately printed

Tamayo Herrera J (2000) Historia general del Qosqo: una historia regional desde el periodo litico hasta el ano 2000 3 vol. Municipalidad del Qosqo, Cusco

Urton G (1981) At the crossroads of the earth and the sky: an Andean cosmology. University of Texas Press, Austin

Valdez LM, Valdez E (1997) Reconsidering the archaeological rarity of guinea pig bones in the central Andes. Curr Anthropol 38(5):896–897

van den Berghe P (1992) From the Popocatepl to the Limpopo. In: Berger B (ed) Authors of our own lives: intellectual autobiographies by twenty American sociologists. University of California Press, Berkeley, pp 410–431

Villanueva U, Macera H, Macera P (eds) (1986) Cusco 1689: informes de los párrocos al Obispo Mollinedo: economía y sociedad en el sur andino. Centro de Estudios Rurales Andinos "Bartolome de las Casas", Cusco

Vitebsky P (2012) Repeated return and special friends: from mythic encounter to shared history. In: Howell S (ed) Returns to the field: multitemporal research and contemporary anthropology. Indiana University Press, Bloomington, pp 180–202

Yépez Valdez W (2005) Qhapaq-Nan del Tahuantinsuyo. Dirección Regional de Cultura Cusco, Cusco

Chapter 2
Urubamba Travelers as Generators of Knowledge

Abstract Three kinds of travelers—tourists, sojourners and scholars—have contributed to understanding the Urubamba Valley. Generations of tourists have captured reproducible images and impressions they share with others in a variety of formats. As a category, sojourners have more deeply penetrated the valley than tourists and many have left valuable accounts. Among them are E. de Sartigues, C. Markham, L. Saint-Cricq (aka P. Marcoy), E. Squier, A. Raimondi, J. Samanezy Ocampo, C. Wiener, E. Middendorf, J. Bryce, H. Chalmers Adams and H. Franck. Researchers, the most intellectually focused of Urubamba visitors, have narrower objectives than sojourners have. Prominent among them are archaeologists who have retrieved the valley past and cultural anthropologists who have done the same for the present. Several deceased scientists in different fields receive in this chapter particular discussion: I. Bowman, American geographer; C. Bües, agronomist and mapmaker of German origin; C. Vargas, Peruvian botanist; C. Sauer, American geographer; and F. Chapman, American ornithologist. Observations by travelers with different objectives explain how and why certain kinds of knowledge have accrued, but in all cases they have provided a baseline for the study of change.

2.1 Tourists in the Region

Millions of people have come for a few days to see for themselves the region's well-advertised attractions. The vogue of romantic primitivism that in the nineteenth century sparked a lot of travel continues to be a motive to go to the Urubamba today. By the standards of Europe and North America, the valley's population and livelihoods belong to an earlier age. Tourism in the Urubamba began after the railroad from the coast arrived in the early twentieth century. With the arrival of air transport in the 1950s, organized tour groups of foreign visitors on tight schedules started to come to Cusco in substantial numbers. Young backpackers later constituted another tourist element; they typically stayed longer, had more flexible travel schedules, sought minimalist accommodations to stay within tight budgets and functioned independently. Backpackers became a familiar presence in the late 1960s and early 1970s when the counterculture movement in North America and Western Europe staked out the Andes as a congenial and cheap and interesting place.

In the first decade of the twenty-first century close to a million visitors went to the valley. Machu Picchu, usually considered to be the highlight of a trip to Peru,

D.W. Gade, *Spell of the Urubamba*, DOI 10.1007/978-3-319-20849-7_2

normally has involved a visit of fewer than 4 h. Fewer than 10 % spend a night near the ruins. Those tourists who spend more time in the Urubamba include those seeking adventure sports or folkloric encounters. Whatever their interests, tourists typically tell about their trips in letters, by cellphone pictures, e-mail messages and personal websites. Their reports frequently relate what happened to them on their trip, suggesting that, for most visitors, personal incidents are more memorable than observations of places or people. In most cases, short-term visitors, seen for a few days at a hotel and then gone, have no lasting impact on the local inhabitants.

Accounts of travelers who come for the sights in the Urubamba are numerous. Their reports in tone and perceptiveness, but generally fail to get below surface observations. For example, Ronald Wright, who in 1984 wrote one of the better travel descriptions of Peru published in English, described Marangani as a "town of eucalyptus trees and textile mills where blankets of sheep and alpaca wool are made". But every town in the valley has clumps of eucalyptus trees and the blanket factory (now closed) was actually located in a neighboring town. The reference to the factory takes on real meaning only when placed in the context of just how unusual it was for this kind of entrepreneurship to take root there. Patrick Fermor conveyed a sense of the past in his 1991 travel book, but his human portrayals carried a note of condescension. At Ollantaytambo he recounted that "all round our rickety chairs squaws were squatting by their baskets amid their mounds of humble wares. Blank eyes watched us, snot fossilized in infant nostrils, and stinging dust blew across the stone slabs in whiffs of grapeshot" (Fermor 1991:53).

Another British author, John Ridgeway, oblivious to the dramatic landscape outside his window focused his attention in his 1986 book on the theft of a tourist's backpack while on the train trip from La Convención to Cusco (Ridgeway 1986). Many of these travel writers reported little reliable knowledge about the localities they reached. The American writer Peter Matthiessen made multiple errors of fact in his book about the Urubamba, yet his evocative descriptions are beyond reproach. He described peasant children as "small bundles of dirty colors, like rag dolls in elfin hats" and at the Pisac market, he commented on "…men lined up against an adobe wall, [who] drowned ancient sorrows in the thick, pasty maize beer known as chicha" (Matthiessen 1961). On the rail trip to Machu Picchu Matthiessen waxed poetic: "Long shreds of cloud settled in the ravines, and the overhanging walls climbed higher, until the train hurried along the shadow, like an ant in a deep rock" (Matthiessen 1961). The number of books published by travelers to the Urubamba and the Andes in general has declined since hundreds of web sites posted by people around the world now offer commentaries on one or more aspects of the valley. More up to date than most books, these sites are often helpful to prospective travelers.

2.2 Sojourners in the Urubamba

Another kind of traveler to the Valley is the one who penetrates more deeply into learning about it. Sojourners move about, but, unlike a tourist, they may also settle for a time. Not all sojourners publish their experiences. One who did was Ethan

Hubbard, a young Vermont man who spent several months in Ollantaytambo in the late 1980s, primarily to get to know its people. His lucid text, tone of respect for the inhabitants, and fine photographs captured the life of the Peruvian peasant (Hubbard 1990). It was early sojourners, those of the nineteenth and early twentieth centuries, who not infrequently left a precious record of great historical interest. With the passage of time, their commentaries and observations acquire value about just how much a place or a culture has changed. Some sojourners make important discoveries.

2.2.1 Eugène de Sartigues a.k.a. E.S. de Lavandais (1809–1892)

One such sojourner was Eugène de Sartigues of the minor French nobility born in Gannat, France (de Sartigues 1851). He is also known under his pseudonym of E. S. de Lavandais. While he was Secretary at the French Embassy in Rio de Janeiro, he traveled to Peru in 1834–1835. After several months on the coast, Sartigues came to the Urubamba Valley by way of Mollepata on a path that led between the two majestic peaks of Salcantay and Soray. The cliffside trail was so narrow muleteers claimed priority by calling out to warn oncoming traffic, in that way avoiding collisions that would have hurtled people or animals into the gorge below. In the annals of exploration, Sartigues is notable mostly for having reached the Inca site of Choqquequirao and written about it. At that time the ruins were on land owned by Hacienda Huadquiña. That estate produced not only wool, hides, horsehair, potatoes, wheat, sugar, coffee, and coca, but also lead, silver and gold from mines on the property. From Huadquiña the Count walked 9 h down the valley to Santa Ana (Quillabamba). From there he went to Cocabambilla where, for the first time, he saw “savages.” Sartigues’s diplomatic career included posting between 1850 and 1860 as Minister Plenipotentiary of France in the United States.

2.2.2 Clements R. Markham (1830–1916)

Son of a cleric near York, England, Clements Markham is usually described as a geographer though his interests were broad. A cadet in the Royal Navy, the work of William Prescott stimulated his imagination and in 1852 he came to Peru setting out “to search the interior for remains” (Markham 1991:ix). In 1853, Clements Markham traveled primarily in the Urubamba zone known as the Sacred Valley, but which he called the “Vale of Vilcamayu.” He described it as having a delicious climate, picturesque farms with “maize towers,” (temporary constructions once used to remove kernels from the cob) and “little forests”—actually orchards—of fruit trees. He took up residence in Urubamba to study the Quechua language and the few writings in that language that existed. On one excursion he went up the lateral valley of the Chicón River through a snowy pass near a glacial lake and saw the puna where alpacas were being raised. Markham traveled down valley to Ollantaytambo and its "eminently picturesque scenery" where he saw a fortress he considered to be the most astonishing monument of antiquity in Peru. His interest in and knowledge of Quechua enabled him to have contact with the common folk. He mentioned that, in Sicuani, he was "mixing much with the people, visiting and sleeping in their huts" (Markham 1991). Returning to England, he served as a geographer in the India Office of the British government. In that capacity he returned to Peru in 1860–1861 to take charge of the seed collection of Peruvian species of *Cinchona* and then transported this material to India for propagation as an anti-malarial. Markham published three books on Peru and was considered to be the nineteenth-century Englishman most knowledgeable about Peru. He dominated the Royal Geographical Society from 1863 to around 1914, serving as secretary for 25 years and president for 12 subsequent years. In 1921 the Peruvian government erected a bust of his likeness in London. The Anglo-Peruvian school in Lima, Colegio Markham, was named after Sir Clements.

2.2.3 Laurent Saint-Cricq, a.k.a. Paul Marcoy (1815–1888)

Born in Bordeaux, as a teenager Laurent Saint-Cricq went to the Antilles for 3 years and on his return to France started to write and paint. Between 1840 and 1846 he traveled to Peru and worked his way down the Urubamba Valley from its source. In Peru he adopted the pen name of Paul Marcoy. Out of that journey came his witty book, published in French in 1869, some content of which had previously been published in journals starting in 1853. Since he made several return trips to Peru between 1847 and 1874, it is not clear whether some of his observations were made on later trips. His account is useful for understanding Peru of the mid-nineteenth century (Marcoy 1869, 1874; Poole 2008). Like most travelers at that time, he entered through La Raya Pass and noted a rustic inn (*tambo*) at Aguas Calientes before he reached Sicuani, then a town of about 3000 people. He described San Pablo and San Pedro de Cacha as miserable twin towns comprised of wretched huts. Saint-Cricq made observations of the nearby Inca temple of Viracocha. At Combapata, a big chicha-making town, guinea pigs ran over his face when he was trying to get to sleep. At Checacupe, he left the valley and ended up in Cusco via a round-about trail, partly through the puna.

After several weeks in Cusco, he reentered the Urubamba Valley at Calca and from there proceeded to Yucay and Urubamba. At Ollantaytambo he moved up the trail to a pass on the puna and from there down into the tropical Occobamba Valley and eventually into the main Urubamba depression. He reached Echarati, known then especially for its cacao, and went further down river into the territory of the Matsigenka. His itinerary from the Urubamba into the Ucayali was chosen as being the "shortest route to Paris from Cusco." Though Saint-Cricq was condescending toward native people, he nevertheless was able to relate to people of all social classes. The engravings of Peruvians in his book are so grotesque that one reviewer was moved to remark that they "…are without comparison the ugliest specimens of humanity we have ever seen" (Cook et al. 1869). Saint-Cricq acquired an interest in plants on his South American journeys and perhaps for that reason he became the director of the botanical garden in Bordeaux between 1876 and 1887 (Chaumeil 1994:273).

2.2.4 *Ephraim George Squier (1821–1888)*

The classic example of an astute and focused sojourner was Ephraim Squier, born in Bethelehem, New York. He taught school, studied engineering and became seriously interested in antiquities. Together with Edwin H. Davis, Squier wrote a book on the prehistoric mound builders of North America. Archaeology became his passion and he pursued it in Central America before he came to the Andes in 1863 to settle claims between the United States and Peru relating to mining and guano exports. He then undertook archaeological explorations at his own expense until he returned home in 1865. He deciphered with clarity and exactitude the tangible remnants of Inca civilization (Squier 1877). Though he assigned no dates to any of the artifacts he studied, he advanced the notion that the constructions he observed had developed in stages over a long time period. He hinted at a pre-Inca occupation.

In the Urubamba, he studied Raq'che in Canchis and the ruins at Pisac, Yucay and Ollantaytambo. He made his trips to those places from Cusco where he set himself up for several months. Squier was the first scholarly traveler to carry a camera in Peru (McElroy 1986). One of his daguerreotypes captured the charm of Ollantaytambo better than any image before or since (Fig. 2.1).

In the field he wore boots, a broad brimmed hat, scarf and poncho and carried a long knife attached to his belt (Squier 1877). Before coming to Cusco, Squier spent 3 weeks exploring the eight islands of Lake Titicaca with Antonio Raimondi whom he described as "a gentleman of high scientific attainments" (Barnhart 2005:254). Squier also recorded observations of landforms and peasant life. Mariana Mould de Pease analyzed his work in terms of the image he presented of Peru (Mould de Pease 1986).

2.2.5 *Antonio Raimondi (1826–1890)*

Antonio Raimondi was Peru's most exceptional explorer (Fig. 2.2). His interests were wide-ranging, he traveled to many out-of-the-way places, and he had the self-discipline to write down his observations and then to use them in articles, books and maps (Raimondi 1874; Yacher 1995). Given the difficulties of travel at the time, Raimondi's explorations were extraordinary achievements.

From Milan, Italy, his birthplace, Raimondi came to Peru in 1850 at the age of 24. His broad education prepared him to notice plants, animals, minerals, rocks, and fossils, as well as geography, hydrography, archaeology, anthropology, diseases,

Fig. 2.1 E. George Squier (1821–1888) in Highland Peru. He made careful descriptions of Inca structures

Fig. 2.2 Antonio Raimondi (1826–1890) came to the Urubamba in 1865. Born in Italy, Lima was his base of operations. Raimondi was Peru's most assiduous and polymathic explorer

and history. In his trip logs, Raimondi always indicated the distance and direction from his stated point of origin, which allowed the reader to follow his routes even without maps. The slow-motion forms of walking and horseback travel permitted him to make a vast array of observations that would not have been possible from a moving vehicle. Understandably, he often commented on bridges and the state of the mule trails. Accounts of human encounters are rare and even hacendados received surprisingly little mention, considering how crucial their hospitality was in his wanderings. When read today, the minutiae of his travel often give insights into the kind and degree of change that has occurred since he was there. Between excursions, he managed to carry on a family life in Lima and to work as a university teacher of chemistry. His 1877 map of Peru was a major contribution to the cartography of his adopted country. Beginning in 1883, he also published 34 sheets of his map of Peru at a scale of 1:500,000.

In 1865 Raimondi traveled through the Urubamba, starting in Puno and moving through La Raya Pass and from there to Cusco (Raimondi 1898b). He often commented on unseemly aspects of what he saw. At Tinta, Raimondi was appalled to see pigs wallowing in mud holes in the streets and patios. He noted there *Opuntia* species planted on top of adobe walls to deter anyone from climbing over, but then stated that its main value was the red dye contained in its seeds. He mentioned maize chicha as being available throughout the upper Urubamba. His notes included such diverse information as the architecture of the Quiquijana bridge and the presence of invasive plants near Urcos. From Cusco he went to Calca, down to Lares and the Yanatile River, and then worked his way up the "valle de Santa Ana" to Huadquiña (Raimondi 1898a). He attributed the dry atmosphere found there to its lack of dense forests, perhaps unaware that the period from May to September receives little rainfall in that part of the valley and that this moisture seasonality resulted in a dry forest. Raimondi recorded that Hacienda Huadquiña produced wool, hides, horsehair, potatoes, wood, maize, sugar, coffee, cacao and coca. Like Sartigues, Raimondi mentioned the practice of *arrendatarios* being required to work for the estate from which they rented agricultural plots. He said not one word about the hacendado.

2.2.6 *José Benigno Samanez y Ocampo (1838–1887)*

A totally different kind of traveler was José Samanez, a Peruvian born in Andahuaylas who, together with several companions, explored the little known north-flowing rivers of Amazon tributaries (Samanez y Ocampo 1885). In 1884, the Samanez team set out on the Apurimac moving from its hazardous mid-reaches to calmer waters below, which eventually flows into the Eni River and, in turn, into the Tambo River. Where the Tambo meets the Urubamba River, the two streams become the mighty Ucayali. Samanez and his companions moved up the Urubamba by canoe against a strong current. Quite aware that they lacked the skills required to negotiate the rapids and other "mal pasos," the expedition depended on Matsigenka and Piro boatmen to paddle them in several stages to the head of Urubamaba navigation at Cocabambilla. Such an upriver trip, called *la surcada*, is impossible from December to April when large volumes of water make upstream paddling too difficult.

Two years after he published a report of his Urubamba trip, Samaenz died of yellow fever in Iquitos. His work remains one of the best nineteenth-century accounts of river travel on the Urubamba, and includes many short though valuable observations of settlements, Indians, locations of tributaries and human activities. His son, David Samanez y Ocampo, became interim president of Peru in 1930–1931.

2.2.7 Charles Wiener (1851–1913)

Born in the Vienna of the Austrian-Hungarian Empire, Charles Wiener was a professor of German at Lycée Condorcet in Paris. He came to Peru near the end of 1876 with the aim of traveling, but also to collect artifacts and antiquities. To Paris he sent from Peru 86 crates, eight of which contained pre-Columbian pots. His interest was in objects he could take back to Europe, which explains why his book included little about people, landscapes or economic activities of the places where he traveled (Wiener 2010). From Cusco he went to Urubamba and from there into the tropical zone by way of the Málaga Pass. From Hacienda Santa Ana (Quillabamba) Wiener went into Matsigenka country. Near the Pongo de Mainique he came across two dead Matsigenka males, whom he proceeded to decapitate so that he could add those heads to his ethnographic collection. In August 1877, he returned to Paris where he was lionized as an intrepid explorer. An exposition there of 4000 objects he brought back from his trip attracted many people. His experience in Peru opened a diplomatic career for him in various Latin American countries.

2.2.8 E.W. Middendorf (1830–1908)

Born in Thuringia, Middendorf is considered to be one of the precursors of scientific archaeology in Peru. He came to Peru from Germany and stayed 25 years. His wide scholarly interests included the Quechua language, pre-Columbian ruins, culture and landscapes. He was an astute observer and his accurate observations are much valued today. His three-volume work on Peru contained photography and was his most unspecialized publication (Middendorf 1893–1895). Almost a century passed before his three-volume work was translated from German to Spanish (Middendorf 1973). In 1887 Middendorf traveled by horseback through the upper part of the Urubamba Valley. He found no evidence of the Inca temple at La Raya Pass, but did comment that the remains of a tambo at that site had walls that looked Inca. Farther down valley, he spent the night at Aguas Calientes, which was the site of another functioning tambo. He made agricultural observations, noting, for example, that wheat was the main crop at Marangani, which remained true in 1963. At Sicuani, he found that the church had a sheet metal roof; today the entire city of Sicuani is largely roofed in that material. In most small places in the Andes, dwellings were thatched much longer than were public buildings. Middendorf visited the Temple of Viracocha and described it in detail. At Tinta he had to deal with drunken officials, a common occurrence in the Sierra of the nineteenth century. Middendorf had much to say about haciendas where he usually spent the night; perhaps out of gratitude to his hosts, he avoided unflattering comments. He traveled in the Urubamba only as far north as Ollantaytambo. Lack of enough time prevented a trip to Santa Ana (Quillabamba); in that era of the mule, a trip there and back required 10 days.

2.2.9 James Bryce (1838–1922)

James Bryce, born in Belfast into the British elite, was a respected jurist, historian, politician and diplomat. He arrived in the Urubamba Valley while on a trip through seven South American countries (Bryce 1913). From 1907 to 1912, Bryce had been the British ambassador the United States and his trip to South America came after the end of that diplomatic assignment. In 1913, he met Hiram Bingham, the self-proclaimed discoverer of Machu Picchu, at a dinner at the National Geographic Society. That encounter may have encouraged Bryce to make the trip to Peru. Bingham had named "Bryce Glacier" in the Cordillera de Vilcabamba in his honor (Bingham 1913:550).

At the age of 74 and as befitted his status as an establishment figure, he assumed no risks in his travels. His nationality gave him free transportation on the recently completed railroad of the Peruvian Corporation constructed by British engineers with British capital. Starting on the coast at Mollendo, Bryce went to Arequipa and on to the high plateau at Puno on Lake Titicaca. From there the train moved north through the Altiplano, over the La Raya Pass and down into the Urubamba corridor. Bryce offered lucid and intelligent observations based on background knowledge acquired from carefully reading the works of Clements Markham (1856). Like many Britishers, he was partial to the works of his own countrymen. His comments about the Urubamba started at La Raya Pass where he noted that a small gorge filled with melt water from the glacier was the river's ("Vilcamayu") source. His reading of Markham prompted Bryce to look for a wall that Markham asserted that the Incas had constructed at the pass as a defense against the Colla people. Finding no evidence of it in the landscape, he reasoned that peasants and herders had carried away ashlars from the wall to build stone huts. In fact, railroad construction through the pass about two decades earlier may have been responsible for the removal of the stones. That building, as is now better understood, was a religious shrine, not a fortress.

Bryce also commented on the mineral springs, then as now known as Aguas Calientes, 10 km north of the Pass, where "station masters cured themselves of rheumatism" in a wooden hut. No such hut stood there in 1963; by then, Peruvians had replaced all the British stationmasters. However, the train continued to stop for a few minutes so that passengers could leave their coaches to sample the water. Today Aguas Calientes (not to be confused with the town below the Machu Picchu ruins) has several small buildings, a pool, an artisan market and a miniature Inca-style rope bridge stretched over a small brook. As Bryce's train moved down the valley toward Marangani, he noted the warming of the air and the beginning of grain and potato cultivation. The window of a moving train allowed him to observe that Indians farmed the steep slopes, whereas non-Indians had taken possession of the valley floor. Bryce's earlier life as a leading liberal in British politics put that socio-geographical commentary in perspective. In Sicuani, the train made a long halt, allowing passengers to visit the big Sunday market then in full swing. Bryce got his first up-close look at the physical characteristics of native people who had come to that market to buy, sell, and trade. Like so many travelers, he could not resist commenting on the great variety of hats. He failed, however, to mention that Andean headgear was a legacy of the colonial period. Each community had its own hat style, a holdover from a similar Inca practice whereby people of different origins wore distinctive hats.

Below Sicuani, Bryce noted the presence of eucalyptus, and mused that this introduced plant had "overspread the world." Bryce compared warmth, color, and variety of form of the snow-capped peaks of the Vilcanota Range to those of the Italian Alps. His observations on the Urubamba Valley ended at Urcos, where the train headed up the Huatanay tributary to the city of Cusco. The spirit was willing to depart from the rail track, but the body was not. He wrote "Gladly would I have followed it down the valley into scenery even more beautiful than that of the upper levels, where luxuriant forests along the stream contrast with the snowy summits of the Eastern Cordillera towering above. But from this point [Urcos], there were only mule paths and travel is so slow that a week would have been needed to reach the finest part of the scenery. Renunciation is the hardest part of travelling" (Bryce 1913). Bryce may have been thinking about his earlier life when, as a mountain climber, he ascended Mount Ararat in 1876.

Later in the chapter, Bryce alluded to the fact that he had heard about "striking ruins, not far off, such as those at Ollantaytambo and Pisac and lower down the Vilcamayu Valley at Machu Picchu and Rosas Pata…." (Bryce 1913). This remark had come from information Hiram Bingham provided Bryce at the National Geographic Society dinner mentioned earlier. When Bryce arrived in Cusco he made a fine description of Sacsahuayman fortress. Although in Peru Bryce functioned in the conventional tourist mode, he recorded thoughtful observations, showing his understanding of the changes that had occurred.

2.2.10 *Harriet Chalmers Adams (1875–1937)*

Another early twentieth century sojourner to the valley was Harriet Chalmers Adams, a professional explorer, writer and photographer from the United States. Born in Stockton, California, she was educated by private tutors. In 1899 she

Fig. 2.3 Born in California, world traveler Harriet Chalmers Adams (1875–1937) came to the Urubamba in 1904 when modern technology was making its first appearance

married Franklin Pierce Adams and in 1904 she and her husband took a trip around South America (Fig. 2.3). In that year the couple came from the coast on the train to Sicuani, then the rail terminus, and from there traveled on horseback to Cusco. After exploring the Inca capital, Adams went to the Sacred Valley (which she called the Yucay Valley) and described its landscape and economy. She noted the practice, now gone, of bringing tropical produce by mule train from the lower valley (Adams 1908). Though technically not very good, the photographic images in her article are especially valuable in conveying how traditional the cultural landscapes and livelihoods were early in the twentieth century. Harriet Chalmers Adams wrote 28 articles for the National Geographic in the period 1907–1935. In spite of her accomplishments, the National Geographic Society did not allow a woman to become a full member. Thus, in 1925, she turned her attention to founding the Society of Woman Geographers of which she was the first president. Adams traveled widely and was a keen observer. A biographer noted that "although she enjoyed widespread fame during her lifetime, she is virtually unknown in the history of geography" (Davis 2009). Such is often the fate of those who popularize their geographical findings.

2.2.11 Harry A. Franck (1881–1962)

Born in Munger, Michigan, Harry Franck's father, an immigrant from Mecklenberg, had joined the middle class. While still a university student in Michigan, Harry took a steamer to Europe, an experience that set him on his way to becoming an intrepid world peregrinator. Shortly thereafter, Franck made a trip around the world about

Fig. 2.4 Harry Franck (1881–1962) walked part of the way through the Andes from Bogotá, Colombia to Buenos Aires. In Peru he acquired the donkey ("Chusquito"), which carried some of his belongings through much of the country. Born in Michigan, Franck traveled to many places in the world. (From Franck 1917:facing p. 229)

which he wrote a book. In 1915–1916, he traveled extensively in South America; his trip down the spine of the Andes from Bogota to Buenos Aires took more than a year (Driever 2011). Thirty-five years old at the time, he walked part of the way and slept often in hovels (Fig. 2.4). The book *Vagabonding Down the Andes* that resulted from his journey captured many Andean traits. Franck made note of the constant yet limp hand shaking habit, frequent use of diminutives in speech, and the poor fit of doors in their frames. To anyone who has spent much time in these highlands, Franck's account still resonates for its authenticity about the place and the people. The book remains an exemplary portrayal for understanding what has changed about the region and what has not.

Moral judgments pepper every page, which, when read today, reflect much on Franck's Midwestern small town values. A sardonic turn of phrase sometimes redeemed the judgmental tone. When told that a local official was at home sleeping off a hangover, Franck commented "far be it from me to blame any man for whiling away an Andean existence in the only available fashion" (Franck 1917). He over generalized when every person he encountered was classified as a type. Indians were feckless and mestizos duplicitous by his definition. Franck, who also had made up his mind about the low quality of the Peruvian clergy, stereotyped the village priest as a less than admirable character. Yet his book contained many astute remarks. Insightful observations about Quechua reflected his early stint as a language teacher, war-time interpreter and translator. Only someone who had penetrated the structure of Quechua would have been able to write that it was a "real language with a complete grammar and all the flexibility and shade of expression of our classical tongues" (Franck 1917:437).

Franck found on his visit to Cusco that most residents were monolingual Quechua speakers. Quechua was also used by the *gente decente* of Hispanic origin who had learned it from their indigenous nursemaids. They used Quechua to keep lower castes in their place. Aspects of that same pattern held true in 1963. In 1916, most cusqueños of full or partial Hispanic heritage knew Quechua and spoke it to Indians and to those Indians hispanicized enough to fall into the cholo category. In the food market cusqueños spoke Quechua, whereas among members of their own class they spoke Spanish. By 1963, the lower stratum in the city also spoke Spanish, although they could not always read it. In addition, the city also had a floating population of country people, many of whom did not speak Spanish. Franck described three kinds of Quechua transporters when he was there: mule drivers (*arrieros*), llama drivers (*llameros*) and *quepiris*, or coca carriers. At that time, Indians still carried coca on their backs from the hot valleys, a survival from the colonial period. About 800 quepiris arrived in Cusco every week from the hot valleys, where they loaded the leaf on their back and brought it on a long trip to Cusco for redistribution elsewhere.

On the Cusco segment of his trip, "Harry the Intrepid" had as his major aim to visit Machu Picchu, a trip which then required "time and disregard for roughing it." He had originally planned to travel there on foot, but then decided on an expedition with two unnamed traveling companions, one an American professor and the other a Peruvian of Italian background who had been in the United States. It seems probable that the professor was Albert Giesecke, then the rector of the University of Cusco, who had first gone to Machu Picchu in 1912. In addition, the government selected a soldier-servant—indeed the same Sargento Carrasco who had accompanied Hiram Bingham to the ruins in 1911—to accompany them. Some nine hours after leaving Cusco, they got to the rim of the Urubamba Valley. Franck described the scene as follows:

> At three we came suddenly to a vast split in the earth, into which we began to go down and ever down by acute zigzags and stony *cuestas* that grew so steep we had to dismount and lead our animals. Before and below us spread the magnificent canyon of the Urubamba, that river of many rivers, which, rising near Titicaca, at length adds its bit to the giant Amazon. Spring plowing was in progression on the valley floor, walled by mountains as far as the eye could reach in either direction. Over this rampart, the sun still peered when we reached the level of the river at last and, picking up the road from up the valley, jogged down along it. (Franck 1917:458).

After spending the night in Ollantaytambo, they proceeded down the valley on the same mule trail that Hiram Bingham had taken in 1911. Franck wrote:

> As we descended still deeper into the vastness of the Andes, the solid granite precipices, rising sheer thousands of feet from the foaming rapids to the clouds, remained at the same height; but the valley of the river continued to descend, and gave us the curious effect of seeming to see the mountains that shut us in rise every higher into the sky. The cañon of the Urubamba had shrunk to a resounding gorge of sharp V-shape, with virtually no room left for cultivation, so that even the hardy *andenes* of the ancients were crowded out of existence, and only the imperious river forced its way through the mountains, permitting the narrow road to follow the precarious footholds blasted for it along one of the towering granite walls (Franck 1917:462).

The four men bivouacked at Mandorpampa, the same place where Bingham and his crew members had camped, and even encountered the same Sr. Arteaga—though Franck did not name him—who had led Bingham to the ruins 6 years before. Just as Bingham had paid Arteaga a coin to show him where to go, the latter apparently had the same expectation of those who came later. Arteaga offered to guide Franck and his companions there in return for a "consideration." Bingham's 1914–1915 expedition laborers had cleared most of the ruins of the tangle of jungle vegetation only 2 years before, greatly easing Frank's exploration of the site. Still, Franck lamented that he did not see the ruins in the same mysteriously overgrown state that Bingham had first experienced in 1911.

Franck returned to Cusco and from there took the train southward through the upper valley. Sparse observations on that trip segment suggested he either fell asleep or became disengaged through the numbing effect of locomotion. A train encases the passenger in a cocoon that greatly affects his perception of place. Any observation one makes is so fleeting since no possibility exists of checking anything that catches one's attention. In a car, one can stop and get out; in a train, one is a prisoner of the iron horse relentlessly plunging ahead. How different were Franck's numerous and incisive observations when trekking on the dusty trail from Ayacucho to Cusco with his donkey in tow. Walking focuses the mind and the people one meets along the way are really encountered. Close up, facial expressions and demeanor can be studied and exchanges can occur, whereas from a moving vehicle, the people seen outside are a blur. Moreover, traveling by foot, one cannot fall asleep. What is seen on foot or horseback becomes worthy of comment; what is seen from a train window is simply an element in a passing, shifting tableau.

The best parts of Harry Franck's *Vagabonding Down the Andes* provide crafted descriptions of place and people. In contrast to James Bryce, the conventional British traveler, Franck represented the non-elitist, outspoken American who got deeper into rural Andean culture than perhaps any other early twentieth century foreign traveler. Use of the word 'vagabond' was meant to convey his willingness to stay in the most minimalist accommodation in order to experience Andean life. It did not mean that he solicited people for money. Franck's evocative descriptions remain valuable as a baseline for pondering how the conditions of Andean land and life have changed or not changed as the case may be. Franck's book contained truths about the conditions of these highlands in the early twentieth century.

2.3 Researchers in the Valley

Modern scholars and scientists usually do not think of themselves as travelers, but displacement for fieldwork nevertheless places them in that category. Researchers differ from other travelers by having a specific objective with a preplanned topical agenda. Unlike sojourners, researchers see themselves professionally as investigators of particular phenomena. Their narrowly defined focus removes from their consciousness a lot of compelling phenomena right under their noses.

Over the last seven decades, the Urubamba has attracted hundreds of specialists in the natural and social sciences. Their publications usually do not make clear why they chose this valley in which to make these studies. Some researchers left no publications to attest to their presence. Cusqueño scholars and scientists have considered the Urubamba Valley to be their backyard, as it were. Research by cusquenos in their own regional surroundings began near the end of the nineteenth century. The Cusco elite, many of whom were hacendados, founded the Centro Cientifico del Cusco. In the decade that the journal, *Boletin del Centro Cientifico del Cusco*, existed, it published articles on alcoholism, Indians and depopulation of Cusco (Rénique 1980).

The arrival of Albert A. Giesecke (1878–1968) in 1910 as rector of the University Nacional of Cusco had a profound impact on turning students and faculty toward investigating their own pre-history, history, geography, culture and economy of their own region (Gade 2006). This American in his early 30s accomplished his objective by frequent field trips and, in one instance, conducting in 1912 a census of all the people in the city of Cusco, using students as census takers. In the three categories constructed, 10,813 were classified as mestizos, 4433 as whites, and 4362 as Indians (Giesecke 1912). Although Giesecke's Ph.D. was in economics, he also had serious interests in demography, geography, archaeology and anthropology. Much that was of great scholarly interest in Cusco Department had never been studied before. Giesecke communicated this enthusiasm for researching the local and regional to his students, many of whom came from towns in the Department.

2.3.1 Culture Historians and Others

The pre-Columbian past has drawn generations of archaeologists to the valley, mainly to study the remains of Inca civilization. Not until much more recently did the Wari civilization that predated the Incas by hundreds of years receive attention. Among the first foreign scholars to come to the valley to delve into its still then mysterious past was Hiram Bingham, an American academic interested in exploration. Though not an archaeologist, Bingham made a spectacular find in 1911 thanks to prompting from Albert Giesecke. Over the years many foreign archaeologists have excavated in the Urubamba Valley. No one made more of an impact than John H. Rowe (1918–2004), a native New Englander who first came to Peru as a Harvard graduate student in 1944 to start Cusco's first archaeology program. That appointment fulfilled the decades-long vision of Albert A. Giesecke, by then in Lima, of having a trained archaeologist come and teach students in Cusco about how to interpret the pre-Columbian past of their region. Although Rowe himself did not undertake a major excavation in the valley itself, he built a foundation of knowledge that included many observations made in the Urubamba. Rowe also contributed much to ethnohistory when he became the first scholar to totally vet the chronicles of the early colonial period about the Inca. His long chapter on the Incas in the *Handbook of South American Indians* published in 1950 still has considerable relevance (Rowe 1946). Rowe, who laid out "the basic structure of Andean archaeology's explanatory framework," became the world's most respected Andeanist scholar (Tantaleán 2014:86).

Other archaeologists have contributed significant work on the Urubamba pre-Columbian past. Jean Protzen of the University of California, Berkeley, made a major contribution by his work on the Inca construction of Ollantaytambo (Protzen 1993). Ian Farrington from Australia and Alan Covey from the United States have combined excavation and ethnohistory in their reconstructions of the Sacred Valley past. Susan Niles has worked on retrieving knowledge about the buildings of the Inca kings. Ann Kendall, a British archaeologist, spent a major period in her career reconstructing the pre-Hispanic presence in the Cusichaca Valley below Ollantaytambo. American archaeologist Gordon McEwan has devoted years reconstructing the Wari occupation of Pikillacta, long a puzzling archaeological site.

The Urubamba has also attracted scholars of folk life and peasant culture, a specialty that mostly emerged after World War II. For decades non-Peruvians dominated research in the Andes prompting two Peruvian anthropologists to note that Peru has been a country where foreigners were in charge of exploration, while the children of the Creole elite tended to act as foreigners in a strange country (de Gregori and Sandoval 2007). Benjamin Orlove studied the commercialization of animal fibers in Southern Highland Peru and Inge Bolin researched the community of Chillihuani above Cusipata in Quispicanchi. French ethnologist and ethnohistorian Antoinette Molinié-Fiorivanti worked at Yucay, and the American Michael Sallnow (1939–1990) prepared a wide-ranging ethnographic study on manifestations of syncretic religiosity partly in the District of San Salvador. Wesley Craig, in the 1960s a North American graduate student, was one of the first to study land reform in La Convención. Jorge Flores Ochoa, who grew up in the Urubamba, has provided anthropological perspectives on many aspects of the Sierra. Two historical studies stand out: one is on Ollantaytambo by the Peruvian scholars Luis Glave and Isabel Remy, and the other on Chanchis and Quispicanchi by the late U.S. historian Ward Stavig.

2.3.2 *Isaiah Bowman (1878–1950)*

Few geographers have done research on the Urubamba Valley, but the best known one was also the first. Isaiah Bowman came in his 30s to the valley as a member of the Yale Peruvian Expedition of 1911 led by Hiram Bingham. On the lookout for

Fig. 2.5 Andeanist geographer Isaiah Bowman (1878–1950) stands to the left of the taller Hiram Bingham. Bowman, a Yale professor, later became director of the American Geographical Society, president of the Johns Hopkins University, and an advisor to Woodrow Wilson. (*Source*: Bingham 1922)

underwriters of his exploratory venture, Bingham chose Bowman for the scientific credibility he represented in the line-up of participants. Bowman, the first real Latin Americanist in North American geography, had written a field-based doctoral dissertation on the Andes of Bolivia and Chile. A professor at Yale, Bowman's course on the geography of Latin America was the first such offering in the United States. Bowman had been to Cusco 2 years before Bingham's first trip there and his knowledge of the Andes exceeded that of other members. The touchy relationship between Bowman and Bingham, kept barely beneath the surface, explained why Bingham schemed to exclude Bowman from accompanying him to "discover" Machu Picchu (Fig. 2.5) To ensure that, Bingham sent him on ahead to make surveys farther down the valley. Unbeknownst to Bowman, Bingham had knowledge that would enable him to cast himself as the Machu Picchu discoverer. Bowman, quite content at the time to go his own way, wanted to learn about the Urubamba jungle. Bowman (1912:884) wrote "We wanted to know what secrets might there be gathered, what people dwelt along its banks, and if the various tales of ruined cities at the top of inaccessible cliffs had any basis in fact". The remark about ruined cities put Bowman in obvious competition with expedition leader Bingham. However, no such archaeological site was found in that lower zone.

Bowman's publications from his 1911 trip started with a potboiler on the "cañon of the Urubamba" (Bowman 1912). A poorly crafted hodgepodge of impressions, the real objective of the article may have been to beat Bingham in print. When Bingham found the article, he reprimanded Bowman for not following his protocol. The contract signed by expedition members required all manuscripts to pass through Bingham before being submitted for publication. Bowman knew enough not to make any allusion in that article to Machu Picchu, which Bingham jealously guarded as his own for publications. The reference in Bowman's title to the "cañon" refers not to the gorge cut through the batholith around Machu Picchu, but rather the Pongo de Mainique, much farther downriver. Bowman discussed navigating the 25 km of unbroken rapids through the 800-m deep gorge of the Pongo. He made an original topographic map of that part of the lower Urubamba (Bowman 1913, 1914). An uncommon usage, "cañon of the Urubamba" has not been used before or since to describe that area (Bingham 1913:419). Bowman also decried the Matsigenka Indians, slave raiders, and rubber traders who were the main occupants when he was there. Bowman couched his article in terms of adventure and courage, purposely upstaging Bingham who promoted himself as the dashing professor-explorer.

Bowman reported on his 1911 fieldwork in the Lower Urubamba in his 1916 book, *The Andes of Southern Peru* (Bowman 1916). Other parts of this book covered studies he made at high elevations. Theodore Roosevelt (1917) reviewed it as "…. one of those uncommon books in which a man who has had the vision to undertake adventure and the hardihood to carry it through sets forth with wisdom what he has seen". Bowman's book inspired Roosevelt, accompanied by his son Kermit, to undertake in 1916 a hazard-filled exploration of a remote part of the Brazilian Amazon. For geographers, Bowman's treatise gained credibility as the essence of field-acquired knowledge. Its audience expanded to Latin America after Carlos Nicholson, a Peruvian geographer, translated the book into Spanish (Bowman 1938). Parts of Bowman's Andes book, especially about the W.M. Davis erosion cycle and certain assertions about environmental control, opened the author to criticism.

The original maps in Bowman's 1916 monograph highlighted the paucity of published topographic maps for the Andes. Reflecting on that deficiency, Bowman conceived of a topographic map series not just of the Andes but of all Latin America. After he became the director of the American Geographical Society in 1915, he crystallized a grand cartographic project, the Million Map of Hispanic America, consisting of 107 sheets at a scale of 1:1,000,000. Completed in 1946, this series of maps contributed much to the geographical knowledge of Latin America. By then, however, Bowman's public service career had overshadowed his South American work. President of The Johns Hopkins University since 1935, he played an important role in the Versailles Treaty Conference, later advised the U.S. State Department on a variety of matters and became a member of several important government commissions on international affairs. Bowman had made his Andean investigations before he was 40, characteristic of most researchers' trajectories in this mountain realm where difficult conditions act as a filter to select those who go there.

2.3.3 *Christian Rudolf August Bües Meislahn (1874–1948)*

A remarkable investigator but hardly known outside Cusco Department was Christian Bües (Cruz Ccorimaya 2009; Bües 1933a, b, 1939). Born in Achiz, Germany near Hannover, Bües studied at the Real Gymnasium in Bremen. He came to the United States as a young man in 1893 and in 1906 received the B.S. degree in agriculture from Cornell University. Liberty Hyde Bailey (1858–1954), a giant of twentieth century American plant science, was one of his professors. After graduation Bües traveled to Peru to become the *mayordomo* of an estate in Junin where he stayed until 1909. In 1910, he worked as the administrator of a company exploiting jungle latex. When the rubber boom ended, Bües moved to the Department of Puno for 3 years where he administered a cattle hacienda. In 1915 he settled in La Convención, patching together a living as a consulting agronomist to neighboring haciendas. Plant pathology was not his specialty, but he knew more about the diseases of valley crops than did local people at that time. Bües also made surveys of property boundaries, an experience that informed him of the history of most large haciendas in La Convención. He traveled on foot and mule-back throughout the region, observing, collecting and recording different phenomena. He earned money by collecting plants for international institutions. Bües discovered pre-Columbian ruins and petroglyphs, studied the material culture of the Matsigenka, and made agronomic experiments.

He wrote several highly regarded scientific papers, including the first scientific article in Peru on coca as a crop (Bües 1916). Much about La Convención that Bües learned was never published. Foreign scientists traveling to the area sought him out. In February 1932 at Hacienda Echarati, Bües met J.W. Gregory, geologist from Edinburgh, and his expedition. They found Bües, who accompanied the group to Rosalina, to be a rich source of information (Coverly-Price and McKinnon-Wood 1933:27). The Scottish expedition without Bües waited 3 weeks at the height of the rainy season before a canoe arrived to take them downriver. At the Pongo de Mainique disaster struck. The canoe, swamped by waves, capsized and two people, Professor Gregory and a Matsigenka, were swept into a whirlpool and drowned.

In 1946, Bües told Arnold Heim when they met in Quillabamba that a secret slave trade involving Matsigenka Indians had lasted on the Urubamba until 1944 (Heim 1957). Anecdotal information about Bües alluded to his fondness for alcoholic spirits of which La Convención, a major rum producing region, had no shortage. In 1942 Bües donated his substantial natural history collections to the Dominican school at La Granja near Quillabamba. He organized that material into a museum, which, after his death, fell into disarray. When I was there in 1963, rats had destroyed several displays containing plants. Its 550 pieces included carved stones and ceramics that provided evidence of Inca and possibly pre-Inca presence in the Urubamba Valley below 1500 m asl. In 1965, the Dominicans rehabilitated the collection, and then donated it to the municipality of Quillabamba, which created a place for it at another site. The Herbario Vargas in Cusco acquired Bües's crytpogam plant collection. Bues and the Cusco plant scientist Cesar Vargas botanized together in the 1930s and 1940s and were good friends.

Christian Bües, thanks to his polymathic interests and focus on La Convención, contributed significantly to knowledge about the Urubamba. His map called "El Señorio de los Incas," displayed natural features, archaeological sites, settlements and trails in Vilcabamba District. Bües had collected information for that map between 1916 and 1928 on the basis of first-hand reconnaissance. Bües was also interested in La Convencion history. He is said to have discovered a 1614 document concerning the area of Machu Picchu. It described the conflict between the Augustinian Order and Francisco Poma Gualpe, a Cañari Indian from Southern Ecuador, who was brought to Cusco by the Inca. Until 1849, Gualpe's descendants considered themselves to be the traditional guardians of Machu Picchu. Hiram Bingham knew nothing of that.

2.3.4 César Vargas Calderón (1903–2002)

Another notable scholar of matters concerning the valley was César Vargas, an eminent botanist. Born in the Province of Paucartambo, he grew up in Cusco where he attended the Colegio Nacional de Ciencias and the National University of Cusco. His botany teacher at the university, Fortunato L. Herrera, inspired him to make a career in plant science.[1] When he was 30 years old, Herrera left for Lima, thus opening a professorial position at the University for which Vargas was appointed. Over his lifetime, Vargas collected thousands of plants in the Urubamba Valley and elsewhere in Southern Peru. Especially in the 1940s and 1950s, César made major additions to taxonomic collections of native plants of Cusco and Apurimac Departments. Plant collecting in the Andes had its hazards. On one occasion, mounted on horseback, Cesar was cutting the stem of a plant when his mount slipped at the edge of the precipice. He threw himself off the horse and saved his life.

In 1938–1939, Professor Thomas Harper Goodspeed from the University of California at Berkeley asked Vargas to join his plant-collecting expedition in the highlands. Goodspeed's main collecting focus was of wild tobacco (*Nicotiana*), a genus on which he was the recognized world authority. In 1940, perhaps as a reward for his good service but also for his scientific promise, Vargas was awarded a scholarship to the University of California at Berkeley. During that year, he studied cytology and systematics and visited herbaria in Cambridge, Chicago, New York and Washington. In 1942, when another Berkeley professor, Carl Sauer, came to Cusco, Vargas acted as guide on a trip to the Urubamba during which the two talked crops and Indian culture incessantly (West 1979:76). Vargas was among several Latin

[1] Fortunato Herrera Garmendia (1873–1945) was born in Cusco and studied at its university where Dr. Antonio Lorena inspired him to become a plant scientist. Herrera began teaching botany there in 1912 and in 1929 became rector of the university. In 1933 he left Cusco to join the Faculty of Science at the Universidad Mayor de San Marcos in Lima.

Americans who impressed Sauer, who was then on a mission for the Rockefeller Foundation. However, there is reason to believe that before he came to Peru Sauer had been impressed by a long article Vargas had written on the potato that had been published 6 years earlier in the *Revista del Museo Nacional* (Vargas 1936). Sauer wrote in a letter: "He [Vargas] has the wit to see that this is the place to study potatoes, wild and domesticated. He is also bright enough to know that the potato is not only a taxonomic but a cultural item" (West 1979). Sauer then wrote that "It wouldn't take long to make [Vargas] into a good cultural geographer as well as a cytologist-taxonomist" (West 1979). Sauer cautioned, however, that "like so many Latin Americans, he is also somewhat unaware as to the difficulty of arriving at generalizations and would require help to keep him from over stimulating himself as to conclusions" (West 1982). With his strong bias for the indigenous, Sauer hinted that Vargas was unlike other Peruvians who "make me tired with the fashion in which they diddle around with wheat and sheep and pay almost no attention to the magnificent cultural acquisitions of their own" (West 1982).

The earlier Goodspeed expedition had awakened Vargas's interest in potato taxonomy, but it was Carl Sauer who, in 1942, gave him encouragement that reinforced his scientific interest and brought him to the attention of the Rockefeller Foundation. In 1944, Vargas applied for a Rockefeller grant to study in Mexico. His initial application contained paragraphs on the "problem of the Peruvian Indian," which the Foundation directors apparently felt overstepped the boundary into the political. Informed of that *faux pas*, Vargas reframed his application in strictly scientific terms and got his grant. His focus on potato taxonomy lasted over two decades during which he developed a collection of tuberous *Solanum* species, both cultivated and wild. In 1950, the Rockefeller Foundation provided funds for facilities at the University of Cusco and for Vargas's two-month study in Colombia with the *Solanum* specialist J.G. Hawkes.

In the same period, the American Consulate in Lima denied a visa for him to visit herbaria and to attend meetings in the United States. Around that time, the U.S. State Department instituted a Cold War policy of refusing visas to ostensible "communist sympathizers." Vargas's youthful support of leftist causes became part of his file at the U.S. Consulate. César had come of age when the *indigenista* movement contested the oligarchy and captured the imagination of many educated cusqueños. But to brand him a Communist was patently ridiculous. He had followed Antonio Raimondi's advice to let go of the politics and devote himself to knowing his country. The irony of his visa denial was that as a professor, he was disdainful of the unceasing and futile political debates at his own university. Leftist students shut down the institution on numerous occasions, and Vargas saw them as distracting learning and inviting violence. As a security measure, Vargas installed his herbarium in his Cusco house, not at the university.

César Vargas contributed much to knowledge of the flora of Southern Peru. By 1962 he had, himself, named 195 new species of the Peruvian flora, some of which had his name as part of the binomial. Vargas's 98 publications between 1935 and 2002 included several dozen reports of new species he had discovered (Tupayachi Herrera et al. 2003). Harper Goodspeed (1961:12–13) referred to

Vargas as an "authority on the vegetation of the Peruvian Andes;" more accurately, Vargas's interest was on the flora.

A pantheistic vision elevating the natural world to the level of a spiritual experience was an aspect of Don César's romantic conception of planet earth. He also loved flowers. At his *casa de campo* in Urubamba, he crossed dahlia varieties, maintained a collection of 130 varieties of roses, and collected more than 200 species of orchids. For Machu Picchu alone, he identified 90 orchid species, some of which he himself was the author of. In response to a request, Vargas (1946:21–28) sent seeds of the red-flowering *pisonay* tree (*Erythrina falcata*) to arboriculturalists in Los Angeles where this species subsequently became a familiar street tree. An essay he wrote on the ornamental flora of Cusco conveyed the sense of the history and traditions of his "*tierra*". Unlike most educated Peruvians of his time, César praised traditional Andean agriculture, knowing before anyone else in Cusco that native crops and cultivars contained priceless germplasm useful in plant breeding. Amid the clamor of voices for Peru to become "modern," Vargas defiantly proclaimed "viva la chaquitaclla, abajo el tractor!" (long live the Andean foot plow and down with the tractor). Although tractors have gained ground since 1963, the indigenous foot plow still has a place in cultivating plots on steep hillsides. The historicist dimension of Don César's romantic science also became apparent in an article he wrote in English identifying the plants represented on molded or painted pre-Columbian pottery (Vargas 1962, 1981).

César Vargas extolled the indigenous and celebrated—particularly to foreigners—the Quechua side of his origins, but in the city he was a *caballero cusqueño* in his comportment and speech (Fig. 2.6). As was the custom of the *gente decente* in Cusco of the 1960s, he usually dressed in a black suit, white shirt and dark tie when he left the house. As a member of the elite Club Cusco, César invited me there when I first arrived to have pisco sours. He was an *eminencia* whose place in the history of science in Peru rests mostly on the many new species of plants he named. Although he was a pioneer in the potato taxonomy of Peru, his interest in it had declined by the 1960s. Another cusqueño, Carlos Ochoa Nieves (1920–2008), had superseded César as Peru's leading potato scientist. Ochoa carried out a program of wild potato exploration throughout the Andes and bred new varieties, some of which became important. Unlike other academic cusqueños who went off to more prestigious positions in Lima, César's professional life remained in Cusco. He identified himself with Cusco and, as a good cusqueno, he was affable, gentle, spiritual and serious. The municipality honored him by naming a new street, Calle César Vargas, in the city.

In retrospect, I realize that the welcome extended to me in 1963 harked back to his personal connection between Carl Sauer. An acquaintance of Sauer's (which I was) could be a friend of Vargas. Sauer had opened the door to Vargas's advanced education, which increased César's self-confidence as a scholar and scientist. In 1942 Sauer wrote that Vargas was not only "the only botanist within a thousand miles of the Andes," but also a "good prospect [who is] bright, clean-cut and looks like a scholar to me" (West 1982). Although Vargas's work on *Solanum* faltered, his other scientific endeavors validated Sauer's judgment of his scholarly and scientific

Fig. 2.6 César Vargas (1904–2003), who saw himself as a proud descendant of the Incas, conceptualized the Inca achievement of working in stone with their adoration of the sun as a backdrop for his own interest in plants. Potatoes, wild and cultivated, had been a decades-long research focus for Vargas. He was also interested in their representation on pre-Columbian pottery. The plant on the left, *Cantua buxifolia*, is the sacred flower of the Incas and national flower of Peru. The plant on the right, *Salvia oppositfolia*, *ñukchu* is part of Cusco folklore. (Source: Vargas 1946:3)

potential. César provided me with information, contacts, and advice without seeking to intervene in my research plan. When I returned to Cusco in 1968, he took my wife, Mary, and me to Urubamba to observe the maize harvest and, on another occasion, to collect plants in the Apurimac Valley below Chinchaypuquio (Fig. 2.7).

2.3.5 Carl O. Sauer (1889–1975)

Carl Sauer, the most famous North American geographer of the twentieth century, spent a short but intellectually significant time in the Urubamba (West 1982; Denevan and Mathewson 2009; Williams et al. 2014). Born in Warrenton, Missouri, Sauer spent most of his adult life as a professor at the University of California at Berkeley. During World War II, the Rockefeller Foundation enlisted him to make a fact-finding trip to the Andean countries to assess the state of science and to encourage promising young scholars to apply for grants to study in the United States. The 1942 trip that took Sauer and his son Jonathan included a segment that began

Fig. 2.7 César Vargas (age 65) and Daniel Gade (age 31) in Urubamba during the maize harvest in 1968. (Photograph by M.K. Gade)

in Antofagasta, Chile. From there, they took the train to La Paz, Bolivia. After a few days in that high-altitude capital, they went by rail to Lake Titicaca, got on the lake steamer, and crossed into Peru. From Puno another train took them through La Raya Pass into the Urubamba Valley. From the window of the coach they observed people in their maturing crop fields and in the towns where the train stopped to pick up passengers. Sauer's claim in a letter that he had seen "the highest corn field in the world" was an observation made from the window of his coach (West 1982). He was particularly interested in crops during that trip (Fig. 2.8). He later was under contract to produce an article on that subject for the *Handbook of South American Indians*.

Among the people Sauer contacted in Cusco was Cesar Vargas, referred to him by his Berkeley colleague Harper Goodspeed. Father Carl, son Jonathan, and Cesar went to the Sunday market in Pisac where they bought samples of maize, beans and squash. The Sauers posed questions to the sellers with Cesar acting as translator to and from Quechua. Sauer was struck by the strongly indigenous character of the Highland Peru he had seen. The process of hispanicization had proceded less in this part of the Central Andes than in Mexico. From Pisac the three men continued down the Sacred Valley to Ollantaytambo, then a zone largely of maize grown under irrigation. Ever on the lookout for clues to origin, Sauer (1950a:493) commented that

Fig. 2.8 Carl Ortwin Sauer (1889–1975) in South America next to a field of quinoa. He came to the Urubamba in 1942, during which time had a special interest in the origin of cultivated plants. (Courtesy of Jonathan D. Sauer)

"it seems reasonable to consider that maize originated sufficiently far away from the Equator to be a well-marked contrast between a warm, rainy summer and a dry, cool fall (but not real winter)." Northward it is difficult to locate an area suitable in climate and soil much short of Guatemala. Southward, however, such a condition exists, south of latitude 10° S in the large eastern valleys of the Andes of Peru, such as the Urubamba. In this valley, incidentally, maize growing is not restricted to irrigation as Mangelsdorf appears to think. There are sections of secure and sufficient rainfall, and the flood plains generally have sufficient flooding and subirrigation.

These cogitations indicate his view that maize may have originated in the Andes. That supposition was before radiocarbon dating and the determination that teosinte, a wild grass of southern Mexico, is the ancestor of maize. Sauer's reflection on the Urubamba led him to assert in a letter that "I know of no other place where agricultural occupation runs continuously through as large an altitude range. And the whole of it is a good valley with a great deal of terrace farming ascending the slopes in many cases from two to three thousand feet, formerly very much more terracing than now" (West 1982:104).

Jonathan Sauer took all the pictures on this trip and several halftones were published in the *HSAI* chapter on the physical geography of South America (Sauer 1950b). These images showed, though not too clearly, the "quebrada of

San Salvador below Cusco", the "Nudo de Vilcanota" at La Raya Pass, "dry cliff vegetation on Rio Urubamba, Ollantaytambo" and "eroded fields on sides of Rio Vilcanota Valley at Tinta south of Cusco". The Sauers saw the southern part of the valley a second time from the opposite direction, when they traveled by train from Cusco to Arequipa. All of Carl Sauer's direct knowledge of the Peruvian Highlands came from his one trip, taken in 1942 to Cusco and the Urubamba Valley and from there to Arequipa. Though Sauer's observations were made over a brief time, his great depth of general knowledge enabled him to make sense of what he saw. Facts for him were a kind of launching pad into speculative inquiry. Much of his research career involved devising original ideas that could not be verified.

2.3.6 Frank M. Chapman (1864–1945)

Born in Teaneck, New Jersey, Frank Chapman never attended a college or university. He started working at a bank, but his growing fascination with birds made him change careers. In 1888, he joined the staff of the American Museum of Natural History where he catalogued and identified bird collections. Chapman wrote many guides and other books about birds and was considered to be the leading American ornithologist of the early twentieth century.

Chapman came to the Urubamba Valley as a member of Bingham's 1915 expedition. While camping at San Miguel Bridge he could make out some of the buildings of Machu Picchu, 500 m above where he was standing. He made the ascent in 2 h and descent in 1 h, and judged that the visit was worth it a thousand times over because of "…the sublimity of its surroundings, the marvel of its site, the character and the mystery of its construction" (Chapman 1933:265). Chapman's (1921) main contribution from that trip was a monograph of bird life in the Urubamba Valley in which he synthesized the findings of his own fieldwork as well as that of several other bird collectors. Chapman organized his monograph according to "life zones" from the puna to the tropics, all together recording 380 species and subspecies of birds. He was the first to think about life zones in the Andes as being different bird habitats. Not surprisingly, the warm part of the valley below 1000 m asl had many more species than the upper valley. Chapman had captured only a small part of that diversity. Between 1974 and 1979, a team from the Louisiana State University made six trips and collected more than 700 species of birds from the valley (Parker and O'Neill 1980).

2.4 Other Field Scientists

Many other field scientists contributed to knowledge of the Urubamba Valley. Ynés Mexia (1870–1938), a Mexican-American who took botany classes at UC Berkeley, collected about 150,000 plants in Alaska, Mexico, Brazil, Ecuador, Peru, Chile and

Argentina. In 1936, at the age of 62, Mexia traveled down the Urubamba Valley to Quillabamba (Anema 2005). Her membership in the Sociedad Geográfica de Lima reflected her wide geographical interest in Peru. An astute ethnobotanist, Timothy C. Plowman (1944–1989) went in the 1970s to the tropical zone around Quillabamba where he collected wild and cultivated coca. He later published an authoritative monograph of the genus *Erythroxylum* (Burger 1989; Plowman 1984; Davis 1996). On the basis of his collections and those of others, Plowman determined that the Urubamba Valley harbored only one cultivated species: *Erythroxylum coca* spp. coca. The cocaine content in the leaves of this subspecies ranged from 0.23 to 0.93 %; unfortunately, he did not correlate that content with the elevation above sea level of the coca fields. Tim Plowman's career as a plant scientist ended at age 45 when he died of AIDS. He had been one of Richard Evans Schultes's most brilliant students at Harvard.

Jeanine Brisseau-Loaiza (1981), a French geographer, prepared an impressive *thèse d'état* on the influence of Cusco to its region which she defined as encompassing the Departments of Cusco, Apurimac and Madre de Dios. Travels to distant districts and relentless interviewing of many different people enabled Brisseau Loaiza to define the varied ways that the city of Cusco economically influences a wide territory. She determined that all parts of the Urubamba Valley except southern Canchis are oriented to Cusco. Arequipa's considerable influence in Canchis Province owes much to its control of the wool market in Peru and to the rail transport that connects it to Sicuani. Her study also showed that the Urubamba Valley in no way comprises an economic unit, partly because exchange and its associated settlement in the Andes occur in overlapping ecological zones. Brisseau-Loaiza's dissertation, assembled over a period of 13 years, brought together an array of information on administrative functions, land tenure entities, trade and transportation. A cartographic feast, the book put Cusco and its region in a spatial perspective never before seen. After her book appeared, Brisseau Loaiza published little, the pattern of many who earned the French *doctorat d'état*. Often described as an endurance test, this kind of doctoral program has passed into history, replaced with a more focused and less ambitious requirement closer to the American Ph.D.

2.5 Conclusion

The Urubamba has been welcoming territory for those who have come to observe different facets of this remarkable valley. Tourism is primarily a money relationship of services rendered and paid for. Most tourists depart with a set of impressions and images that become part of their travel experiences. A small number among these hordes make a point of learning a good deal about Andean landscapes and culture. It can also work the other way. Some tourists have changed the perceptions that locals have of foreigners. Sojourners and researchers develop different sets of impressions than do tourists. They deal with civil authorities and make varying requests of country and town dwellers beyond food and lodging. Tourists have been welcomed as sources of income and, in some places, as sources of amusement. Sojourners and researchers have generally been welcomed in the Urubamba. However, authorization for archaeological excavation has not always been granted, a function in part of the maturing of that profession in Peru itself. Unlike some North American Indian tribes who have protested outsiders seeking to gain knowledge or images of their lives, indigenous folk of the Urubamba have rarely sought to keep scholars out. The result is that the Urubamba is one of Peru's best known valleys and is certainly its most visited.

References

Adams HC (1908) Some wonderful sights in the Andean highlands. Natl Geogr Mag 19(9):597–618
Anema D (2005) Ynés Mexía: botanist and adventurer. Morgan Reynolds Publishing, Greensboro
Barnhart TA (2005) Ephraim George Squier and the development of American anthropology. University of Nebraska Press, Lincoln
Bingham H (1913) On the wonderland of Peru. Natl Geogr Mag 27:387–574
Bingham H (1922) Inca land: explorations in the highlands of Peru. Little Brown and Co, Boston
Bowman I (1912) Cañon of the Urubamba. Bull Am Geogr Soc 44:881–897
Bowman I (1913) First report of Professor Bowman's expedition. Bull Am Geogr Soc 45:750–753
Bowman I (1914) Results of an expedition to the Central Andes. Bull Am Geogr Soc 46:161–183
Bowman I (1916) The Andes of southern Peru. Henry Holt & Co., New York
Bowman I (1938) Los Andes del sur del Peru (trans: Nicholson C). Editorial La Colmena, Arequipa
Brisseau-Loaiza J (1981) Cuzco dans sa région: etude de l'aire d'influence d'une ville andine. Centre d'Etudes de Géographie Tropicale, Bordeaux
Bryce J (1913) South America: observations and impressions. Macmillan Co., New York
Bües C (1916) La coca: apuntes sobre la planta, su beneficio, enfermedades y aplicación. Ministerio de Fomento, Lima
Bües C (1933a) Estudios sobre la flora del Departmaneto del Cusco. Revista Universitaria 64:58–64
Bües C (1933b) Helechos de La Convención: Rio Alto Urubamba. Revista Universitaria 64:150–172
Bües C (1939) El valle del Río San Miguel: Cuenca del Urubamba. Revista Universitaria 77:42–56
Burger W (1989) Timothy Plowman (1941–1989). Taxon 38(2):329–330
Chapman FM (1921) The distribution of bird life in the Urubamba Valley. U.S. National Museum Bulletin 117. Government Printing Office, Washington
Chapman FM (1933) Autobiography of a bird-lover. D. Appleton Century C, New York
Chaumeil J-P (1994) Una visión de la Amazonía a mediados del siglo XIX: el viajero Paul Marcoy. Bulletin de l'Institut Français d'Etudes Andines 23(2):269–295
Cook JD et al (1869) French Literature. Saturday Rev Polit Lit Sci Art 694:192
Coverly-Price AV, McKinnon-Wood M (1933) Professor J.W. Gregory's expedition to Peru, 1932. Geogr J 82(1):16–38
Cruz Ccorimaya N (2009) Christian Bües, biografia y recopilación de obras. Centro Cultural Jose Pio Aza, Lima
Davis W (1996) One river: explorations and discoveries in the Amazon rain forest. Simon and Schuster, New York
Davis K (2009) Harriet Chalmers Adams: remembering an American geographer. Calif Geogr 49:51–70

de Gregori CJ, Sandoval P (2007) La antropología en el Perú. Revista Colombiana de Antropología 43:299–334

de Sartigues E (pseudo. Lavandais ES) (1851) Voyage dans les républiques de l'Améique du Sud. Revue des Deux Mondes 2:1019–1059

Denevan WM, Mathewson K (eds) (2009) Carl Sauer on culture and landscape: readings and commentaries. Louisiana State University Press, Baton Rouge

Driever SL (2011) Geographic narratives in the South American travelogues of Harry A. Franck. 1917–1943. J Lat Am Geogr 10(1):53–70

Fermor PL (1991) Three letters from the Andes. Penguin, New York

Franck HA (1917) Vagabonding down the Andes. Garden City Publishing Co., Inc., Garden City

Gade DW (2006) Albert A. Giesecke (1883–1968)—A Philadelphian in the land of the Incas. Expedition 48(3):27–33

Giesecke AA (1912) Informe sobre el censo levantado en la Provincia del Cuzco el 10 de setiembre de 1912. Revista Universitaria (Cusco) 4:2–45

Goodspeed TA (1961) Plant hunters in the Andes. University of California Press, Berkeley

Heim A (1957) Wunderland Peru:naturerlebnisse. Verlag Hans Huber, Berne

Hubbard E (1990) Journey to Ollantaytambo: in the Sacred Valley of the Incas. Chelsea Green Publishing Co., Post Mills

Marcoy P (pseudo. of Laurent de Saint-Cricq) (1869) Voyage à travers l'Amérique du Sud, de l'Océan Pacifique a l'Océan Atlantique, 2 vol. Hachette, Paris [English edition: Marcoy P (pseudo. Laurent de Saint-Cricq) (1874) A journey across South America, from the Pacific Ocean to the Atlantic Ocean. (trans: Rich E) 2 vol. Blackie & Sons, London]

Markham CR (1856) Cusco: a journey to the ancient capital of Peru. Chapman and Hall, London

Markham CR (1991). In: Blanchard P (ed) Markham in Peru: the travels of Clemens R. Markham, 1852–1853. University of Texas Press, Austin

Matthiessen P (1961) The cloud forest. Viking, New York

McElroy K (1986) Epharaim George Squier: photography and the illustration of Peruvian antiquities. Hist Photography 10:99–129

Middendorf EW (1893–1895) Peru: Beobachtungen und Studien uber das Land und seine Bewohner wahrend einer 25 jahrigen Aufenthalts, 3 vols. Robert Oppenheim, Berlin [Spanish edition: Middendorf EW (1973) Peru: observaciones y estudios del pais y sus habitantes durante una permanencia de 25 años. Universidad National Mayor de San Marcos, Lima]

Mould de Pease M (1986) Observaciones a un observador: hurgando en el tintero de Ephraim George Squier. In: Masuda S (ed) Etnografia e historia del mundo andino: continuidad y cambio. Universidad de Tokio, Tokyo, pp 36–107

Samanez y Ocampo JB (1885) Exploración de los ríos peruanos Apurímac, Eni, Tambo, Ucayali y Urubamba. Imp.de "El Pais," Lima

Parker TA III, O'Neill JP (1980) Notes on little known birds of the Upper Urubamba Valley, Southern Peru. The Auk 97:167–176

Plowman T (1984) The ethnobotany of coca (*Erythroxylum* spp., Erythroxylaceae). In: Prance GT, Kallunki JA (eds) Ethnobotany in the neotropics. New York Botanical Garden, New York, pp 62–111

Poole D (2008) Marcoy P (1815–1888). In: Pillsbury J (ed) Guide to documentary sources for Andean Studies, 1530–1900, vol III. University of Oklahoma Press, Norman, pp 374–380

Protzen J-P (1993) Inca architecture and construction at Ollantaytambo. Oxford University Press, New York

Raimondi A (1874) El Peru 6 vol. Imprenta del Estado, Lima

Raimondi A (1898a) Itinerario de los viajes de Raimondi en el Perú: valle de Lares, Santa Ana y regreso por Mollepata y Limatambo [1865]. Boletin de la Sociedad Geográfica de Lima 8:1–61

Raimondi A (1898b) Itinerario de los viajes de Raimondi en al Perú: Cuzco, Quispicanchi, Lucre, Pisac, etc. y regresó hasta Abancay [1865]. Boletín de la Sociedad Geográfica de Lima 8:361–387

Rénique JL (1980) El Centro Científico del Cusco (1897–1907). Histórica (Lima) 4(1):41–52

Ridgeway J (1986) Road to Osambre: a daring adventure in the high country of Peru. Viking, New York

Roosevelt T (1917) Review of: The Andes of Southern Peru by Isaiah Bowman. Geogr Rev 3:317–322

Rowe JH (1946) Inca culture at the time of the Spanish conquest. In: Stewart JH (ed) Handbook of South American Indians, vol 2. Smithsonian Institution, Washington, pp 183–330

Sauer CO (1950a) Cultivated plants of South and Central America. In: Steward J (ed) The handbook of South American Indians, vol 6, Bureau of American Ethnology. Smithsonian Institute, Washington, pp 487–543

Sauer CO (1950b) Geography of South America. In: Steward JH (ed) Handbook of South American Indians. Smithsonian Institution, Bureau of American Ethnology, Bulletin 143, Washington Government Printing Office, pp 319–344

Squier EG (1877) Peru: Incidents of travel and exploration in the land of the Incas. Henry Holt & Co, New York

Tantaleán H (2014) Peruvian archaeology: a critical history (trans: Stanish C). Left Coast Press, Walnut Creek

Tupayachi Herrera A, Franco Navia F, Moscoso Zambrano D (2003) Bibliografía de Julio Cesar Vargas Calderón, botánico cusqueño (1903–2003). Cantua 12:19–31

Vargas C (1936) El solanum tuberosum atras del desenvolivimiento de las actividades humanas. Revista del Museo Nacional 5(3):193–248

Vargas C (1946) Diez años al servicio de la botánica en la Universidad del Cuzco. Universidad Nacional del Cuzco, Cuzco

Vargas C (1962) Phytomorphic representations of the ancient Peruvians. Econ Bot 16(2):106–115

Vargas C (1981) Plant motifs on Inca ceremonial vases from Peru. Bot J Linn Soc 82:313–325

West RC (1979) Carl Sauer's fieldwork in Latin America. Department of Geography, Syracuse University, Syracuse

West RC (ed) (1982) Andean reflections: letters from Carl O. Sauer while on a South American trip under a grant from the Rockefeller Foundation, 1942. Westview, Boulder

Wiener C (2010) Voyage au Pérou et Bolivie (1875–1877). Ginkgo, Paris

Williams M, Lowenthal D, Denevan WM (2014) To pass on a good earth: the life and work of Carl O. Sauer. University of Virginia Press, Charlottesville

Wright R (1984) Cut stones and crossroads: a journey in the two worlds of Peru. Viking, New York

Yacher LI (1995) Antonio Raimondi del Acqua, 1826–1890. Geogr Biographical Stud 16:80–87

Chapter 3
Urubamba Verticality: Reflections on Crops and Diseases

Abstract First studied elsewhere in the Andes by Alexander von Humboldt and Francisco de Caldas, verticality refers to the highly telescoped climate from glaciers and puna above 5000 m to the tropical rainforest (*selva baja*) below 400 m above sea level (asl). A seasoned perspective on verticality as a phenomenon has grown out of fieldwork on it contained in a 1975 publication. Crop distributions, though controlled by thermal tolerance limits, are primarily about human decisions. Coca and maize are analyzed to grasp that nature-culture duality. Verticality of agriculture influenced the location of points of ecological exchange and account for migration of people between the lower and upper Urubamba. Another set of organisms, tropical parasite diseases, manifest a verticality largely controlled by temperature thresholds of both the pathogen and vector. Occurring in the pre-Columbian Andes, mucocutaneous leishmaniasis played a role in the Inca avoidance of the jungle as a zone of permanent settlement. The other disease, malaria, introduced to Peru in the sixteenth century, also became endemic in the Urubamba below 1800 m asl. In its epidemic form, malaria in the 1930s caused high mortality and major dislocation. Verticality of crops and diseases raise questions about boundary shifts through time brought on by global warming. Advances in plant breeding increase tolerance and thus can change limits of crops, and medical advances can either eliminate or cure disease.

3.1 Elevation and Crop Plants

Verticality refers to the pattern of highly telescoped environments arranged in steep gradients of elevation above sea level. The notion is most applicable to tropical highlands where the gamut of climate and life forms associated with different altitudes is largest. Between the cool or cold conditions at high elevations and warm to hot at low elevations, the climate is temperate in the sense that "cold" and "hot" do not apply. Along this gradient, wild biota broadly sorts itself out spontaneously primarily according to temperatures and secondarily according to precipitation. The 84 ecological zones defined in Peru incorporate these two elements in various environmental combinations (ONERN 1976). However, the vertical distribution of crops and zoonotic diseases does not necessarily conform to these ecological zones. With crops, availability of irrigation water negates the importance of rainfall, making temperature the most salient factor in explaining distributions. Development of

D.W. Gade, *Spell of the Urubamba*, DOI 10.1007/978-3-319-20849-7_3

parasites and vectors depends more specifically on thermal conditions. Scholars of mountain geography have frequently alluded to the concept of verticality without necessarily using that expression. Generally an observer perceives that the spontaneous vegetation in a landscape reflects environmental conditions. But weather stations are few and recorded observations are sometimes unreliable. Detailed assessments of Andean climate will require the establishment of many more stations. Knowledge of verticality has advanced since the eighteenth century when two scientists began thinking in those terms. Alexander von Humboldt (1769–1859) was the first to make a zonation in the Andes in a macrothermal to microthermal scheme from lower to higher elevations. That simple scheme is still used today: *tierra caliente*, *tierra templada*, *tierra fria* and *tierra helada*. Humboldt's so-called Chimborazo map encapsulates "the formative cross-fertilization of cartography, scientific visualization, protoenvironmental science, and nascent environmental social sciences…" (Zimmerer 2011:125).

Humboldt focused on wild flora, whereas Francisco de Caldas (1761–1816), born in Popayán, provided an early discussion of the effects of thermal differences on crops in a vertical slope (Appel 1994). This brilliant scientific mind of eighteenth-century Spanish America found that the boiling point of water was different at different altitudes and, from that, constructed a hypsometric thermometer to determine altitude. Caldas turned his attention to the distribution of crops in an area spanning the 500 km from 4° N latitude in Southern Colombia to the equator in Ecuador. He asked why the lower limit of wheat cultivation should be at 2034 m asl: was that boundary simply an unquestioned tradition? In Spain, wheat was cultivated down to sea level. Caldas noted, however, that in his part of South America *polvillo*, a fungus blight, made wheat cultivation unsustainable below the 2034 m asl threshold. The greater humidity at low elevations, not the greater temperature, accounted for the fungus infestation. In contrast, the upper limit of wheat cultivation placed at 2833 m was due to the fact that, above that elevation, the flour made from wheat was "black and bitter." Caldas did not invoke climate directly in either boundary, though ultimately that factor entered into the explanation. He noted that oats grew successfully both below the lower limit of wheat and above its upper limit. One species, maize, had the most vertical range of all grain crops, since its different varieties and landraces tolerated wide ranges of climates. In contrast, although the potato contains several species its ecological amplitude was considerably narrower than maize. Bananas, plantain, and manioc grew from the coast up to 2034 m and sugar cane slightly higher to 2092 m asl.

Altitude limits on crops from place to place are largely incommensurable. Factors other than temperature complicate the identification of boundaries, among them exposure to sunlight, kind of soil, and rainfall. Latitude has a thermal effect as one moves away from the equator. A crop field at 3000 m asl at 20° S will be cooler than one at the same elevation at 12° S. Moreover, the array of certain cultivars available can account for the crops found in any one place. Unusual events and idiosyncratic decisions of particular farmers further affect crop distributions. An investigator's perception can also explain why recorded boundaries differ for a particular plant. Among those who synthesized verticality in the Central Andes was the late French geographer Olivier Dollfus (1981:23) whose altitudinal graphic showed the upper limit of crops as determined by available moisture. Those limits, when assigned to specific places, are not accurate. Dollfus placed the upper limit of maize at 3300 m asl; in the Urubamba Valley, the absolute maize limit has long been very close to 3700 m asl.

Dollfus placed the limit of wheat at 3500 m asl; however, in the Urubamba it is 350 m higher. His upper limit of the vine at 2000 m asl is a valid generalization. Considered only generically, the potato in Southern Peru is grown up to 4400 m asl, whereas its upper limit in Ecuador is 3500 m asl. Unlike Ecuador, Southern Peru has frost-resistant potatoes. Known as *luki* or *ruki*, they fall into two species, *Solanum juzepczukii* and *Solanum curtilobum*. In short, like many broadly-defined vertical boundaries in the literature, this schematic treatment cannot be usefully applied to any particular place. Moreover, Dollfus did not give a definition for "upper limit." A crop boundary can be the difference between what is there and what could be there. However defined, a crop boundary is not easily generalized. In some places, exposure to the sun's rays may be more important than its elevation above sea level.

Uppermost crop boundaries in southern Peru involve two unusually resistant plants. Thomas and Winterhalder (1976:55) compared relative crop hardiness in a limited area on the Altiplano at Nuñoa, distant from any ameliorating effect of Lake Titicaca. There the two most resistant crops raised were the bitter potato ("luki") and canihua (*Chenopodium pallidicaule*) both grown up to 4450 m asl. Below that, tetraploid potatoes, the three small tubers of oca, añu and ullucu, and the two seed crops of quinoa and barley, were harvested up to 4250 m. Wheat in Nuñoa grew up to 4100 m asl. Temporality has to be factored into any discussion of crop limits both for the past and the future. Warming climate, as evidenced by the melting of glaciers, is generally acknowledged for the Central Andes. In Huancavelica Department, Peru, the upper limit for bitter potatoes in 1975 was 4150 m asl. By 2009 that limit had increased to 4300 m asl, opening crop competition with livestock pasture (de Haan et al. 2010). The limits for any one crop is based on local determinants. Microclimate may be the main factor in one place, but in another, a simple idiosyncratic peasant decision may account for the boundary.

3.2 Approaches to Verticality

The geographical trifecta of coast, Sierra, and montaña has been the starting point for making sense of Peru's complicated environment. Jesuit chronicler José de Acosta (1940) made that fundamental distinction early in the colonial period to describe basic thermal and pluviometric differences of a diverse territory. The seventeenth-century chronicler Bernabé Cobo (1956:I:56–90) refined those observations, building upon another century's worth of accumulated knowledge. Alexander von Humboldt introduced a deeper scientific understanding of the differences among these three broad environments. When, in the early twentieth century, Carl Troll constructed a *Staffelsysteme*, the study of climatic verticality became better defined in spite of the paucity of climatic data (Gade 1996). Troll's account also included the perceptions, knowledge, and values of highland people and their economic and political institutions got attention. He introduced environmental determinism in asserting that the Inca Empire could only have developed in the Southern Peruvian Highlands. In that area, freezing nocturnal temperatures in the dry season permitted the elaboration of potato into chuño, which, being largely indestructible, enabled food to be stored indefinitely. Furthermore, llamas, indispensable for transport, were most numerous there. Troll's hypothesis is intriguing, but ignores the fact that other staple foods, including maize, can also be dried needing only sunshine.

The work of Javier Pulgar Vidal (1987), who developed the most widely adopted conceptualization of geographical differences in Peru, captured the essence of verticality. In 1941 this Peruvian geographer presented his first scheme of eight natural regions, six of which were in the highlands at different elevations. Temperature and rainfall differences governed these regions, though Pulgar Vidal's organization was more intuitive than scientific. Very few meteorological records were available and even today the climatic complexity of the Sierra remains far from sorted out. Subsequent editions of his book refined the original outline with more information and illustration, but climate data remains scarce. Pulgar Vidal compensated for that by reasoning that altitudinal differences in temperatures created a series of life zones, each with its own assemblage of wild plants and animals as well as its own crops and human adaptations.

Although most categories had a certain amount of zonal overlap, Pulgar Vidal's classification remained a compelling way to organize fundamental knowledge about the geography of a country as extravagantly diverse as Peru (Córdova and Bernex 2008). By elaborating enough of Peru's diversity without being weighed down with excessive detail, the classification has been widely applied. For example, Donald Dyer (1962) used the Pulgar Vidal organization to show that the largest population concentrations in Peru are located in the coastal zone ("chala") below 500 asl and in the "quechua" between 2500 and 3500 m asl. This last zone is agriculturally the most productive region of the highlands, but no one has as yet convincingly determined why that is the case. French geographer Evelyne Mesclie (2001:546) praised Pulgar Vidal's efforts to put resources into larger contexts. American geographers, Karl Zimmerer and Martha Bell (2013), who examined the utility and limitations of the Pulgar Vidal scheme, also found it to be heuristically useful. By integrating the biophysical with the anthropogenic, the scheme clarifies some of the vertical complexity that characterizes the geography of Peru and other Central Andean countries. The 1976 classification of Peru into 84 ecological zones makes useful and revealing nuances, but its great detail makes it too unwieldy for the human mind to grasp without extended study (ONERN 1976).

Andeanist and anthropologist John Murra contributed a third perspective to verticality, one based on ethnohistory. In citing Troll's 1931 paper, Murra (1989:205) mentioned the German geoecologist's influence on his environmental conceptualization of the Andes. He found evidence in early colonial documents that an Inca ayllu normally had rights to a range of ecological zones in which inhabitants of that ayllu raised the crops and/or animals considered to be most suitable for each zone. The notion that a community held vertical "islands" ensured livelihoods by expanding dietary choices and reducing subsistence risks. Murra explained that this arrangement precluded the need for much market exchange in the pre-Columbian Andes. As a result of Viceroy Toledo's reducción program, the archipelago notion of land use at different elevations was largely effaced. One isolated community east of Cusco, Q'eros, has maintained the kind of verticality that once prevailed in many places (Webster 1971; Flores Ochoa and Fries 1989). The people of Q'eros today still deploy their livelihood over a vertical spread of 1200 m, raising llamas and sheep in the puna (4000–4600 m asl); growing potatoes in the quechua/suni (3200–3800 m asl); and producing maize in the "monte" (1400–2000 m asl). To reach that last zone requires a 9-h walk from the main town which is in the quechua zone. Even though the land they occupied was in the hands of a large estate, Hacienda Cusipata owned by Luis Angel Yábar, much of their land use persisted without outside interference. In 1955, when it received its first scholarly visitors, Q'ero inhabitants were monolingual Quechua speakers, used the knotted string counting device (*quipu*) of the Incas, and wore the sleeveless tunic (*uncu*) known to the Inca (Escobar Moscoso 1958).

Verticality research of agricultural land use has emphasized production zones characterized by one dominant activity. The raising of llamas, alpacas and sheep is characteristic of the puna and potato cultivation characterizes the zone below that. Deemphasized is the fact that sheep and llamas are also in the potato zone and certain kinds of hardy potatoes grow in the puna. Thus in that zonal scheme, boundaries are not an overriding concern. Anthropologist Enrique Mayer (1985), working in the Mantaro Valley, described verticality in terms of production zones, but he emphasized the social organization of the villages more than the different zones.

Other verticality studies demonstrate the variations in perception of environmental zones. In the 1970s American anthropologist Stephen Brush (1977) studied a corporate peasant community, Utcubamba in Cajamarca Department, which encompassed an entire valley (Brush 1977). Brush defined seven production zones running the gamut from *temple* (800–1500 asl), *kichua fuerte* (1500–1900), where firewood was collected; *kichwa* (1900–1450 m asl), the main grain zone; *templado* (1450–3100 asl), *jalka* (3100–3500 asl), the potato zone; and *jalka fuerte* (3500–4300 m asl) or paramo. Brush understood that this diversity provided agricultural security. A crop disaster in one zone could be offset by crop success in other zones. Also in the 1970s, Ethnologist Harald Sklar (1982), working in the Pincos Valley of Apurímac Department, decided that the *chaupiyunga*, *ca* 2300 m asl was the "ecological floor" around which the economy revolved (Sklar 1982). Before the agrarian reform, haciendas growing sugar cane for rum completely dominated that land. Their cane cutters came from hamlets above the floor. The chaupiyunga occupied a particular ecological niche whose place in the culture history of the Central Andes has received little attention (Gade 1973). Pulgar Vidal subsumed the chaupiyunga into the yunga, which obscured the character of an environmentally important threshold. Boundary work is still rare in Andean studies, perhaps because biophysical parameters are more difficult to discern than social processes. Geographers who consider themselves political ecologists are less interested in natural history than social issues.

Still other perspectives on verticality deserve mention. A study launched in the 1960s by the French ethnologist Antoinette Fiorivanti-Molinié (1975) conceptualized verticality in terms of local zonation. She considered the Urubamba Valley floor to be one zone, the plateau to the west (as around Chinchero) another, the cordillera to the east a third and, as an outlying component, the tropical part of the Urubamba Valley 100 km to the north. Many people from highland Urubamba went to the tropical part of the same valley to farm in a kind of pendular migration. Fiorivante-Molinié argued that the population of each zone manifested not only economic and social differences, but also cultural and ideological ones. Other conceptualizations include that of Geographer Karl Zimmerer (1999) who saw verticality as an "overlapping patchwork," of agricultural landscapes that are the historical product of human ecological processes, not only of thermal differences related to elevation. Likewise geographer Kenneth Young (2009) about the same time conceptualized Andean landscapes as mosaics more than in elevated gradients. Closest to my approach in the Urubamba was that of the Japanese ethnographer Yamamoto (1981) in the 1970s who used the upper and mid-section of the Marcapata Valley to study the arrangement of crops according to their elevation above sea level.

3.2.1 Urubamba Verticality

The Urubamba manifests an extended type of verticality: its valley floor drops in altitude from the puna to the selva. However, the vertical ideal of one community holding land at various elevations on which to grow different crops apparently did not develop here. The Inca appropriation of the Sacred Valley and coca lands farther down the valley precluded dispersed archipelagos to be used by separate communities. When the Spaniards arrived, they imposed the encomienda system that later opened the door to the creation of haciendas. Over time these estates took over most of the valley floor, thus removing the best arable land from native control. Following this land grab, Viceroy Toledo's *reducción* program in the early 1570s grouped previously semi-dispersed Indian populations into tightly clustered villages (Gade and Escobar 1982). This forced nucleation took the native farmers away from much of their more distant agricultural and pastoral lands. Too remote from population clusters to be cultivated, those fields fell into the hands of Spaniards.

Andean people have to a greater or lesser degree always used multiple ecological zones in a wide variety of arrangements. Many native communities work a valley floor and the slope land above. If not, they trade with communities in other zones of different elevations. The Incas imposed a vertical land use pattern that survived into the early colonial period. Maize-growing estate peons in the Yucay area were obligated to pick coca for specified periods in the Yungas of Paucartambo. The arrangement of obligatory corvée connected to imperial lands was not the typical example of verticality. Likewise the practice that started in the colonial period of highlanders working on haciendas in the lower valley and returning to their home villages represents a kind of verticality since they also farmed plots for their own use in both places.

3.2.2 Personal Interest in Verticality

My fascination with the verticality of mountain environments grew out of a 1961 travel experience in Mexico. In the course of one day I moved from near sea level on the Gulf Coastal Plain to the Central Plateau along a gradient from Veracruz to Córdoba and Orizaba and then up to Tehuacán. A succession of biotic and agronomic landscapes that went from hot to cool and wet to dry presented themselves. My immediate grasp of what I was seeing came from a keen appreciation of climate as the key variable in making sense of the world's geography. Although Köppen's version of world climates has since receded among geographers, a course in climatology forever inscribed this classification in my brain. Mountains telescope these thermal differences, making them the most complex of all climatic realms, though that is not apparent on small-scale maps. Wide reading provided more striking examples of verticality than that of Highland Mexico. Although I no longer remember the exact source, Japanese research on the Himalaya described the habitats that range from the tropical Terai region to the iconic snowy peaks. But it was the Peruvian Andes that particularly enchanted me. Over a distance of only 100 km, highland valleys draining into the Amazon Basin have a stunning vertical change in land use and a more diverse range of crops than that of Asia.

Two readings particularly sparked my geographical interest in the Andes. O.F. Cook, a USDA plant scientist, went to Peru in 1915 as the chief botanist on the last expedition led by Hiram Bingham that was funded by the National Geographic Society. Cook (1916) published an excellent article on Andean agriculture in a 1916 issue of the *National Geographic Magazine* that focused on the Urubamba Valley as an Inca legacy. Cook described paleotechnic ways of working the land without patronizing Andean peasants by advocating "rural development." My affinity for Cook's mode of thinking extended to his whole published corpus, which stimulated me to write an article about his scientific contributions (Gade 1970) Though meant for semi-popular consumption, Cook's *National Geographic* article presented the Urubamba Valley as an intriguing place in which to study verticality. The article alluded to maize growing up to 11,000 ft (3353 m asl), potatoes up to 14,000 ft (4267 m asl), and placed the "general limit" of agriculture at 13,200 ft (4023 m asl). He mentioned wheat as the chief crop grown in the Urubamba between 10,000 and 12,000 ft (3048–3657 m asl). His description of crop boundaries made me realize that every crop and even landrace had its own particular boundary and that field observation could tell me what many of those limits were. It also opened questions about temporal changes in boundaries reflecting climatic cooling or warming. Population pressure also could explain boundary changes. The agronomic and geographic literature revealed that no one had yet given crop verticality much thought. Boundaries at different latitudes are often not comparable. In some cases, inductions of limits were totally incorrect (Moseley 1992:30; 2001:31).[1] By comparison, natural vegetation boundaries in mountains, such as alpine tundra and the krummholz, were topics of innumerable articles.

The other compelling piece of writing for me at the time was Carl Sauer's (1950) chapter on native crops in the *Handbook of South American Indians*. Sauer's articles showed him to be the kind of scholar I admired. In that chapter, he fetchingly described Andean crop diversity that was then still poorly known, by combining his 1942 Peruvian and Ecuadorian observations, first-hand knowledge of Mexico, and a firm grasp of the literature on New World crop plants, but minimalist attention to boundaries challenged me to make that the centerpiece of a field study.

The notion of vertical crop limits incorporated the multifaceted concept of boundedness (Hastorf 1993:25). Crop boundaries were one of several kinds of limits found in the Andes. District limits, village confines, communally-owned lands that everyone knows the limits of and, remarkably, also named fields, all were bounded entities that had to be taken into account. The idea that a field would have its own toponym suggested generations of personal connections to the land. Since boundaries can either provide or close access, they are important in peasant land-use strategies and in shaping the landscape.

3.2.3 Defining and Defending the Field Site

Anthropologists typically choose one village to be their field site. An example was Stephen Brush's choice of Utcubamba, a northern Peruvian community in possession of several ecological zones. In a large, long valley such as the Urubamba, communities had no such diversity. In conceptualizing my project within this site stretching from 4343 m asl down to the confluence of the Urubamba River with an important tributary at 723 m asl, I was most interested in the biophysical changes. My dissertation proposal was to study the spatial, temporal and ecological patterns of agricultural plants in this Peruvian valley. Though poorly organized and infelicitously written, not one of the four members of my dissertation committee commented on those deficiencies. Dissertation proposals, of course, are provisional in-house documents with no independent existence, yet an intense textual criticism at that stage would have been helpful. Graduate programs perhaps fail most when professors do not respond to the need for hard criticism.

[1]The table in his book of altitudinal limits for the principal native Andean crops contains a bundle of errors that placed, for example, the upper limits of arracacha at between 830 and 956 m asl (the normal limit is 2400 m asl); coca's upper limit is given as 1200 m asl (when, in fact, it extends more than 800 m above that), and the limit of oca, placed at between 850 and 1700 m asl. This hardy tuber is rarely grown below 3200 m asl and most often above 3600 m asl.

My three main objectives in that proposal were to provide essential information on every useful plant in the valley, work out the distribution of each crop as a function of ecological amplitude and human decisions, and prepare a series of maps showing the distribution of the 20 most important crops in each successive century since the fifteenth. That last objective turned out to be quixotic, for a historical record of crop distributions simply does not exist. Documents in archival repositories dating from the colonial period do not include this kind of agricultural information. The processes of introduction, expansion, displacement, contraction and disappearance involve temporalities never considered worthy of a written record and so, in most cases, can only be inferred. My project also did not lay out a theoretical framework that would have made my study applicable to other parts of the Andes and to mountainous regions elsewhere in the world. Nevertheless the proposal passed muster with the dissertation committee.

The methodology I proposed for collecting field data proved to be inoperable in the field. A line traverse across the valley in which each crop could be recorded with its elevation above sea level seemed feasible enough. In fact, however, the inaccessibility of many steep-sided slopes made it impossible to identify with any certitude crops in fields high above the valley floor. Moreover, in many places, the river barred access to both sides of the valley. I did, however, determine altitude by using a pocket altimeter. I wondered if I would have been so obsessed if I had had to carry with me the kind of bulky barometer used by Alexander Von Humboldt on Chimborazo.

3.2.4 *Thoughts on Crop Boundaries in the Urubamba*

A crop boundary is, in part. a response to thermal limits beyond which a plant cannot successfully grow, but it is not absolute, for some farmers take more risks than others. Depending on the crop, low temperatures halt plant growth. Above 3000 m asl elevation, frosts become inevitable in the three dry months of June, July and August. Above 4000 m asl, freezes occur in every month of the year. Knowledge about frost occurrence is locally shared and may determine planting practices of those not affected. Low elevation limits are set less by high temperatures than by excessive transpiration or the presence of plant disease or pests. Of the two kinds of limits, temperature and moisture, cold affects the germination, growth and reproduction of most crop plants. Freezing injury occurs primarily in the cellular membranes with the loss of electrolytes from plant tissue (Lyons et al. 1979:7). Andean people, perhaps more than any other people in the world, developed crops with considerable cold tolerance. Where irrigation is not available too little rain can also exclude certain crops. Yet it is easier to control water availability than to control temperature. Ultimately, however, a crop and a crop distribution reflect a human decision.

3.3 Verticality of Selected Crop Plants

3.3.1 Highland Seed Crops

In the Urubamba, amylaceous maize heavily dominates the crop inventory between 3200 and 2500 m asl. This crop is most productive in that zone and the risks of growing it here are smaller than elsewhere in the valley. Growth requirements of flour maize differ from the flinty maize grown in tropical climate. Germination requires a minimum temperature of 10 °C (Guerrero Barrantes 2012). Growth and maturation of the maize plant is best achieved between 15 and 20 °C. Freezes affect maize most seriously during flowering by damaging the pollination mechanism that functions over a period of 13 days in the life of the plant. A freeze also shrivels maize leaves impeding the development of new leaves needed for photosynthesis. Maize reaches its upper limit of cultivation slightly above Maranganí at 3700 m (Fig. 3.1). Freezes here are so common during the growing season and maize, as a plant, is unusually sensitive to low temperatures. Cold stunts the height of the plant and the size of the ear (Fig. 3.2). Above 3400 m asl, freezing temperatures late in the growing season make it a risky crop to grow. Yet many farmers around Combapata assumed that risk, for maize could be profitably sold or traded to the plateau communities to the west, notably Marcaconga, Pomacanchi and Sangarará. Combapata has long been one of the big chichi making towns in the valley because people from surrounding higher elevations came there to buy chicha. Only 13 km up valley, in Tinta, maize was uncommon a difference accounted for less by differences in freeze frequency than by its lack of Combapata's market for chicha. Above 3400 m asl, maize, when grown, is frequently mixed with crops like quinoa in case maize does not yield. In Calca and Urubamba the few freezes that occur descend in June or July after the maize harvest. Since it has been widely cultivated there for several millennia, it is a strong part of valley agriculture.

A second highland crop, less sensitive to low temperatures than maize, is broad bean (*Vicia faba*). Known in Spanish as haba, this is a hardy crop well suited to southern Canchis. A temperature of 5–6 °C is required for broad bean to germinate and its development proceeds at temperatures as low as 12 °C which is also the threshold temperature for flowering. Seed formation requires 16 °C. These low temperature thresholds account for successful cultivation of broad bean up to almost 4000 m asl (Fig. 3.3) in the Urubamba and Calca provinces, Broad bean successfully also grows in the Urubamba down to 2700 m asl as long as temperatures do not exceed 25 °C. Another seed crop, quinoa (Chenopodium quinoa), also tolerates low temperatures but grows as well where the risk of freeze is low (Fig. 3.4).

3.3.2 Comments on Highland Root Crops

Root crops occupy a wide ecological range. Potato, especially, has considerable amplitude, for, like maize, human selection has occurred over long periods enabling it to fit into a wide range of climatic niches. For Andean peasants living above

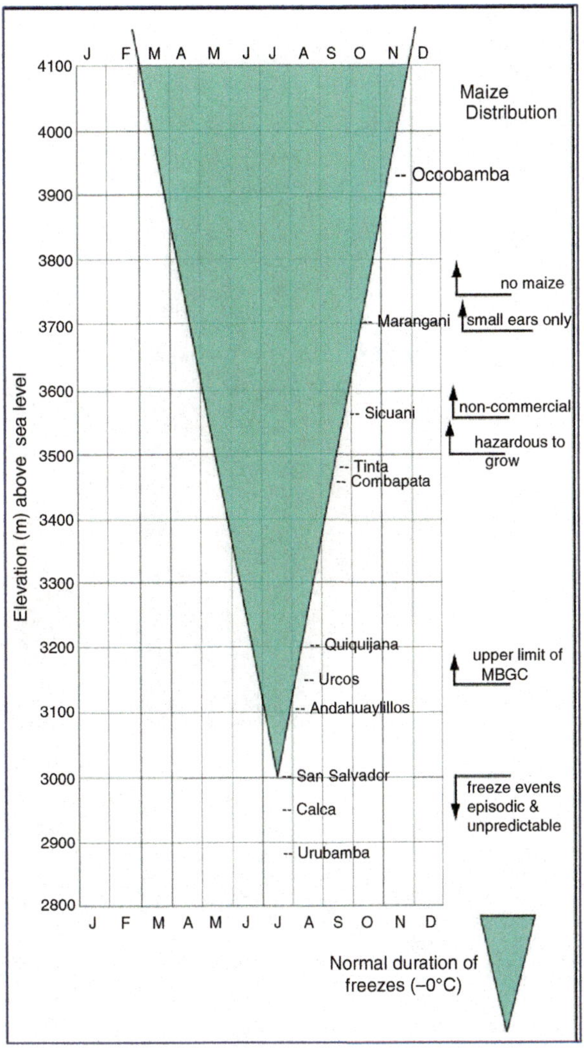

Fig. 3.1 Freeze frequency in the Upper Urubamba Valley. Freezes increase with increasing elevation. Above 3400 m asl, maize is vulnerable and above 3750 it is so vulnerable that it is no longer cultivated. Episodic freezes in the Sacred Valley farther north occur after the crop has been harvested

3400 m elevation, potatoes are the safest crop to grow (Fig. 3.5). Pests and disease, not the ability of the plant to grow or bear tubers, determined the lower limit of potatoes in the valley. Weevils, nematodes and tuber moths multiply to such an extent in warmer places that farmers considered the potato not worth the bother. Most devastating is late blight (*Phytophthora infestans*). Without hard freezes to kill them, the airborne spores of this fungus spread quickly. At high elevations, the risk of late blight is considerably less. But resistance to late blight varies with the kind of potato. *Solanum phureja*, a diploid, yielding less-esteemed tubers, is better adapted to elevations where frosts do not normally occur.

Fig. 3.2 Maize differences with elevation. On the left yellow flour maize grown in the Sacred Valley and on the right a maize ear near the upper limit of its cultivation

Fig. 3.3 Broad bean harvest at 3890 m asl at Occobamba in southern Canchis. Known in Spanish as haba (*Vicia faba*), this Old World crop plant introduced in the sixteenth century to the Andes has a highland niche between 3000 and 4000 m above sea level

Fig. 3.4 Quinoa harvest. Quinoa is a cold-tolerant crop that once only indigenous people grew for their own food needs. It has since become an export product in high demand. Landraces correspond to color differences of the husk

Fig. 3.5 Potato harvest above 3500 m asl. In native Andean agriculture, both men and women participate in the harvest. Tubers are dug up with a tool called the lampa

Fig. 3.6 Añu (or mashua) harvest in Southern Canchis. A member of the nasturtium family, this plant has strong Indian associations

The selection of which potato species or cultivar was grown at a particular site depended also upon the locale's storage conditions. *Solanum tuberosum* stores poorly in warm humid conditions and was rarely grown in those places. At the upper limit of cultivation, luki (or ruki) potatoes (*Solanum juzepczukii* and *S. curtilobum*) grow in a climate of frequent freezes that they withstand down to temperatures of −4 °C. During the flowering phase, ruki potatoes require temperatures of at least 8 °C. Potatoes actually tolerate lower limits than the literature implies, which has tended to assign them to the Andean high country. Compared to the potato, oca (*Oxalis tuberosa* and añu (*Tropaeolum tuberosum*) are rarely grown below 500 m asl at least in the Urubamba (Fig. 3.6).

3.3.3 Warm Climate Crops

3.3.3.1 Coca Cultivation

In the northern, lower elevation part of the valley grow crops sensitive to low temperatures. The historical changes in crop distributions that have occurred there are not, however, necessarily related to temperatures. One temporal shift that can be extrapolated from historical documents is the upper limit of cocoa cultivation (Fig. 3.7). In the early 1960s, coca's effective upper limit as an important crop was at Hacienda Ccolpani Grande (1600–1700 m asl). Beyond that, small *cocales* were

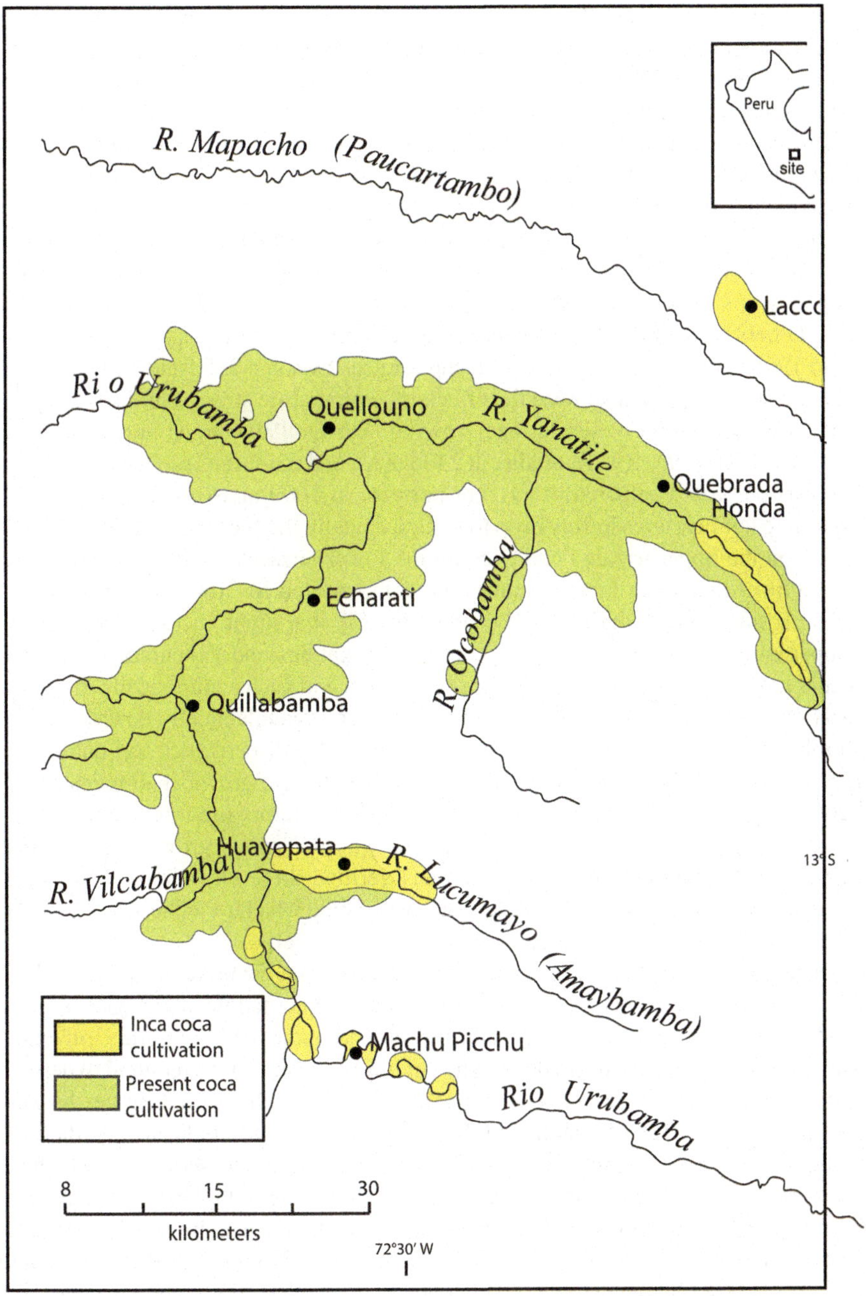

Fig. 3.7 Coca distribution past and present in the Urubamba Valley. For a time in the late Inca and early colonial periods, coca was cultivated up to ca. 2400 m asl. It is now cultivated from ca. 1800 to 650 m asl

planted up to 1950 m asl. The shrubs bore darkened leaves indicating damage from temperatures below 10 °C. In the late Inca and early colonial periods, coca was cultivated up to 2400 m asl (Julien 2001:232). Pachacutec, whose imperial juggernaut included expanding into the upper fringes of the hot country, ordered the plantation of coca fields. This Inca king also ordered the transplantation of workers (*mitmaqkuna*) from Chachapoyas in northern Peru to the lateral valleys of Amaybamba and up the Urubamba to Torontoy. Sixteenth-century references to the "indios de Picchu" point to the colonial continuation of the late Inca locations and methods of cultivating coca. Some coca-growing sites were on the valley floor or on narrow strips of useable land placed near stone faces that radiated microclimatic heat. Others were 300–400 m higher on stone-faced bank terraces.

In 1968 coca planted on Machu Picchu terraces between 2350 and 2400 m asl as a demonstration garden showed the effects of temperature stress. Shrubs were only half the size of those of similar age growing near Quillabamba at an elevation of 1000 m asl. The leaves were smaller at 2400 m asl and the shrubs produced only two harvests per year compared to the four per year at 1000 m asl. At Machu Picchu, nighttime temperatures in July drop to 8 °C. Logically the Incas would have planted maize on the upper terraces and coca on the lower terraces (White 1989:16). The microclimate on stone terraces somewhat compensated for low ambient temperatures. Absorbing heat during the day and releasing it at night, the terraces increase temperature by 2 to 3 °C. A higher content of alkaloids and flavonoids in the leaf would have been a big motivation for growing coca in such a climatically marginal location. Content of the coca leaf could have been organoleptically determined by simple taste. As in the Inca period, coca farmers today discern coca leaf quality by taste and flavor. Those who know coca realize that the leaf grown at altitudes above 1200 m asl represents a higher quality. Coca for masticatory use rarely comes from elevations lower than 800 m asl and explains why it is rarely grown in that hotter part of the Urubamba Valley. Research has verified these elevational differences (Acock et al. 1996:20). The coca once grown around Machu Picchu was probably recognized as superior and its use confined to the Inca elite.

After Hernando Pizarro in 1539 received the Amaybama Valley and "Picho" (or "Picchu") as his encomienda, the *mitmaqkuna* who had been engaged in coca production for the Inca around Machu Picchu and about 10 km up the Urubamba Valley from there continued to do so. The "indios of Picchu" were required to deliver to him tribute of 105 baskets (*cestos*) of coca leaf, which, at 10.4 kg per basket, amounted to only 1092 kg (Julien 2001). This small quantity indicates the thermal marginality of the area, the small amount of appropriate land surface on which to grow the shrubs, the difficulty of drying the leaf in a zone of high humidity, and the limited number of people (15–20 households) in that zone. In 1550, Spaniards reduced the tribute demand to 35 baskets of coca, half of it delivered to Cusco and the other half to Ollantaytambo, but included two other exactions, chile peppers and fruit. In 1560, the encomienda reduced the coca tribute to 15 baskets (about 174 kg of leaf), along with unspecified amounts of maize, eggs, chickens, firewood, fish, chile pepper and fruit. All that changed in 1572 when, with the death of Manco Inca, the remaining workers fled the settlements to which the Inca had assigned them. In that decade, the coca plantations became privately owned.

Fig. 3.8 Coca harvest in La Convención. Most coca is grown on slope land and traditionally women have been the pickers (palladoras). A plant is normally stripped of its leaves four times a year. In the Urubamba, only one cultivated species of coca (*Coca erythroxylyn* var. *coca*) is grown, though the valley holds many wild species of the genus

Demise of the encomienda system brought abandonment of the small cultivable patches in the canyon. That geographical extension of coca cultivation southward disappeared (Fig. 3.8). That land subsequently passed to the Augustinian religious order who rented it to individuals (Glave and Remy 1983:191). In 1650, the earthquake that devastated the city of Cusco also triggered landslides that buried cultivable land and destroyed the Inca-constructed trail through the canyon. Viewed as a cul-de-sac rather than a passageway to the hot country farther down the valley, Spaniards rarely entered that rainy zone covered by a tangle of impenetrable bamboo (*Chusquea* spp.) and small trees, exuberant vegetation. To reach the lower Urubamba Valley with its haciendas of tropical products required a steep climb over the Málaga Pass and down through the Amaybamba Valley.

Historically coca production in the Urubamba below 1600 m asl has been far lower than in the valleys that drain into the Rio Alto Madre de Dios, collectively known as Kosñipata, located to the east of Cusco. An early colonial document mentioned that the coca fields of Francisco Pizarro had belonged to the Incas in "Tono" (i.e. along the banks of the Rio Tono below Paucartambo), "Canavire" (unknown), and "Yanati" (i.e. Yanatile Valley of Calca) (Covey and González 2008:101). Around 1750 the productivity of these valleys declined as a result of insect plagues, belligerence of Indian tribes and lack of labor. Yet that zone remained much more important for coca than the Urubamba well into the nineteenth century. Nevertheless, it was an important crop in nearly all the estates in the tropical Urubamba; in the

1840s at Huyro in the Amaybamba, of the 230 workers on the property, 150 were women who picked the leaf (de Castelnau 1850–1859). After the chemical makeup of coca became known in the nineteenth century, an hacendado, Juan B. Samanez, established in Quillabamba a factory to extract cocaine hydrochloride to sell as a local anesthetic and for use in making a health drink (Tamayo Herrera 1992:II:542). César Lomellini, Cusco's biggest entrepreneur, later took over ownership of that then-legal factory.

Even before the agrarian reform of the 1970s broke up the larger land holdings of haciendas, peasant migrants seasonally came and went in pendular fashion. They rented parcels from the hacendado and took over most of the coca production. Hacendados demanded that his workers (*arrendires*) supply for harvesting the hacienda's own coca crop one female coca picker (*palladora*) for each coca harvest. In the 1950s, discontent of renters over this and other aspects of the rental agreements mounted. Arrendires at some point refused in unison to meet the requirements imposed by the haciendas. In 1959, Hacienda Santa Rosa had 13 ha in estate coca production, but 675 ha in arrendire coca production. By 1963 all the coca (701 ha) produced at Santa Rosa was in the hands of arrendires and their families picked the leaves. When haciendas lost the workers who picked and dried the leaf, whose salaries represented 80 % of the cost of production, they could no longer grow the plant as part of estate production.

Although the haciendas that once dominated coca production are gone, the plant remains important. Many cocales owned by ex-arrendire families were once coca land of the haciendas. Coca's special advantage over other crops is its ability to produce despite poor soil nutrients. Lower moisture requirements of the roots of this perennial make coca well suited to slope land and quite impoverished soils. One ha of coca yields 350 kg of leaf, and four harvests per year are expected. How much coca is now grown in the Urubamba Valley can only be a gross estimate. Official coca statistics cover only legal producers who must be registered by the ENACO state monopoly (Empresa Nacional de la Coca). Since 1978, ENACO has purchased all the coca produced under registration. Specific coca merchants, of which the Department of Cusco has about 800, are authorized to retail the leaf. The actual areal extent and production may be several times the official figures. Illegal growers, if caught, have their harvest seized and cocales destroyed. If they can successfully evade the authorities, private buyers pay a better price for the leaf than does the state monopoly. Rogue growers often use herbicides forbidden by ENACO. From whatever source, La Convención coca mostly ends up in traditional markets, not in cocaine processing.

3.3.3.2 Verticality of Coffee, Cacao and Tropical Root Crops

Coffea arabica can withstand a light freeze, but to play it safe, commercial cafetales are placed below 1600 m asl. Within protected patios enclosed with mud walls, coffee trees are grown as high as 2850 m asl in the town of Urubamba. Less coffee is grown below 700 m where hot temperatures lower the quality of the arabica bean.

Coffee growers at higher elevations realize their marketing advantage. The cooperative that today occupies the lands of the Hacienda Huadquiña (1500 m asl) earned a controlled appellation for its coffee. Since good coffee is defined in the international trade as that having a mild flavor, the zone between 1400 and 1600 m asl fetches the highest price from wholesalers.

In the Urubamba Valley, from 15,000 to 20,000 ha of land are in cacao. More susceptible to low temperatures than is coffee, cacao is normally injured by temperatures below 10 °C (Hall 2001:61). Its cultivation in the Urubamba centers on two distinct cultivars, virtually subspecies, neither of which have had any Inca associations. The oldest cultivar, South American in origin, and placed in the so-called forastero group by cacao taxonomists, is known as cacao chuncho. As suggested by its name, it may have originally been a trade item brought up the valley by Piro Indians in the eighteenth century from where it was first and mainly cultivated at Hacienda Echarati (800 m asl). This cultivar has 30 or so white seeds per pod. The thick green or yellow pods hold astringent beans yielding an intensely rich flavor and floral aroma. Criollo cacao, also known as cacao blanco, is a Central American cultigen introduced to the Urubamba by an hacendado in the late nineteenth century from Colombia. Criollo cacao has white, ivory or sometimes pale purple pods and about 20 red seeds ("beans"). In the Urubamba, intolerance of criollo cacao to low temperatures has set its uppermost cultivation near 800 m asl, whereas chuncho grows up to 1000 m asl. Chuncho and criollo have hybridized and there is good reason why that is not lamented. Chuncho cacao is not very productive compared to the hybrid. Marketing cacao from La Convención, 90 % of which is exported, as chuncho is a profitable deception. Much is actually hybrid.

A taxonomically diverse set of domesticates used for their underground parts also has also been part of the crop inventory. Arracacha (*Arracacia xanthorrhiza*), known in the valley as *virraca*, had a narrow niche between 2300 and 1800 m asl, though it could grow higher. Its low commercial value may best explain its normal upper limit. Uncucha (*Xanthosoma sagittifolium*) thrived between 1800 and 1400 m, mainly because it was most appreciated there—not because it cannot survive in hotter climates (Fig. 3.9). Likewise, manioc (*Manihot esculenta*) was an important subsistence crop below 1100 m asl, and was more sparingly grown up to 1800 m asl. No altitudinal zonation could be determined for two minor root crops, yacon (*Polymnia sonchifolia*) and achira (*Canna edulis*). They appeared as sporadic agricultural plants in the lower valley, but also sparingly in the Sacred Valley. Such a taxonomically diverse set of domesticates used for their underground parts cannot be found anywhere else in the world.

3.4 Ecological Exchange

Slim evidence does not foreclose the possibility that the Incas were involved in some trade with people from different environments. In the colonial period and after, exchange of products between the temperate and the tropical linked these zones either by the movement of peoples or the use of market trade.

Fig. 3.9 Uncucha, with an edible underground root, is the New World equivalent of taro. It is grown in the Urubamba especially between 1600 and 1300 m asl

3.4.1 Movement Between Environments

Before 1950 traders from certain highland towns traveled with llamas and mules to the coca country to exchange their dried beef, chuño and flour maize for coca leaf, rum and coffee. Known in Quechua as *q'epirkuna*, these itinerant folk came especially from Acos, Pillpinto and Sangarará which all lie above and west of the Urubamba depression. Another group of traders (*huasaq'epes*) carried to La Convención merchandise on their back, mainly manufactured clothing suitable for the hot climate. Another mode of vertical exchange occurred thanks to the pendular migration of highland peasants. These highlanders maintained houses and farmed plots in their home communities but seasonally moved to the tropical valley to work for large land owners in exchange for the use of several hectares of arable land on which they could grow their own crops for sale and consumption. Over time these renters (*arrendires*) became well enough established to sublet small plots to peasants (*allegados*) poorer than themselves, in return for their labor. Frequent displacement of both groups between the tropical valley and their home communities brought tropical products into the highlands. Large bundles characterized these peripatetic peasants when they move in both directions. The same peasant who carried bananas and citrus to the highlands moved months later in the opposite direction, loaded down with highland wheat, broad beans and potatoes.

3.4.2 *Markets and Fairs*

Large amounts of vertical exchange in the valley also occurred at fixed points. The main periodic markets in the valley have taken place on Sundays. On the day peasants gathered not only to buy and sell, they also came to hear mass in the church. Pisac, Urcos, Quiquijana and Sicuani all had big Sunday markets at which high country products were exchanged for those from the valley floor. Several annual fairs served as religious-themed harvest festivals whose main function was the exchange of goods from a wide array of ecological zones. The fair at Tiobamba, near Maras, on the western rim of the Urubamba Valley is held on August 15, the feast day of the Virgin of the Assumption ("Virgen Assunta"). Colonial authorities reportedly established it in 1601, but its location suggests an event with pre-Columbian origins. The grand panoply of tropical, temperate, and puna products bought and sold there has long attracted thousands to buy and sell. Calca also has its Virgen Assunta fair on August 15, attracting people especially from the tropical and puna zones of Calca Province itself. The third big fair in the Urubamba is held on September 14 at Huanca near San Salvador. Many who buy or sell also perform religious devotions associated with the shrine of the Señor de Huanca, a Christ-figure with many devotees. Ceramic pots, chuño, and dried meat from Puno are offered there in exchange for maize. This fair has traditionally set the price of maize for the entire Cusco region. Formerly the Altiplano traders, known in the valley as Qolla, came to Huanca with their llamas through La Raya Pass; by the 1960s they came mostly in trucks. That the fair may predate not only the Conquest, but also the Incas is suggested by the fact that the Christian shrine appears to have been built on a site of Wari origins.

Elsewhere in the valley annual festivals at different times of the year have been occasions to exchange products with those brought down from the adjacent high country. The San Bartolomé festival at Tinta, Virgin of Carmen fiesta at Checcacupe, and the Señor de Pampacucho at Sicuani have been major events in the upper valley in the dry period of the year. The town of Cusipata also has its own August 15 event, called there the "feria de la Emerita" (hermit fair), but nevertheless associated with Virgen Assunta. Several events during the rainy season include two fairs associated with the Festival of the Kings on January 6, one at San Pablo and the other at Pampamarca; the San Hilarion fiesta in Cusipata on January 14; the Virgin of Candelaria fiesta in Urcos and, near the rim of the Urubamba depression above Urcos, the Virgin of Canincunca festival on the second of February.

3.5 Writing About Verticality

Retrospection of a scholarly work by its author provides an opportunity to reexamine its impact and shortcomings. Although informed readers, not the author, should most appropriately be the ones to evaluate it, authors do. The book, *Plants, Man, and the Land in the Vilcanota Valley of Peru*, was substantially more than the 1967 dissertation (Gade 1967, 1975). Research trips to the Department of Cusco in 1968 and 1970 had added information and ideas beyond those recorded in 1963 and 1964.

The published work was not complete. Since at the time I could not determine their uses, I omitted several plant species. Observing actual use of plants often required being in the right place at the right time. A complete plant use study of such a large area would have required 3–4 years of research. Thus the book did not measure up to being a guide to all the useful plants in the Urubamba Valley. Although I never made that claim, some readers assumed completeness. A second lacuna was my failure to elicit more deeply the insider's view of valley verticality, that is, the environmental distinctions made by rural folk themselves. They distinguished only puna, quebrada (to refer to the mesothermal part of the upper valley) and "valle," by which they meant the tropical zone. I also did not succeed in determining which plants were the best indicators of particular climates. It would have been the way in which people in the colonial period visualized zonal distinctions. Thirdly I did not develop a grand theory about verticality. When a theory becomes passé, as all theories do, then the book loses part of its usefulness.

3.5.1 *Reception of the Work*

Specialized scientific monographs attract a limited readership. Few bookstores carry such monographs and the media do not report on them, but over the years authors learn about their readers. They write or email; others cite the work, sometimes in ways not predicted. *Plants, Man and the Land in the Vilcanota Valley of Peru* has had several audiences. Some readers sought specific information on the valley; others were interested in the details of Andean agriculture; and yet others saw it as an example of a field-based study in the Sauerian tradition of cultural-historical geography.

According to the World Catalogue, most large university libraries in the United States accessioned the book. In California 22 repositories catalogued it as did five in Washington, D.C. In other countries accession was uneven: 17 in Germany, 10 in Australia, eight in Canada and six in the Netherlands, but none in France, Spain, Italy, China or Argentina. In all, the Web of Science and Google Scholar listed more than 300 different citations and numerous other citations appeared in the literature before the initiation of these two citation indices or for other reasons were not flagged by them. Many journals are not indexed in the Web of Science.

Because a Spanish translation of the work never materialized, the book's impact in Peru except on specialists was modest. Even for Peruvians who could read English, two other barriers existed. The European publisher was not easily accessible, and the cost, then about four or five times the price of a book published in Peru, hindered its acquisition even by institutions. I delivered gift copies to several Peruvian friends and libraries. The Library of the Catholic University (PUCP) in Lima immediately accessioned my donation in its collection. In an in-house misappropriation, the Biblioteca Nacional in Lima never catalogued its personally-delivered copy.

The National University of Cusco (UNSAAC) did not catalogue it for more than a decade; at some point in the 1990s it was finally entered into the system. Since then, pages were cut out of the book. One Peruvian author plagiarized two of the book's maps (Gonzales Mires 1993).

Besides the brief book notices that appeared in the *IUCN Bulletin*, *Angewante Botanik*, *Field Crops Abstracts*. and *Handbook of Latin American Studies*, eight reviews of the book appeared between 1976 and 1978. Plant scientists made up one category of reviewers. Robert M. Bird (1976), a maize specialist, wrote in *Science* that "the description this study provides of a part of Andean ethnobotany and cultural geography is good and it should be widely used." His two major criticisms were that "more discussion of the interaction of factors and the establishment of smaller, more inclusive, zonal units could have clarified matters" and that "distribution of crops might have been better presented by zonal diagrams than by the many maps the book includes." Bird was correct on both counts. *Science* included as part of its review of the book one of Guaman Poma's panels and the long caption I had made about it. The review by Jack Harlan (1976), a notable expert on plant domestication, provided an elegant assessment. He wrote "this is a competent ethnobotanical study of agriculture and plant usage of the Vilcanota Valley as of the early 1960s." Harlan correctly noted the time frame, the only reviewer who did so. Temporality governs everything that humans do.

The economic botanist Hardy Eshbaugh (1977) noted in his review that this book manifested a growing interest in ethnobotany, especially as it related to "cultural investigations detailing the impact of plants on the very lives of various tribal groups." Eshbaugh opined that "Gade's book is certainly one of the best of these studies to appear in recent years" and went on to state that "unlike so many similar publications it shows strength both anthropologically and botanically." C. Ronald Carroll (1977:463) in the *Quarterly Review of Biology* noted that it is "a fascinating historical treatment of the agricultural development of the valley." Carroll was the only reviewer who put the whole book in historical terms. M. J. A. Werger (1976), in his review in *Vegetatio*, wrote that "the present ethnobotanic account is invaluable because of its integrated approach to the subject," which contradicts to some degree Bird's previous assessment that more was needed in the interaction of factors. Werger summarized it as a "very useful and interesting book."

Another European reviewer, Hans Richter (1977:156), a biologist in East Germany, wrote in German that "The book has a lot of weight as a contemporary discussion of the topic," which contradicted Carroll's review that the work was historical. In my view, both remarks fit the content. Richter was the only reviewer to invoke the name of Carl Sauer, noting that I was Sauer's *Schüler* (disciple). He was also the only one to describe the work as Kulturgeographie (cultural geography). He further stated that the book was a "careful field study and critical assessment of the literature …on the special problems of plant use," and "recommending it to geographers and biologists in equal measure." Biologist R. Schubert (1976) wrote that the book was "interesting and informative and contributes very stimulating ideas for geoecology." A third German reviewer, G. Schwabe (1976), wrote rather enigmatically that the "richly illustrated work offers on the whole more concrete information than might be expected from its topic."

Antoinette Fiorivanti-Molinié (1976), a French ethnologist and specialist in this part of the Peruvian Andes, provided an anthropological perspective. She was the only reviewer who had done intensive research on part of my study area and produced a considerable publication record related to it. Mme Fiorivanti-Molinié wrote that the book contained an "overly generalized presentation of the history of Andean agriculture," but that "on the Urubamba it was precise about the natural conditions of the valley." She went on to say that "It is a detailed study and very useful manual of consultation for the ethnologist. D. Gade shows a good ethnographic knowledge of the region. However, it would have been appropriate to have also included the use of plants in magic and ritual." Fiorivanti-Molinié wrote that its theme, method and preciseness make it a work of great interest to the ethnologist for two reasons. Its central subject was human use, not the valley flora, and secondly, each plant entry had an historical dimension. She concluded by noting that "the author succeeds in explaining in a balanced way his own observations, perception of native inhabitants, and the agronomist's point of view all at the same time."

Geographers who reviewed the book were less concerned with details about plants and more with people on the land. In the *Annals of the Association of American Geographers*, Charles Bennett (1976:636–637), unfamiliar with the specifics of the valley, confined his criticisms to the generalizations I had made. He took issue with a remark I made in passing that the domestication process may have been influenced by the need for protein and my suggestion that high insolation and sharp temperature fluctuations in the Andean Highlands may have triggered mutations in crop plants. Crop diversity is greatest in mountains and that occurred through human selection of naturally occurring mutations. N. Vavilov had made that suggestion to explain why so many of the world's crop plants had their origins in highland areas. Bennett wrote that "Gade has presented a valuable contribution to human ecology and he has also provided us with a lode of problems to be researched in greater detail…I congratulate Gade for his solid field work well presented to the reader…" British Latin Americanist Harold Blakemore discerned that "…the author's sympathies were fully engaged with the cultivators he studied, and that engagement, together with his industry, has produced a valuable micro-study, of interest, not only to geographers and botanists, but also to historians and specialists in the social sciences willing to study traditional peoples whose ways of life have been, and are continuously eroded by other cultures."

Michael Wilson (1976), a North American geographer who had apparently thought a good deal about the notion of scale, made the insightful remark that "Gade covers some 500 square kilometers through 400 years in 240 pages…which makes the volume of detail so much more impressive." Wilson commented that "new options opened by the introduction of plants to the valley have been carefully documented, too, and the consequences followed in detail. Although Gade does not place the theme ahead of others, I would say that this record shows, in concrete ways, the interactions of all parts of a plant-man-culture symbiotic system." French geographer

Pierre Usselmann (1977:506–507) balanced positive with critical comments, notably a failure to quantify crop production, a "parsimonious" spatial representation of phenomena, and the lack of citations to two French Andeanists.

3.5.2 *Retrospective Personal Assessment*

The book had its greatest impact in Andean studies by providing an empirical account of plant use and agriculture in an important valley in the Cusco region. Anthropologists, who have done much fieldwork in this part of the world, cited it more than geographers. Its temporal context appealed to archaeologists and historians. The Swedish historian Magnus Mörner (1985:271) described it as "an excellent study of the river valley of Cuzco, also known as the Urubamba." Environmental writer Wendell Berry (1981), interested in agricultural sustainability, referred to my book and what it might be able to teach the world when he went to Peru to study native agriculture. My correspondence included exchanges with those seeking material for plant breeding, additional information about a particular plant use, and, and in one case, why I omitted this or that plant from my discussion. The work has not been much cited for its findings on the vertical boundaries of crops, the concept of the boundary or as a model to examine the specifics elsewhere of vertical crop boundaries.

On four return visits to the valley, I determined that much of the book's content is still broadly valid. Inhabitants continue to cultivate the same crops and use many of the same wild plants. Commercial-type agriculture has increased and people eat more packaged foods. Folk medicinal plant use survives, though with more competition from packaged pharmaceuticals. Towns depend less on firewood than in the 1960s. Thatched roofs have largely disappeared. These shifts have occurred along with a major change in land tenure, acculturation toward Hispanic language and culture, and less extreme poverty. These trends, assuming they continue, will eventually make obsolete the information in this book. The state of knowledge itself also advances. Already some scientific binomials have changed as taxonomists make their revisions. For example, a dozen *Cassia* species for the valley have now been described whereas in the 1960s there was only one. Within just the borders of Machu Picchu Sanctuary, five species of *Polylepis* have been identified. In the 1960s, there was only one.

After ca. 2030, the book's main value will be as a baseline to evaluate change. In that way, new invasive species that have entered the valley flora can be flagged by knowing what was there in the 1960s. Changes in crop boundaries can be deduced, the result of global warming, reduction in water supplies, or acculturation of peasant farmers. Decline in agrobiodiversity in the future may or may not occur as market sales increase (Zimmerer 2010). Absence increased my appreciation of the Urubamba. By 1973, I had reached an existential crossroads in my research program. My intellectual energy about the Andes began to flag as other parts of the world beckoned. My attention in the decades of the 1970s turned to France, Ethiopia and Paraguay, and in the 1980s to France again, Italy, Bolivia, Madagascar and the tiny island of Rodrigues. Sendero Luminoso kept me away from doing research in Peru, but once that movement disintegrated, the seductive power of that country prompted a return. The changes observed brought me a deeper understanding of the country and the valley which I had studied decades before.

3.6 Disease Verticality in the Urubamba

Disease manifests verticality when it is carried by a vector and the parasite is subject to thermal control. Contagions—diseases spread person to person—do not have that kind of spatiality. For example, the contagious character of tuberculosis enabled it to occur at all elevations. Likewise, the contagious nature of smallpox, measles, and scarlet fever caused such high mortality that the native population fell sharply in the colonial period. Typhus spreads from person to person but only when infected body lice move to other victims. The disastrous unidentified epidemic of 1719–1720, possibly measles, raged in the upper Urubamba for 6 months. The disease killed so many native people that families disintegrated and farmland was abandoned. The colonial government sold the vacant parcels to non-Indians, which further changed the valley's cultural character (Stavig 1999). Disease outbreaks were long part of the valley's history, but by the late twentieth century, the availability of vaccines and a government-funded health care infrastructure made contagious epidemics less frequent and less deadly.

Several vector-borne diseases have environmental constraints. In the Urubamba below 2000 m asl, warm temperatures enable vectors and parasites to reproduce. In La Convención sylvatic yellow fever, a viral disease spread by mosquitoes of the genus *Haemagogus*, has occurred in small waves. The mosquito breeds in tree holes and monkeys are its main reservoir. Waves of yellow fever may have episodically occurred far back in time, though not until recent decades did medical scientists analyze the outbreaks of 1994, 1998, and 2007 in those terms. Several unvaccinated people of highland origin living between 800 and 1100 m asl died in each of those above years. Another kind of yellow fever, which spreads in cities by *Aedes aegypti*, has no known history in the lower valley.[2]

Two important vector-borne maladies, leishmaniasis and malaria, are classic examples of diseases with environmental thresholds that explain contraction. They each left their morbid mark in La Convención, setting back population growth, mobility, and economic improvement. Both the parasite and the vector of each disease require warm temperatures to complete their life cycles. For that reason no record of contraction of either disease is recorded above 1800 m asl. However, since weather variations can shift that altitudinal threshold up or down, the possibility

[2] Description of an epidemic between 1854 and 1856 in the city of Cusco suggested yellow fever, even though the mosquito vector cannot normally survive the low temperatures that descend over the Cusco Valley during most of the year. A warm period in October—usually the warmest month of the year—possibly permitted the mosquito to survive and spread the disease.

exists that elevation limits of disease contraction have also moved from year to year. Taking those variations into account, the "safe" upper limit to avoid contraction is 2000 m asl. Defining the upper limit of leishmaniasis at 800 m asl is incorrect (Lumbreras and Guerra 1985).

The two diseases appeared at different times in history. Leishmaniasis posed a serious health problem both before and after the Spanish Conquest right up to the recent past. By contrast, malaria, in this valley and elsewhere in the Andes arrived with the Conquest. Spaniards brought in their blood or liver at least one form of that disease and African slaves who disembarked later brought another form of it. Since neither malaria nor leishmaniasis is contagious, by avoiding the hot country contraction was also avoided. However, agricultural laborers, both before the Conquest and after, were conscripted and sent to this problematic environment. Agents called *enganchadores* moved Indians to lower elevation to work on haciendas. Sometimes the labor need was so dire that goons known as *guatucos* kidnapped workers. Some entrepreneurial highland peasants had enough economic motivation to risk entering the zone. There they acquired coca leaf and brought it back on mules to their home villages. These pendular movements between environments had disease ramifications that affected the population of the entire Urubamba Valley.

3.6.1 *Characteristics of Leishmaniasis*

This zoonotic malady, caused by the protozoan *Leishmania braziliensis*, is spread through the bite of a sand fly popularly called in La Convención *manta blanca* or *q'hete*. Other species implicated are *L. peruviana* and *L. lainseni* (Grimaldi et al. 1989; Lucas et al. 1998). Two different genera and dozens of species of sand flies, only some of them anthropophilic, occur in the Andes. The disease they spread is present in the Urubamba in two clinical forms: cutaneous leishmaniasis (CL) and mucosal or muco-cutaneous leishmaniasis (MCL). Peasants group both CL and MCL as *jucuya*. Migrants to the Urubamba from Apurimac where leishmaniasis is also known, used the term *tiac araña*. Still other names were *espundia*, *quepo* and *llaga*. Most historic descriptions of it refer to MCL, for it was this form that prompted horror and disgust. One colonial description was of an affliction combining syphilis and leprosy in which the flesh is eaten away (Villanueva Urteaga and Macera 1982:262). Another characterized it as a "cancer that eats the nostrils" (Vega 1896:VI:128). de Lizárraga (1987:170) mentioned the "trompa gruesa y colorada," i.e. the elephant-like nose that is an intermediate stage in the development of MCL.

3.6.1.1 Pre-Conquest Leishmaniasis

As a protozoan infection, MCL has a drawn-out sequel which probably explains why its etiology was not clarified until the twentieth century. It starts with an infected sand fly bite on the arm or leg. The bite slowly enlarges into a scabby nodule.

One to six months later, the parasite, spreading through the lymph or blood, starts to invade the mucous membrane of the nose, mouth and throat (Walton 1987). Over succeeding months, sometimes years, the nasal septum and the cartilage of the palate, pharynx, and epiglottis of the vocal chords, larynx and trachea are eaten away. Death from starvation occurs when necrosis of these tissues makes the victim unable to swallow. Before that happens, death frequently results from pneumonia brought on by blockage of the airways.

The gruesome facial disfiguration of MCL in its advanced stage was a powerful reminder of the connection between disease and environment. MCL was a geographic metaphor acknowledged in Andean culture history of the dangers lurking in the hot valleys. This manifestation of the disease has long been present in coastal valleys, intermontane depressions in the highlands, and on the eastern slope. Long before the Inca, molded ceramics dating from the Mochica period (100–700 AD) displayed the characteristic facial disfiguration that left no doubt about which disease was represented (Gantzer 1972; Altimirano et al. 2001). Paleopathologist Marvin Allison (1993) interpreted a skull of Tiwanaku age (500–1000 AD) retrieved from the Atacama as having a deformation caused by mucosal leishmaniasis. It is reasonable to believe that the Incas were aware of this disease and knew where, but not why, it was contracted.

So striking was the association between the hot valleys and this insidious affliction that it elicits a powerful reason for why the Incas chose not to permanently settle this part of the valley. The scarcity of Inca ruins below 1500 m asl is, I believe, strong evidence of that choice (Fig. 3.10). The early colonial documents comment on the relative lack of permanent residents in the late Inca period below the 1800 m asl threshold. To minimize time spent in this zone, the people who set out and cared for the coca plants were workers (*cocamayoq*) brought from elsewhere. If present In the hot country over the long term, they may have become acclimatized enough to minimize disease contraction. Coca workers established their dwellings just above the zone of disease contraction, but still within walking distance of coca fields. In peridomestic settings, sandflies bite at night; to avoid them meant locating dwellings above the normal distribution of the insect. Another arrangement involved pendular movement, in which coca workers were exposed for only several weeks. Though not a failsafe solution, it nevertheless became institutionalized as a way to cope with an endemic disease. The most permanent coca workers were criminals who had been exiled to the coca zone to work there as punishment.

3.6.1.2 Post-Conquest Leishmaniasis

After the Conquest, accounts linked MCL with coca growing when Spaniards became involved in coca production and commerce. With no interest in using coca themselves, Spaniards understood the large profits that coca represented. They took over *cocales* the Incas had planted and had workers clear forested land to make more plantings. When the *mitmaqkuna* whom the Inca had sent to the lower Urubamba left the zone in the 1550s, labor shortages constituted the biggest constraint to leaf

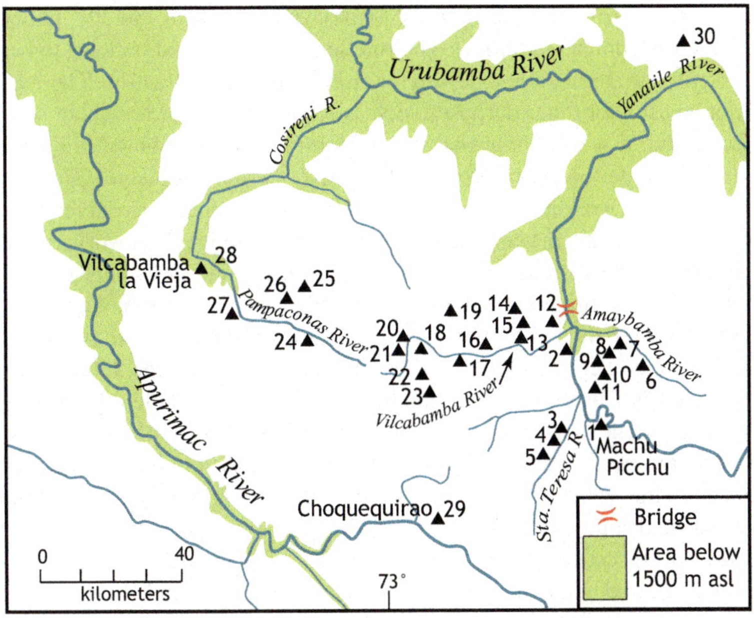

Fig. 3.10 Major Inca sites north of Cusco and their relationship to endemic leishmaniasis which occurs mainly below 1500 m asl. Main Urubamba depression: 1. Machu Picchu (2400 m), 2. Sapamarca (2698 m). Santa Teresa Valley: 3. Chaqui Orcco (3500 m), 4. Plateriayoc (2621 m), 5. Llactapata (2650 m). Amaybamba: 6. Wamanmarka (1889 m), 7. Qochapata (2504 m), 8. Incacárcel (2275 m), 9. Pasto Grande (2275 m), 10. Maqtuyoc (2168 m), 11. Luqmayoc (2161 m). Vilcabamba: 12. Chuquichaka (1250 m), 13. Mesakancha Alta (1565 m), 14. Media Luna (2421 m), 15. Kurkupata (2223 m), 16. Marraniyoq (1746 m), 17. Kinoagarqay (2015 m), 18. Inkahuarkana (2316 m), 19. Inkawasi (3920 m), 20. Pillau (2615 m), 21. Pinkullunka (2960 m), 22. Vitcos (3080 m), 23. Chqupalta aka Yurak Rumi (3050 m). Pampaconas: 24. Ututo (2180 m), 25. Yanakachi (2540 m), 26. Palmapata (2503 m), 27. Tambo (2180). Consevidayoc: 28. Espiritu Pampa aka Vilcabamba la Vieja (1210 m). Apurimac: 29. Choquequirao. Chirumbia: 30. Mosoc Llacta (1800 m). (Map by M.K. Gade)

production. Coerced into going to the valley to do this work, many workers contracted leishmaniasis and died (de Santillán 1968; Loayza 1889; Vega 1896; de Quiroga 1922; de la Bandera 1965:I:178–180). The remarks of Pedro Pizarro in 1571 encapsulated the connection: "…the natives die in this (coca) trade especially those who enter the Andes, for it gives them a sickness of the nose like that of Saint Anthony, and which has no cure, albeit there are some remedies for checking it, yet in the end, it returns and kills them. This sickness attacks those Indians who are not natives born and bred among the Andes, or even touches some of those who are born there, and for this reason there are so few of them" (Pizarro 1965:218).

Unhealthiness of the "Andes del Cuzco" prompted Viceroy Toledo in 1573 to seek protection of coca workers by issuing an "ordenanza de la coca," the aim of which was to impose rules on coca production (Spain 1841:II:288–289). The decree obligated owners of coca plantations to pay their workers directly rather than through their headmen (*caciques*). The rule required work assignments to be made according to a strict schedule to lessen the risk of disease contraction. Their services could not be used for other purposes such as fetching water or firewood. As during the Inca period, it was thought that risk of contraction of leishmaniasis decreased when the work period was restricted to a few weeks. The decree also directed owners to provide workers with beds off the floor and to make sure, given the heavy rain in the valley, that workers had a dry change of clothing. Coca workers and their families were to be supplied with adequate food. Plantation owners were to hire salaried physicians, surgeons and pharmacists to protect the health of coca workers. Many victims were treated for the disease after they had left the valley. The chronicler Bartolomé de Vega (1896:VI:128) wrote that at any one time 200 Indians were in the hospital in Cusco—the only medical facility in the whole region—with the disease. The extent to which plantation owners complied with the points of this Toledan decree in a zone where law enforcement was minimal is not recorded. Nevertheless, the decree informs scholars how the problem was viewed at the time and its indicated solutions.

In the tropical section of the Urubamba and its tributary valleys, coca growing in the colonial period gradually, though discontinuously, extended from Ccollpani down to below Quillabamba and up the tributary valleys of the Amaybamba, Vilcabamba, Yanatile, and Occobamba. This distribution overlapped between 700 and 1500 m asl the zone where the disease was most virulent. As an analogy, in the Yungas of Bolivia, the Desjeux (1976) team found that of 113 documented case of leishmaniasis, 80 % of infecged people were from altitudes between 1000 and 1700 m asl, 15 % from 400 to 1000 m asl, and 5 % from 1700 to 2500 m asl. In the valley below 700 m elevation, some evidence suggests that leishmaniasis was less of a problem. The Matsigenka have a word, *cachinori*, for MCL, but its lower level of incidence in their hotter environment may reflect the ecological tolerance limits of the organism. In one study, the minimum temperature threshold of the protozoan *Leishmania* survival was found to be 12 °C and the optimal temperature was between 20 and 28 °C. Above 33 °C, growth of the parasite stops (do Amaral 1940).

Both before and after the Conquest, this disease was a malignant symbol of a stark ecological reality. Entry into the tropical realm of the Urubamba and other eastern valleys exposed one to the risk of a fearsome though unpredictable sequel. Weeks, months or sometimes even years after migrants had returned to their highland communities, their lesions migrated from the leg or arm to the nasopharynx. These individuals were symbols of the mutilation awaiting anyone who ventured into the jungle (Fig. 3.11).

The connection between disease and environment was apparent even without having fathomed its etiology. A person exhibiting metastatic mucosal lesions in the mouth and nose was immediately identified as someone who had spent time in the hot country. It was recognized that the malady was not passed from person to person. To Spaniards, the facial disfiguration reminded them of the transi, the sculpture of human decay adorning tombs of the late Middle Ages, designed to trigger instant revulsion and fear of death.

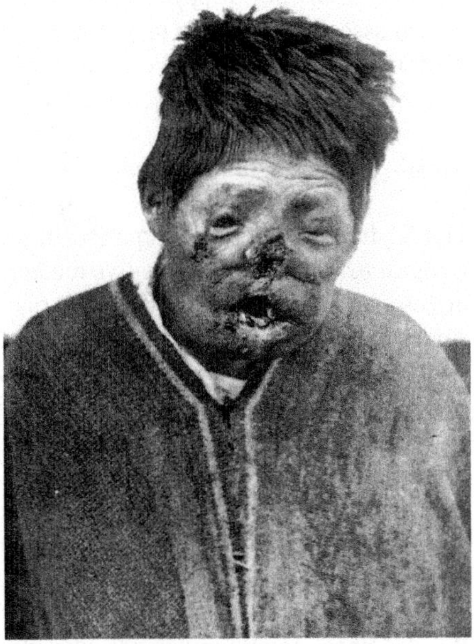

Fig. 3.11 Known as juk'uya in the Urubamba, muco-cutaneous leishmaniasis is a necrotic disease that affects the mucous membrane. Grotesque lesions develop on the face of those who contract it. (Source: Tamayo 1909)

How the Inca might have treated *jucuya* is suggested by medicines available to cauterize the lesions. Several native Amazonian plants, *Plumbago cerulea*, *Plumeria rubra*, and *Hura crepitans*, exude caustic milky juice used in the colonial period to treat lesions and possibly also before the Conquest. Spaniards introduced ceruse (P_6CO_3), the medieval European remedy for the bubos of bubonic plague. Called *solimán* in Spanish, ceruse consisted of highly caustic white crystals which were applied to necrotic tissue. The name "arbol de solimán" suggests that the milky sap of *Hura crepitans* became a substitute for ceruse. In 1776, a Cusco physician, Eugenio Texeyra, stated in a sworn testimony that "leishmaniasis, called by the tropical valley people *hucuya*, was incurable" (Lastres 1956:191). Another eighteenth-century observer wrote that these ulcerous wounds that destroy the face caused by this disease were difficult to cure. He also wrote that the sores derived "from the bites of certain animals too small to see" (Oricaín 1907:350). Sand flies are not imperceptible, yet in that description he correctly detected the zoonotic link.

Since the 1970s, medical intervention has made CL and MCL less of a devastating problem than it was in the past. Pentavalent antimonial marketed as Glucantine®, became the therapeutic mainstay of those with CL or MCL. However, the emergence of drug-resistant strains of the parasite indicates that the disease will not easily be eliminated. Even in the 1990s, health authorities in the Andean countries annually reported almost 13,000 new cases of leishmaniasis in its various manifestations (Davies et al. 2000:927). Those exposed to the disease are increasingly better informed. Treatment has become more common (Camino and Anderson 1993). The folklore that MCL is caused by dirty water, lack of hygiene, dietary change or alcoholism has not disappeared, but it is now less often believed now. Self-help patient associations set up in highland communities, whose members descended to work in the hot country, have lobbied the government for access to antimonials.

3.6.1.3 Critique of an Article

The leishmaniasis story I reported in a 1979 issue of the *Journal of Historical Geography* contained an attentive reading of the full range of chronicles about this disease, but also my knowledge of coca and of the tropical environment east of the Andes acquired largely through fieldwork in 1963–1964 and in 1968. Eight years after publication, an archaeology professor and his then graduate student at the University of Calgary, appeared in print in the same journal with their reply (Gade 1979). They disputed my linking the paucity of permanent Inca settlement in the tropical forest east of the Andes to fear of mucocutaneous leishmaniasis. Their assertion was that social, political and economic considerations accounted for the relative absence of settlement. For both political and economic reasons, the Incas had much to gain from settlement. Reached in 4 days, the Urubamba at Quillabamba had no forest Indians, nor were there any along the way to hinder movement. Besides coca, sumptuary goods like wood and bird feathers could be acquired. Genevieve Le Moine and Scott Raymond, the authors of that article, did not elaborate on either the trade networks or sociopolitical alliances that they claimed foreclosed Inca forest settlements. Their reply also made a series of assertions that contained errors or misconceptions. It also conflicted with Raymond's (1992) own find of pottery remains identified as Wari in the warm tropical section of the Apurimac Valley below 1500 m asl where coca was grown. Significantly for my argument, no pre-Conquest settlement remains were found at that site.

The authors failed to grasp the heterogeneity in the transmission ecology of CL and MCL. Their comment that "no sand flies carrying the parasite have been recorded from houses" is an egregious error (Le Moine and Raymond 1987:116). Contrary to their assertion, evidence is found in Herrer et al. (1980), Lumbreras and Guerra (1985), Grimaldi et al. (1989), Ogusuku et al. (1993), and Villaseca et al. (1993). The vectors are both sylvatic and peridomestic. Le Moine and Raymond assumed people contracted leishmaniasis only while clearing forests. Their sandfly vector was identified as only *Psychodopygus wellcomei*, when, in fact, two particularly anthropophilic species in the genus *Lutzomyia*, *L. peruvianem* and *L. verracarum*, are the dominant vectors. These sand flies bite humans who intrude into the cycle between the vector and the *Leishmania* reservoir. In the wild the reservoirs for the parasites are usually wild mammals, but in peridomestic settings, they are the family dog. Le Moine and Raymond failed to describe the cycle found in peridomestic settings in which sand flies enter rural dwellings and bite people late at night as they sleep.

Secondly, the article ignored a critical point of my argument: that the Incas had instituted a regime of limited work periods in the hot lands specifically to lessen exposure to disease. In the high rainfall and humidity between December and April, the sand fly vector population multiplied (Villaseca et al. 1993). The decrease in disease incidence during the May to October period meshed with the highland cycle whereby in June the major crops had been harvested and peasants were free to pick coca. After a maximum of 30 days in the coca harvest, the workers returned to their ayllus in the highlands from whence they came. The coca shrub yields four harvests a year, but if labor is not available, the leaves do not have to be picked. After the Conquest, the Spaniards, to maximize profits, replaced the Inca traditional short term migration with permanently transplanted coca workers. Sick workers went back to their communities and others were forcibly brought from the highlands to take their place. Disregard for seasonal migration led to widespread contraction of the disease.

Thirdly, Le Moine and Raymond asserted that MCL was "a rare disease." It certainly was not rare in the early colonial period among those who went to lower elevations, which explains why it received so much notice. If, in the Inca period, *jucuya* was less common, the explanations rested on the precautions taken. The chronicler de Loayza (1889) wrote that out of ten who went to work in coca cultivation, only five returned. Two other chroniclers, Pedro de Quiroga (1922:103) and Bartolomé Vega (1896:VI:103), made the observation that most people who lived in the coca country had the disease. Quiroga proclaimed that more Indians died in coca growing than in the civil wars between the Pizarro and Almagro factions. Another Spanish chronicler, Reginaldo de Lizárraga (1987:179), conscious of the devastating effect of this disease, viewed forced displacements to coca plantations as a death sentence ("*llevarlos a la casa de la muerte*").

Travelers in the republican period mentioned the disease. Antonio Raimondi (1898), grand explorer of the Andean backcountry, traveled in the valley in 1865 where he noted that around Huadquiña (1500 m) MCL was quite common ("bastante común"). In the early twentieth century, eight people of an expedition to the Urubamba contracted the disease (Tamayo 1909:29). A 1968 survey I conducted of highland colonists then living in the jungles north and east of Cusco reported leishmaniasis to be a serious health problem. Peridomestic leishmaniasis decreased when DDT was sprayed against malaria; but away from dwellings, the sand fly population was unaffected, thus people clearing forest lands remained vulnerable.

In the fourth place, the authors claimed that the Inca could not have understood causation: "If the Incas did connect espundia with the Montaña, they had a much more sophisticated ontology and etiology of leishmaniasis than did modern medicine until very recently" (Le Moine and Raymond 1987:125). In fact, etiological knowledge about the parasite or vector is not required to make a linkage between disease and environment. The prima facie evidence said it all: when people went to the hot valleys and came back with facial lesions, the connection was obvious to both the victim and those who knew the victim. Pedro de Quiroga (1922:101) wrote that "one cannot look without great horror and fear at such ugliness." Vega (1896:VI:102) made a quite specific observation that mal de los Andes was not passed from person to person ("*no se pega*"), but that the disease was acquired when one actually went to the tropical valleys. One historic Quechua name for leishmanisais was *anti-onccoy*, meaning "disease of the eastern tropical valleys". By extension, *anti* in Quechua referred to the forest-dwelling inhabitants on the lower eastern side of the cordillera. The Spaniards glossed the word Anti to derive Andes, which in the early colonial period specifically designated the eastern slope in the Amazon drainage. Thus, in Spanish usage of the colonial period, "*mal de los Andes*" referred to diseases of the tropical valleys. When contextualized with other kinds of information, the expression most commonly referred to leishmaniasis (Basto Girón 1977:49). Only in the nineteenth century did the word Andes take on its present denotation as the highland region of Western South America.

As a last point in their commentary, Le Moine and Raymond failed to come to grips with the correlation between verticality of the disease and the distribution of pre-Conquest settlement. Sorting out that linkage requires a field-based knowledge of settlement sites in southern Peru and their environmental settings. Normally, the "safe" upper limit in southern Peru is 1500 m asl, though in any one year the weather can bring higher or lower temperatures, thus changing somewhat the upper limit of contraction. Where leishmaniasis is endemic, evidence of permanent pre-Hispanic presence is rare. The most important exception to the above generalization is Vilcabamba la Vieja, also called Espiritu Pampa. First brought to world attention by Hiram Bingham, this site received much greater publicity in the early 1960s when Gene Savoy explored the zone. Located between 1200 and 1300 m asl, Vilcabamba la Vieja lies well within the leishmaniasis contraction zone. However, it represents a special case of settlement under duress: it was Manco Inca's refuge, a place safe from Spanish attack after Pizarro had conquered Cusco.

Since the 1970s, further explorations in the Urubamba drainage north of Cusco have documented that Inca settlement was overwhelmingly above 1500 m asl. The settlement associated with Choquechaca (1200 m) suggests that the need to control ingress and egress to the bridge overrode other considerations. Between 2500 and 2750 m asl on the side of Cerro Yanatin, a series of early Inca settlements existed that historically were known as Sapamarca. They housed people who then may have worked at lower elevations. In the Vilcabamba Valley proper, five Inca sites were above 1800 m asl and only one, Maranniyuq (1740 m asl) fell below that altitudinal threshold.

In the Santa Teresa Valley, which joins the Urubamba at 1500 m asl, the three important Inca sites were all above the zone of leishmaniasis contraction: Chaqui Orcco (2506 m asl), Colpapampa/Platerioyoc (265 l m asl) and Llactapata at 2650 m asl. In the Pampaconas Valley, except for Vilcabamba la Vieja, the three known archaeological sites are all above 2000 m asl. Many Inca settlements not in a macrothermal climate were, however, accessible to coca fields. If viewed as an intentional Inca health measure, that separation of work site and place of habitation cannot be dismissed. Likewise in the Lucumayo (Amaybama) Valley that joins the Urubamba on its east side, *mitmaqkuna* brought from other warm climates cultivated the coca fields on a permanent basis. But there too, living sites and coca fields appear to have been separated. Coca grew on the valley floor and slopes below 1900 m asl but the Inca located their dwelling sites between 1889 and 2712 m asl. The three largest sites, Llacta Pata (2656 m asl) and Incacarcel (2326 m) and Incatambo (2127 m asl), were areas of coca cultivation ordinarily above the limit of disease contraction. At Huamanmarca (1889 m asl), 11 km from the town of Huyro, Pachacutec built a palace.

Other Inca sites, Luqmayuq (2161 m asl) and Nust'ayuq (2168 m asl) and Maqt'ayuq (2391 m asl) were above the normal disease limit. After the Conquest, when this valley became the encomienda of Hernando Pizarro, mitmaq workers remained to harvest coca there. With the assassination of Manco Inca in 1544, they returned to their home communities and new coca workers had to be found.

Le Moine and Raymond argued that social, economic and cultural factors, not leishmaniasis or any other disease, inhibited the Incas from settling the hot forested country. The archaeological circle of affinity to which Raymond belonged followed the line of thinking of its main protagonist, Donald Lathrop (1927–1990) of the University of Illinois, who held to an interpretive agenda that the Western Amazon Basin had developed a high cultural level. The Lathrop school of thought was not prepared to view environmentally-related disease as a hindrance to cultural development (Raymond 1992:31). Yet nothing in my remarks in that article was meant to suggest that disease in the rainforest of the Amazon Basin precluded high cultural levels.

3.6.2 Malaria as the Post-Conquest Scourge

Like leishmaniasis, the risk of contracting malaria is affected by temperatures related to altitude above sea level. Malaria arrived in Peru in the sixteenth century with Spaniards who carried the *Plasmodium vivax* parasite in their blood. Spread of the disease essentially depended on one native species of anopheline mosquito. Although more than 40 species of anophelines have been described for Peru, the only one known to have transmitted malaria in the coastal and intermontane valleys up to 1700 m asl is *Anopheles pseudopunctipennis*. This markedly anthropophilic species bites humans in and around the domicile in the 6-h period leading up to midnight (Calderón et al. 1995). Much more than infected mosquitoes, infected humans carried the disease organism far and wide within and between macrothermal environments. Uninfected mosquitoes acquired the parasite by feeding on the blood of infected humans. *P. vivax* had established itself in the Iberian homeland, but if a Spaniard sailed to the Indies free of it, he could still have easily contracted malaria in the steamy Panamanian Isthmus, a strip of land widely considered to be the most malignant zone of the New World. Colonists going to Peru from Spain had to cross this unhealthy 50-km stretch. The *vivax* parasite has a dormant cycle by which it persists for years in the livers of comparatively healthy humans, thus facilitating contraction. With Spaniards as carriers and local mosquitoes as vectors, the argument offered by Linda Newson (1993:1190) that the Andean highlands acted as a barrier to the spread of malaria from the coast to the eastern valleys is invalid. An infected human sufficed, for the mosquito vector was already present. Muleteers, who regularly moved through several environmental zones and so brought the parasites in their blood to places where the disease was previously unknown, were major human spreaders of the disease in the Andes. The other form of malaria, *Plasmodium falciparum*, most plausibly arrived in the blood of African slaves. Both kinds of

Fig. 3.12 Fiestas in Cusco Department often included performers who symbolized the connection between malaria and the "yungas" i.e. the hot country. Here a nineteenth-century rendering of a festival in Cusco feature a man of the jungle carrying a cinchona tree. Source: Marcoy (1874)

malaria had lethal or seriously morbid consequences on native people. Their lack of resistance to the parasites made Indians "prisoners of the disease of the lowlands" (Basto and Luis 1977:49). When malaria did not kill, it rendered its victims feeble, disconsolate, and generally unable to work or even perform simple tasks. Thus malaria made them vulnerable to other diseases, especially tuberculosis. Indians from the high country who went to work on estates both on the coast and in low intermontane valleys more often than not contracted malaria. All highlanders associated malaria, called *chucchu* in Quechua, with contraction in the "valle," meaning the quasi-tropical zone. At highland fiestas, choreographic troops wearing yellowish face masks performed the "dance of the chucchu," associating malaria sufferers with the hot country (Fig. 3.12). As with leishmaniasis, people did not have to understand the etiology of the disease in order to make the environmental connection between hot climate and disease contraction.

The same vector, *A. pseudopunctipennis*, carried both species of parasite. Together these two *Plasmodia* decimated many native people on the coast and in intermontane valleys below 2500 m asl between 1550 and 1950. Children were especially vulnerable to the disease. Contraction rendered many who survived incapable of productive work. In the New World, epidemic malaria of *P. vivax* and *P. falciparum*, both transmitted by just one vector, *A. pseudopunctipennis*, triggered a much higher mortality rate than the normal endemic pattern.

3.6.2.1 Malaria in the Urubamba

For four centuries malaria prevailed in Peru. An early twentieth century survey of every province in each department recorded endemicity of the disease in all valleys of the coast, intermontane depressions below 2000 m asl, and throughout the forested montaña to the east (Lorente and Cordova 1925). The physician Erving (1912:338) erroneously believed that malaria was endemic only below Rosalina (600 m asl), but that epidemics of it reached as far upstream as Echarati (750 m asl). In fact, endemic malaria occurred up to 1500 m from the colonial period until ca. 1960. In stark contrast to that pattern of endemicity, the highlands above 2500 m asl were free of it. Temperature thresholds, since they affected the life cycles of both the mosquito vector and the parasite, rendered malaria a vertically controlled disease. *Plasmodium* has a cycle of sporogeny that extends from the time of gametocyte ingestion to the time when sporozoites are found in the salivary glands of the mosquito (Boyd 1949). With *Plasmodium vivax*, that cycle occurs within the temperature range of 15–30 °C (Stratman-Thomas 1940). At 15 °C it takes 38 days to complete the cycle; at 28° it takes only 8 days. *Plasmodium falciparum* requires more heat to develop than *P. vivax*. Temperatures below 18 °C abort the development of *falciparum* and prevent completion of its sporogenic cycle.

In the upper valley, some parts of the depression have an annual average temperature of 15 °C, but at night temperatures fall below that in every month making it impossible for *Plasmodium* to complete its sporogenic cycle. Thus the boundary of malaria contraction normally lay at about 1500 m asl; there the daily temperatures rarely fall below 18 °C. Between 1500 and 1000 m asl, only *P. vivax* could develop, whereas below 1000 m, the two kinds of malaria overlapped. From 1500 to 2000 m asl, *P. vivax* may have episodically occurred when the annual temperature and rainfall pattern in a year deviated from the normal.

No colonial document unambiguously identifies the early presence of malaria in the Urubamba, a measure of the imprecision attached to naming this disease. Use of malaria, a term of Italian origin, does not in the Andes predate the twentieth century. The Spanish word *paludismo* does not seem to predate the late nineteenth century. In the eighteenth century, the common Andean word for the disease was *tercianas*. In the seventeenth century, the term *mal del valle* ("disease of the valley") sometimes applied to malaria, sometimes to leishmaniasis. *Fiebre* ("fever") and *camara* were two other disease terms in Spanish, that in certain contexts and places, referred to malaria.

Besides the confusing nomenclature for the disease, paucity of inhabitants, remoteness, and neglect of the authorities kept the several pathologies of the area hidden. The presence of malaria can be deduced, for Spaniards would not have brought in black slaves to the lower valley if disease had not decimated the Indian labor force. Many Indians knew what was in store for them in the hot valleys and refused to go there of their own free will. High incidence of malaria among Indian workers forced estate owners to substitute African slaves as the main labor force. Adult resistance to *P. vivax* was related, as medicine has come to learn, to the lack of Duffy proteins on the surfaces of their red blood cells to which *vivax* parasites attach. From 30 to 40 % of Africans were also immune to the effects of *falciparum* malaria as a result of the deficiency of the enzyme glucose-6-phosphate-dehydrogenase. Resistance to malaria made black slaves valuable as laborers in the lower valley. Aware of their health advantage without knowing why, the Jesuits brought African slaves to work on their sugar estates in the Urubamba as they did on many of their properties where they grew sugar cane (de Ocampo 1906:336). The order had learned that black workers were superior to Indians in warm climates. No one then understood what was verified in the late twentieth century, namely, that Africans hold genetically-linked polymorphisms in their hemoglobin.

That linkage, the result of millions of years of humans and the malaria parasite evolving together, provided the insight that Africa was the parasite's place of origin. In the Andes of South America below 2000 m asl, it explained why blacks stayed healthy when Indians died or got so sick they could not work. Since slaves were expensive and Indians cheap, only the fact that Africans as a group rarely contracted malaria as adults did that substitution make economic sense. Blacks, however, since they were deprived of liberty, were also prone to revolt. In 1603, a massive uprising of slaves led by a Piro Indian burned three of the ten sugar mills then operating in the valley (de Ocampo 1906:185–187). The rioters also destroyed many houses and killed 24 Indians, including one headman (*curaca*), suggesting that highland natives had not joined the uprising. After that event, slaves were viewed as a problematic solution to hacienda labor. The Jesuit departure from the valley in 1767 brought further decline of the black presence in the Urubamba. Even before the nineteenth-century abolition of slavery in republican Peru, blacks had largely disappeared in the Urubamba. No prima facie evidence of negroid phenotypes remained in the area in the 1960s.

Fear of malaria accounted for the perpetual labor shortage in the valley estates. In order to entice enough highlanders to work on their properties, hacendados tried to develop a relationship with certain labor sources. Hacienda Sahuayaco, for example, got most of its migrants from Urubamba and Calca, whereas Hacienda Potrero had contacts with communities in Apurimac Department. Some estates worked through *enganchadores* who, for a fee, persuaded or entrapped highland Indians to come to the valley. A certain number came of their own free will, but with no immunity to disease, they suffered accordingly. A glimpse into the stark connection of climate and malaria came from the comments of two North American travelers. In 1911 Isaiah Bowman (1912:881–882), a member of Bingham's expedition to Peru, met a priest on the trail into the Urubamba tropical zone. When Bowman asked him why more Indians did not live in the "warm and pleasant lower portions of the valley," he was told that the "savages" (i.e. uncivilized forest tribes) lived down there. Reasoning that the small number of those forest-dwelling "salvajes" could not have been a major hindrance to settlement, Bowman pressed his interlocutor and received a more satisfactory second answer posed as a rhetorical question: "doesn't the stranger [sic: foreigner] know that fever-spirits live in the valley?" Five years later, another American, Harry Franck (1917:461), on his cavalcade through the same canyon, encountered "yellow, fever-eyed walking skeletons, straggling languidly up from the tropical valleys." These gaunt and hollow-eyed Indians, were of a "lifeless caste of countenance." Pendular migrants had little time to acquire immunity, whereas those who had established themselves permanently over years had a better chance of surviving the disease.

In the Urubamba, malarial mosquito populations peaked not in the rainiest month, but in May and June when evaporating pools of water formed green algae mats (*Spirogira* spp.) on which *A. pseudopunctipennis* laid her eggs. Thus decreased rain increased the number of breeding places until those pools dried up between July and November. Malaria contraction was misunderstood in La Convención and throughout much of Western South America until the peculiar habits of this species of mosquito were clarified in the 1940s (Shannon 1930).

3.6.2.2 Epidemic Malaria

Epidemic malaria may have occurred episodically in the colonial period, but the tropical zone of the valley was so isolated that epidemics in that zone went unrecorded in Cusco. The earliest recorded malaria epidemic, in 1878, broke out first in Matsigenka Indian territory and spread up the Urubamba as far as Echarati. Many people who died in that year had come without immunity from the highlands to the selva to collect rubber. In 1897, another malaria epidemic forced abandonment of haciendas as far up valley as Cocabambilla. In both those epidemics, the Matsigenka moved to higher elevations in the hills, away from the river to escape the disease. A pattern of upward displacement seems to have been a survival tactic against malaria.

The most devastating paludic outbreak ever known in the valley occurred three decades later. In 1932 and 1933, malaria devastated the whole zone below 1500 m asl. Just as in 1878 and 1897, the epidemic spread upstream from Koribeni Mission as far as Huadquiña (1500 m asl). *Anopheles pseudopunctipennis* was the only vector species implicated, but different mosquitoes carried either *P. vivax* or *P. falciparum*. An estimated 7000 people died; another calculation increased the number of victims to 16,000 (Paz Soldan 1933; Cueto 2001:88). Virtually everyone who did not die either got sick or left the zone. Whole families died in their huts without medical attention. Abandoned dwellings came to be seen as burial chambers. Since the malaria victims were mainly pendular migrants, their home communities high above the zone of contraction also suffered population loss. Having fled the hot valley before they became delirious, many died in those highland communities (Chiriboga and Donayre 1947). A 1916 Peruvian law requiring haciendas to give their workers free quinine and to destroy mosquito larvae had been ineffective. The small number of living souls who remained in the tropical hacienda zone languished, sick and hungry, unable to perform any productive activity. Compounding their misery, many of them also suffered from tuberculosis. Hacendados left their estates for safe-havens in Cusco. Though their properties desperately needed attention, estate owners feared returning to the valley long after the epidemic subsided. Almost all workers vanished; coca fields had no pickers and sugar cane no cutters. At that time, about 40 % of the cultivated land between 800 and 1500 m was in coca; 30 % in sugar, and 30 % in food crops.

For a decade between 1933 and 1942 the tropical valley from Huadquiña to Rosalina was a scene of total desolation (Cornejo-Bouroncle 1975). Death or flight accounted for the absence of people. Commerce came to a standstill. Middlemen (*vallunos*) who had long been the intermediaries between the highlands and lowlands refused to go into the malarial zone. As a human disaster, the La Convención malaria epidemic stands as a transcendental event in the annals of public health in Peru. Administrative disorganization of the state and department was total. Belated response from the government in Lima worsened the tragedy. Although national authorities provided funds to construct a hospital in Quillabamba, that facility was inadequate. The epidemic itself came to an end in spite of, rather than because of, public health interventions. Endemic malaria, however, remained a health problem in La Convención into the 1950s. Starting in 1945, the first campaign of spraying DDT on dwellings had a startling effect on mosquito populations. Elimination of breeding sites did not occur until the anti-malaria service learned about the particular habit of this anopheline vector. Spraying pools of standing water was ineffective; only later did they learn that the mosquito bred on algae mats. The World Health Organization, Rockefeller Foundation, and the Peruvian Ministry of Public Health developed programs to spray all dwellings several times a year, spray breeding areas, and dispense malaria medication to inhabitants. So successful were these three interventions that public health officials frequently spoke and wrote in terms of malaria eradication.

3.6.2.3 Post-Epidemic Urubamba

In 1947 migrants began to trickle back to the valley, lured by the prospect of growing coffee. World coffee prices, which sharply increased through the 1950s and 1960s, became a profitable cash crop everywhere in Latin America. The pre-epidemic pattern of haciendas renting land to arrendires continued, though now the renters wanted to grow coffee trees. In-migration increased the population. In 1940 the census of the province counted 27,243 people, half of whom lived in the highlands above 3000 m asl, beyond the reach of malaria; by 1960, population had increased more than 100 % (Perú 1944, 1966). Thus, very few families there in 1960 had histories in the valley that predated the twentieth century. With continued control of malaria, the population surged to 166,000 by 2007. Yet, during that time agencies diverted money away from spraying programs and malaria was not eliminated. Complacency had led to deterioration in the malaria control program, allowing the disease to again appear between 1969 and 1973. The number of cases rose more after 1987 when DDT was declared to be environmentally dangerous and not to be used inside buildings. Moreover, anti-malarials that once were effective became less so as *Plasmodium* rapidly evolved to outwit the synthetic compounds brought onto the drug market. By 1993, the malaria parasite index had increased more than 90 % (Roberts et al. 1997:300). One hypothesis is that decades of malaria eradication had led to separate allopatric parasite refuges where distinct clonal lineages evolved through inbreeding and genetic drift. Anopheline vectors also move as a result of deforestation. Long the major malaria carrier in the Brazilian Amazon, *A. darlingi*, has now become the primary vector of *falciparum* malaria in Loreto Department around Iquitos (Aramburu Guarda et al. 1999). Deforestation of the Amazon tributaries, including of the lower Urubamba Valley, may facilitate the migration of *A. darlingi* to La Convención. Its biting behavior is known to be more aggressive than A. *pseudopunctipennis*.

The greater epizootic health problem today is dengue, spread by an introduced peridomestic mosquito, *Aedes aegypti*. This vector entered Peru from Brazil in about 1984 and by 2011 had reached La Convención. Another vector-driven malady, recent in the Urubamba but ancient elsewhere in the Andes, is bartonellosis, also known as Carrion's disease, which occurs in two clinical phases. An acute septic phase of high fever is followed from 2 weeks to several years later by the eruption of nodular skin lesions ("verruga peruana") on the head. The pathogen, *Bartonella baciliformis*, is an intracellular bacterium spread by the same sand fly, *Lutzomyia peruensis*, that transmits leishmaniasis. Reservoirs of the bacterium include domestic guinea pigs and chickens. Its distribution in the Urubamba, only documented in 1997, is at elevations between 2600 and 3200 m asl (Castillo-Schilder et al. 2008). Whether the vector has always been there or is a recent arrival is not known, but the disease is still rare in the valley.

3.7 Human Social Verticality

Up to a point, altitude can be studied as social distributions. In Western South America, people differ in certain respects according to elevation above sea level. Dress is an obvious difference; at elevations above 3500 m asl, people wear wool whereas in the lower Urubamba, cotton is the textile of choice. The logic cannot be pushed too far, however, for dress, in the end, reflects culture not environment. Native people in the far south of South America mostly went naked. In the microthermal portions of the Urubamba Valley, soups and stews that warm the body are common food preparations. But even in the jungle, those foods are common, reflecting, in part, many people's highland origins. Other factors such a religion have less of an obvious environmental connection. In general people at higher elevations are more religious than those in hot climates. Is it that divinities come to the fore when life is more tenuous? High elevations require more subsistence effort than does the jungle where fruit and fish are abundant. On another front, my own experience is that violence is more common at high elevations. Canchis, for example, has many murders even in indigenous communities (Paponnet-Cantat 1991). Connecting violence to excessive intake of alcohol is a useful hypothesis, for high elevations have favored binge drinking. But if temperatures play some explanatory role, what about oxygen, which also varies with elevation above sea level?

3.8 Conclusion

Duality suffuses Andean culture history and is conceptualized in terms of "upper and "lower." The Andes and, in particular, the Urubamba Valley, encourage me to think how verticality functions in an intricate interplay of space, time, culture, and ecology. Although biophysical forces constrain human decisions, people are capable of outwitting or attenuating these same forces. With cultivated plants, humans voluntarily decide which and where cultigens are grown within generally understood thermal tolerance limits. Climatic changes may have moved crop boundaries in the past and will likely do so in the future. Change through time is best documented for the sixteenth century European introductions. Though competition with Old World plants marginalized some native cultigens, details on how that actually happened can only be inferred (Gade 1969). Of the two leguminous species, tarwi (*Lupinus mutabilis*) and broad bean (*Vicia faba*), in competition at the same elevation range, both are resistant to disease and to low temperatures. Tarwi, however, requires processing to remove bitter alkaloids in the seeds (Fig. 3.13). Broad bean gained ascendancy though tarwi has survived and, if the alkaloid can be bred out of the seed, it may have a revival. In recent decades land tenure changes have occurred that have changed agriculture though one cannot infer that crop boundaries will change.

No evidence yet exists that agronomic advances have increased crop hardiness enough to shift crop boundaries. Greater productivity has always been the prime

Fig. 3.13 A field of tarwi (*Lupinus mutabilis*) grown by a indigenous farmer. It grows well at high elevations, but requires processing to remove the alkaloid in the seed. Community of Chiara, Department of Canchis. (Photo by DW Gade 1968)

objective. A crop boundary can move when humans provide special protection to plants, allowing them to grow where otherwise they would not. Advances in understanding and treating plant and zoonotic infections affect crop and disease boundaries. Malaria transmission has decreased to a point that the risk factor no longer enters into a migration decision. If malaria is contracted, modern drug therapy is available to cure it. Effective new cures are available for leishmaniasis. The public health system is also now much better able to respond in times of need. Low disease contraction clouds evidence that vector-borne disease boundaries have changed. Neither disease, however, has been eliminated and other vector-borne diseases have been recorded in the valley. How dengue will develop in the future remains a concern. With yet no way to control it, a new disease verticality factor has entered the Urubamba.

References

Acock MC, Lydon J, Johnso E, Collins R (1996) Effects of temperature and light levels on leaf yield and cocaine content in two *Erythroxylum* species. Ann Bot 78:40–53

Allison MJ (1993) Leishmaniasis. In: Kiple KE (ed) The Cambridge world history of disease. Cambridge University Press, New York, pp 832–834

Altimirano MJ, Moreira JS, Marzuchi MCA (2001) Lesión lítica craniana por leishmaniasis in Makat-Tampu durante el Imperio Inca, siglox XV-XVI, Valle del Bajo Rimac, Peru. Revista del Museo Arqueológico e Etnológico 11:227–242

Appel JW (1994) Francisco José de Caldas: a scientist at work in nueva granada. American Philosophical Society, Philadelphia

Aramburu Guarda J, Asayag CR, Nitzig R (1999) Malaria reemergence in the Peruvian Amazon region. Emerg Infect Dis 5(2):209–215

Basto Girón LJ (1977) Salud y enfermedad en el campesino peruano del siglo XVII. Seminario de Historia Rural Andina. Universidad Nacional de San Marcos, Lima

Bennett CF (1976) Review of plants, man and the land in the Vilcanota valley of Peru. Annals of the Association of American Geographers 66(4):636–637

Berry W (1981) The gift of good land: further essays cultural and agricultural. North Point Press, San Francisco

Bird RMcK (1976) Vegetation and culture: review of plants, man and the land in the Vilcanota valley of Peru. Science 193(4254):669–670

Bowman I (1912) The canon of the Urubamba. Bull Am Geogr Soc 44:811–897

Boyd MF (1949) Malariology. Saunders, Philadelphia

Brush SB (1977) Mountain, field and family: the economy and human ecology of an Andean valley. University of Pennsylvania Press, Philadelphia

Calderón GR, Valle F, Valle J (1995) Especies de la faune anofelina, su distribución y algunas consideraciones sobre su abundancia e infectividad en el Perú. Revista Peruana de Epidemiologia 8(2):1–20

Camino L, Anderson A (1993) Pastores altoandinos:lenguaje y conocimiento de la leishmaniasis una approximación empirica. Revisita Peruana de Epidemiologia 6(1):1–5

Carroll CR (1977) Review of plants, man and the land in the Vilcanota valley of Peru. Q Rev Biol 52 (4): 462–463

Castillo-Schilder R, Campos CT, Bedoya DY, López PV (2008) Conocimientos, actitudes y prácticas respecto a la bartonelosis aguda [fiebre de la Oroya] en los pobladores del districto de

Ollantaytambo, Provincia de Urubamba en el Valle Sagrado de los Incas. Cusco, Peru. Acta Médica Peruana 25(2):58–62

Chiriboga J, Donayre R (1947) Epidemiologia en los Andes y el problema de las migraciónes humanas. Revista Médica Peruana 20(239):701–780

Cobo B (1956) Historia del Nuevo Mundo [1653], 2 vol. Ediciones Atlas, Madrid

Cook OF (1916) Staircase farming of the ancients. National Geographic Magazine 29(5): 495–534

Córdova Aguilar H, Bernex N (2008) El amauta Javier Pulgar Vidal. http:ciga.pucp.edu.pe:1–6 (accessioned 20 September 2014)

Cornejo-Bouroncle J (1975) Tierras ajenas. Ediciones Inca, Cusco

Covey RA, González DA (eds) (2008) Imperial transformations in sixteenth-century Yucay, Peru (Memoir No. 44) Ann Arbor Museum of Anthropology, University of Michigan

Cueto M (2001) The return of epidemics: health and society in Peru during the twentieth century. Ashgate, Burlington

Davies CR et al (2000) The epidemiology and control of leishmaniasis in Andean countries. Cadernos de Saúde Pública (Rio de Janeiro) 16(4):925–950

de Acosta J (1940) Historia natural de las Indias (1590). Fondo de Cultura Económica, Mexico City

de Castelnau F (1850–1859) Expéditions dans les parties centrales de l'Amérique du Sud, 15 vol. Bertrand, Paris

de Haan S, Núñez J, Bonierbale M, Ghislain M (2010) Multilevel agrobiodiversity and conservation of Andean potatoes in central Peru. Mountain Res Dev 30(3):222–231

de la Bandera D (1965) Relación general de la disposició y calidad de la provincial de Huamanaga-año 1557. In: de la Espada MJ (ed) Relaciones geográficas de Indias, vol 1/3. Ediciones Atlas, Madrid, pp 178–180

de Lizárraga R (1987) Descripción del Perú, Tucumán, Rio de la Plata y Chile (1609). Historia 16, Madrid

de Loayza R (1889) Memorial de las cosas de Perú tocantes a los Indios (1586). Colección de documentos inéditos para la Historia de España, vol 94. Imprenta de Miguel Ginestra, Madrid, pp 554–605

de Ocampo CB (1906) Descripción y sucesos de la historia de Vilcabamba (1610). In: Maurtua VM (ed) Prueba peruana: Juicio de límites entre Perú y Boliviatua, vol 7. Editorial Rodrigues, Barcelona, pp 306–344

de Quiroga P (1922) In: Cuevas JZ (ed) Libro intitutlado coloquios de la verdad, trata de las causas e inconvenientes que impeden la doctrina e conversión de los indios de los reinos del Perú (1563). Tip Zarzuela, Seville

de Santillán H (1968) Relacion del origin, descendencia, politica y gobierno de los Incas (1563–1576). In: Barba FE (ed) Crónicas peruanas de interés indígena 970150. Ediciones Atlas, Madrid

Desjeux P (1976) Relations leishmaniose et altitude en Bolivie. In: Ruffiė J, Quilici JC, Lacoste MC (eds) Anthropologie des populations andines. INSERM, Paris, pp 247–258

do Amaral ADF (1940) Observações sobre a resistência das culturas de *Leishmania braziliensis* a varias temperaturas. Arquivo do Instituto Biológico de São Paulo, pp 115–119

Dollfus O (1981) El reto del espacio andino. Instituto de Estudios Peruanos, Lima

Dyer DR (1962) Population of the Qeuchua region of Peru. Geogr Rev 52(3):337–345

Erving WG (1912) Medical report of the Yale Peruvian expedition. Yale Med J 18:325–335

Escobar Moscoso M (1958) Reconocimiento geográfico de Q'ero. Revista Universitaria-UNSAAC 47 (No. 115):159–188

Eshbaugh WH (1977) Review of plants, man and the land in the Vilcanota valley of Peru. Econ Bot 31(3):35–60

Fiorivanti-Molinié A (1975) Contribution á l'étude des sociétés étagés des Andes: la vallée de Yucay, Pérou. Etudes Rurales 57:45–60

Fiorivanti-Molinié A (1976) Compte rendu du Plants, man and the land in the Vilcanota valley of Peru. Journal d'agriculture tropicale et de botanique appliquée 23(1-2-3):59–60

Flores Ochoa J, Fries AM (1989) Puna, qheswa, yunga: el hombre yu medio en Q'ero. Fondo Editorial BCR, Lima
Franck HA (1917) Vagabonding down the Andes. Henry Holt, New York
Gade DW (1967) Plant use and folk agriculture in the Vilcanota Valley of Peru: a cultural-historical geography of plant resources. PhD dissertation, University of Wisconsin, Madison
Gade DW (1969) Vanishing crops of traditional agriculture: the case of Tarwi (*Lupinus mutabilis*) in the Andes. Proc Assoc Am Geogr 1:47–51
Gade DW (1970) Contributions of O.F. Cook to cultural geography. Prof Geogr 2(4):206–209
Gade DW (1973) Environment and disease in the land use and settlement of Apurimac department, Peru. Geoforum 16:37–45
Gade DW (1975) Plants, man and the land in the Vilcanota Valley of Peru. Dr. W. Junk B.V. Publishers, The Hague
Gade DW (1979) Inca and colonial settlement, coca cultivation and endemic disease in the tropical forest. J Hist Geogr 5(3):263–279
Gade DW (1996) Carl Troll on nature and culture in the Andes. Erdkunde: Archiv für wissenschaftliche Geographie 50(4):301–316
Gade DW, Escobar M (1982) Village settlement and the colonial legacy in Southern Peru. Geogr Rev 72(4):430–449
Gantzer J (1972) Die Gesichtsverstümmelungen auf den Keramiken der Mochica-Kultur. Verhandlungen des XXXVIII Internationalen Amerikanistenkongresses. Stuttgart München 1968. Band IV, pp 367–374
Glave LM, Remy MI (1983) Estructura agraria y vida rural en una región andina: Ollantaytambo entre los siglos XCVI-XIX. Centro de Estudios Rurales Andinos "Bartolomé de las Casas", Cusco
Gonzales Mires C (1993) Historia y cultura del agro andino. Municipalidad del Qosqo, Cusco
Grimaldi JR, Gabriel RB, McMahon-Pratt T, McMahon-Pratt D (1989) A review of the geographic distribution and epidemiology of leishmaniasis in the New World. Am J Trop Med Hyg 41(6):687–725
Guerrero Barrantes J (2012) Caracterización agroclimática de cultivos priorizados y evaluación de impactos de la variabilidad y cambio climático sobre el desarrollo fenológico de los cultivos y su productividad: medidas actuales de adaptación en las regiones Apurimac y Cusco. http://www.paccperu.org/pe/publicaciones/pdf/39.pdf. Accessed 13 Sept 2014
Hall AE (2001) Crop responses to environment. CRC, Boca Raton
Harlan JR (1976) Review of plants, man, and the land in the Vilcanota Valley of Peru. J Agronomic Educ 5:87
Hastorf C (1993) Agriculture and the onset of political inequality before the Inka. Cambridge University Press, New York
Herrer A, Hidalgo V, Meneses O (1980) Cutaneous leishmaniasis and insecticides in Peru. Revista do Instituto Médico Tropical de São Paulo 22:203–206
Julien C (2001) Inca estates and the encomienda: Hernando Pizarro's holdings in Cusco. Andean Past 6:229–275
Lastres JB (1956) Historia de la medicina en el Perú: La visita médica de las parroquias de la ciudad del Cuzco Siglo XVIII. Revista del Archivo Nacional del Peru 20:175–194
Le Moine G, Raymond JS (1987) Leishmaniasis and Inca settlement in the Peruvian jungle. J Hist Geogr 13(2):113–129
Lorente S, Cordova PF (1925) Estudios sobre geografía médica y patología del Perú. Imprenta Americana, Lima
Lucas CM et al (1998) Geographic distribution and clinical description of leishmaniasis cases in Peru. Am J Trop Med Hyg 59(2):312–317
Lumbreras H, Guerra H (1985) Leishmaniasis in Perú. In: Bray RS, Change KP (eds) Leishmaniasis. Elsevier Science Publishers, New York, pp 297–311
Lyons JM, Raison JK, Steponkos PL (1979) The plant membrane in response to low temperature: an overview. In: Lyons JM, Graham D, Raison JK (eds) Low temperature stress in crop plants. Academic, New York, pp 1–24

Marcoy P (pseud) (1874) A journey across South America from the Pacific ocean to the Atlantic ocean (trans: Rich E). Blackie & Sons, London

Mayer E (1985) Production zones. In: Masuda I, Shimada C, Morris C (eds) Andean ecology and civilization: an interdisciplinary perspective on Andean ecological complementarity. University of Tokyo Press, Tokyo, pp 45–84

Mesclie E (2001) De la complementaridad a la voluntad de "aplanar los Andes:" representaciones de la naturaleza y pensamiento económico y político en el Perú del siglo XX. Bulletin de l'Institut Français d'Études Andines 30(3):541–562

Mörner M (1985) The Andean past: land, societies and conflicts. Columbia University Press, New York

Moseley ME (1992/2001) The Incas and their ancestors: the archaeology of Peru. Thames and Hudson, London

Murra JV (1989) High altitude Andean societies and their economies. In: Genoves ED, Hochberg L (eds) Geographic perspectives in history. B Blackwell, New York, pp 205–214

Newson LA (1993) Highland-lowland contrasts in the impact of old world diseases in early colonial Ecuador. Soc Sci Med 36(9):1187–1195

Ogusuku E et al (1993) El hombre como fuente de sangre para *Lutzomyia* spp. (Diptera: Psychodidae) andinos. *Revista Peruana de Entomologia* 34:56

ONERN (1976) Mapa ecológico del Perú. Guia explicativa. Oficina Nacional de Evaluación de Recursos Naturales, Lima

Oricaín P (1907) Compendio breve de discursos varios sobre diferentes materias y noticias geográficas (1780). In: Maurtua VM (ed) Juicio de límites entre el Perú y Bolivi, vol 5. Imprenta de Heinrich y Compañia, Barcelona, pp 319–377

Paponnet-Cantat C (1991) Violent heights in an Andean community of southern Peru. Can Rev Soc 28(3):340–356

Paz Soldan CE (1933) La pandemia malárica en los valles cálidas del Cuzco. Instituto de Medicina, Lima

Perú (1944) V censo nacional de población de 1940. Dirección Nacional de Estadística, Lima

Perú (1966) VI censo nacional de población de 1961. Dirección Nacional de Estadística y Censos, Lima

Pizarrro P (1965) Relación del descubrimiento y conquista del Perú. (1571). In: de Tudela Buesco JP (ed) Crónicas del Perú, vol 5. Ediciones Atlas, Madrid, pp 167–242

Pulgar Vidal J (1987) Geografía del Perú: Las ocho regions naturales. PEISA, Lima

Raimondi A (1898) Itinerario de los viajes de Raimondi en el Perú: valle de Lares, Santa Anna y regresos por Mollepta y Limatamabo (1865). Boletín de la Sociedad Geográfica de Lima 8:1–81

Raymond JS (1992) Highland colonization of the Peruvian Montana in relation to the political economy of the Huari Empire. J Steward Anthropol Soc 20(1–2):17–36

Richter H (1977) Rezension auf: plants, man and the land in the Vilcanota Valley of Peru. Wissenschaftliche Zeitschrift Universität Halle 26:156

Roberts DK, Laughlin LL, Hsheih P, Legters LJ (1997) DDT global strategies and the malaria control crisis in South America. Emerg Infect Dis 3(3):295–302

Sauer CO (1950) Cultivated plants of South America, handbook of South American Indians, vol 6. Bulletin of the Smithsonian Institution No. 143. Government Printing Office, Washington

Schubert R (1976) Rezension auf: plants, man and the land in the Vilcanota Valley of Peru. Biologische Rundschau 14:400–401

Schwabe GH (1976) Rezension auf: plants, man and the land in the Vilcanota Valley of Peru. Berichte Biochemie und Biologie 432(3):298–299

Shannnon RC (1930) Observations on Anopheles pseudopunctipennis in Peru. American Journal of Hygiene 12(2):442–448

Sklar HO (1982) The warm valley people: duality and land reform among the Quechua Indians of highland Peru. Universitetsforlaget, Oslo

Spain (1841) Recopilación de leyes de los reinos de las Indias (Carlos II), 4 vols, 5th edn. Boix Editor, Madrid

Stavig W (1999) The world of Tupac Amaru: conflict, community and identity in colonial Peru. University of Nebraska Press, Lincoln

Stratman-Thomas W (1940) The influence of temperature in Plasmodium vivax. Am J Trop Med 20(5):703–715

Tamayo MO (1909) La uta en el Perú. Boletin de la Sociedad Geográfica de Lima 25:1–64

Tamayo Herrera J (1992) Historia general del Qosqo 3 vols. Municipalidad de Qosqo, Cusco

Thomas RB, Winterhalder BP (1976) Physical and biotic environment of southern highland Peru. In: Baker PF, Little M (eds) Man in the Andes: a multidisciplinary study of high altitude Quechu. Dowden, Hutchinson and Ross, Stroudsburg, pp 21–59

Usselmann P (1977) L'homme et la terre dans la vallée du Rio Vilcanota (Pérou) [compte rendu]. Annales de Géographie (Paris) 86(476):506–507

Vega B (1896) Memorial....al real Consejo de Indias sobre los agravios que reciben los indios del Perú vol. 6 In: de Zalaburu F, Ragón JS (eds) Nueva coleccóon de documentos inéditos para la historia de España y de sus Indias 6 vols. Imprenta de los Hijos de M.G. Hernández, Madrid, pp 105–130

Villanueva Urteaga H, Macera P (eds) (1982) Cusco 1689: Economia y sociedad en el sur andino. Centro de Estudios "Bartolome de las Casas", Cusco

Villaseca PA, Llanos-Cuentas D, Davies D (1993) A comparative field study of the relative importance of Lutzomyia peruensis and Lutzomyia verrucarum as Vectors of Cutaneous leishmaniasis in the Peruvian Amazon. Am J Trop Med Hyg 49:260–269

Walton BC (1987) American cutaneous and mucocutaneous leishmaniasis. In: Peters W, Killick-Kendrick R (eds) The leishmaniases in biology and medicine, vol 2, climatic aspects and control. Academic, Orlando, pp 637–664

Webster S (1971) An indigenous Quechua community in exploitation of multiple ecological zones. Revista del Museo Nacional (Lima) 34:174–183

Werger MJA (1976) Review of: plants, man and the land in the Vilcanota Valley of Peru. Vegetatio 32(3):188

White PT (1989) Coca—an ancient herb turns deadly. Natl Geogr 175(1):2–47

Wilson MR (1976) Ecological futuristics: review of plants, man and the land in the Vilcanota Valley of Peru. Geographical Survey (Muncie, Indiana) 1: 28–30

Yamamoto N (1981) Investigación preliminar sobfre las activides agro-pastoriles en el Distrito de Marcapata, Departamento del Cuzco, Perú. In: Masuda S (ed) Estudios etnográficos del Perü meridional. Universidad de Tokio, Tokyo, pp 85–137

Young KR (2009) Andean land use and biodiversity: humanized landscapes in a time of change. Ann Mo Bot Gard 96:492–507

Zimmerer KS (1999) Overlapping patchworks of mountain agriculture in Peru and Bolivia: toward a regional global landscape model. Hum Ecol 27(1):115–165

Zimmerer KC (2010) Biological diversity in agriculture and global change. Annu Rev Environ Resour 35:137–166

Zimmerer KS (2011) Mapping mountains. In: Dym J, Offen K (eds) Mapping Latin America: a cartographic reader. University of Chicago Press, Chicago, pp 125–130

Zimmerer KS, Bell MG (2013) An early framework of national land use and geovisualization: policy attributes and application of Pulgar Vidal's state-indigenous vision of Peru (1941–Present). Land Use Policy 30:305–316

Chapter 4
The Sacred Valley as a Zone of Productivity, Privilege and Power

Abstract The Urubamba mid-section, commonly now known as the Sacred Valley, has flat land on the valley floor, temperate climate, abundant irrigation water and scenic attributes. Five Inca kings acquired parts of the Sacred Valley as their own reserve and several of them ordered the construction of earthworks that materially enhanced resource use. For at least a millennium the agricultural mainstay has been flour maize, adapted to moderate temperatures and an unusually long growing season. After the Conquest, Spaniards appropriated the land and labor of this privileged section of the Urubamba. In the twentieth century, haciendas cultivated a large-kernelled, white flour corn that expanded after the breakup of the estate system. Decreasing water supply brought on by global warming and changing transportation adjustments favor crops with less water demands. Overarching the future, however, is land speculation that predicts a shift to tourist-oriented land use. Pollution of the Urubamba River, generalization of alien vegetation and settlement sprawl on rich farm land are Sacred Valley issues needing resolution.

4.1 Sacred Valley Definition and Location

For more than a millennium, the 65 km long section of the Urubamba drainage now known as the Sacred Valley has attracted people for its productive agriculture, benign climate and scenic attributes. This temperate stretch is the heart of a depression spanning a range of ecological zones, population densities, and degrees of human impact. More effort has been expended in humanizing the landscape of this section of the Urubamba than of any other, upstream or down. Pharaonic work projects in the century before the Spanish Conquest changed the population and increased agricultural productivity. Soon after they arrived, Spaniards coveted this neat and sweet little corner of the Andes, finding it analogous to their homeland valley of Aranjuez. Exaltation of the Sacred Valley as a special place continues into the modern period. However, the arrival of mass tourism and the exacerbation of environmental problems have had an impact on its agricultural base and landscape integrity.

As a regional expression, Sacred Valley is a direct translation of the Spanish term "valle sagrado," a nomenclature with an aura of antiquity, but actually coined in its now popular usage only in the 1950s. As defined here, its southern limit is at San

D.W. Gade, *Spell of the Urubamba*, DOI 10.1007/978-3-319-20849-7_4

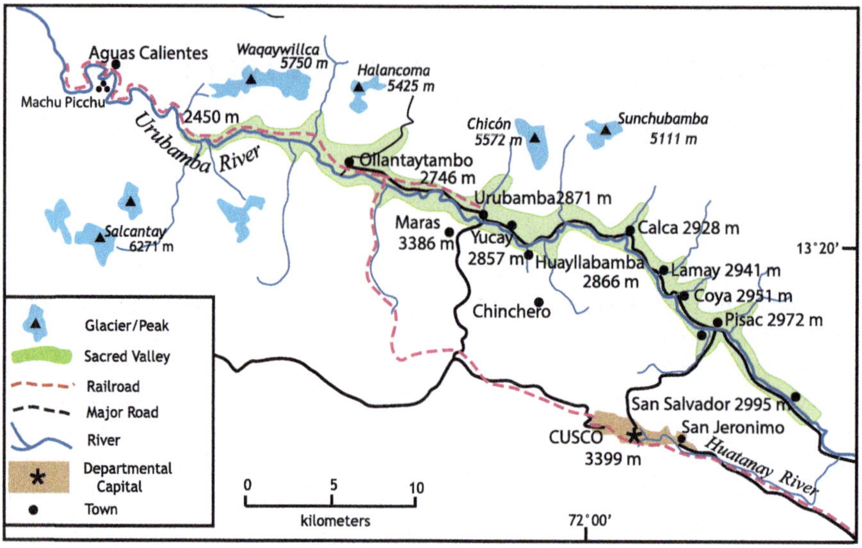

Fig. 4.1 Map of the Sacred Valley section of the larger Urubamba Valley. The northern end of this section extends to within 19 km of Machu Picchu. (Map by M.K. Gade)

Salvador (3050 m asl) and its northern boundary is at Torontoy (2450 m asl) that lies just upstream from where the Urubamba incises to form a deep canyon (Fig. 4.1). In some writings, Machu Picchu is glossed so as to place it within the Sacred Valley, but that is not justified either historically or environmentally. The core of the Sacred Valley encompasses the three towns of Urubamba, Yucay, and Huayllabamba, which hover around 2800 m above sea level (asl). In the colonial period Spaniards referred to this subzone as the Yucay Valley.[1] That name prevailed throughout much of the nineteenth century when the American archaeological explorer Ephraim Squier (1866:483) expanded that Yucay Valley to include the space from San Salvador to about 10 km downstream from Ollantaytambo well below Huayllabamba. Another term, though scarcely used now, is "Valle de Pisac," which referred to that part of the depression in the Province of Calca. If the Inca had their own regional name for the valley, it has now been lost, but it could have been a conflation of local names to make the word Huillcamayu, which is translated as Sacred River. To extend sacredness from the river to the depression is a short step. Sacred Valley as a popular expression came into widespread use when tour operators embraced it as a

[1] The expression "valle de Yucay" had both topographic and political meanings in the late colonial period when it was a corregimiento with 24 repartimientos. That duality of reference explains why Yucay was discussed under "sugar fields." Lack of enough hot weather makes sugar cane cultivation unsuccessful in the Sacred Valley, but in the tropical areas 70 km farther north around Huadquiña and Ccolpani below Machu Picchu, cane cultivation did well. In the early colonial period, that hot country was in the same jurisdiction as Yucay.

seductive designation for the Cusco-based day excursion that visited sites from Pisac to Ollantaytambo.

The Sacred Valley's centrality to Cusco's geographical imagination owes much to the former's proximity and accessibility. In the Inca period, the 49 km trip between Cusco and Urubamba required a day's hard walk from dawn to dusk. No overnight stop along the way was normally required. Spaniards riding horses covered the same distance easily within daylight hours. Beginning in 1924, the train from Cusco, chugging slowly on switchbacks in and out of the Cusco Basin, reached the valley in about 3 h. By the 1930s, trucks and buses from Cusco made that trip on primitive roads during the dry season in two and some hours. Today the paved highway makes the Cusco to Urubamba trip an easy hour's ride. The Sacred Valley's principal identity has been defined in its relationship to Cusco. The valley climate is mild, in contrast to the colder weather in Cusco. Agricultural products and people have traditionally gone between the valley and Cusco, whereas the movement of people and goods through the valley between San Salvador and Ollantaytambo has always been much less important.

4.2 Physical Character of the Sacred Valley

4.2.1 Landforms

Deep and narrow, the flat-floored valley is, on an average, 3 km wide with 70–80 % declivities on both sides. Locals call it the *quebrada*, a deep gash in the landscape. From Cusco, which lies at 3400 m above sea level in the valley of a tributary of the Urubamba River, one moves up and on to a high plateau at 3700 m asl and then descends in zigzags to reach the Urubamba River on the valley floor. The steep valley sides give a feeling of tightly bounded intimacy. At certain angles on a cloudless day the icy peaks of the Urubamba Cordillera, which extends 30 km in a northerly direction, can be glimpsed. Downstream from Ollantaytambo looms the hoary head of Waqawillca (5750 m asl), known also as Verónica (Fig. 4.2). Nearby is the iconic peak of Padre Eterno. Above the town of Ollantaytambo itself stands the Halancoma snowcap (5425 m asl), whose melt waters sustain agriculture in the Patacancha Valley. Above the Urubamba-Yucay area, the Yucay Massif incorporates the three main glaciated peaks: Pumahuanca (5650 m), Chicón (5572 m) and Sirjuani (5400 m). Farther south is Sawasiray Massif (5720 m) with Sunchubamba (5111 m) as the *nevado* closest to the town of Calca. According to local tradition and folk beliefs these mountains are the abode of the spirit gods (*apus*). The Incas made human sacrifices (*capacocha*), especially of children, on mountains to ensure good crops. Though no mummies have so far been recovered on any of those peaks, current glacial retreat may begin to expose them (Reinhard and Ceruti 2010). Before refrigeration, certain individuals climbed up to the glaciers to collect chunks of ice to sell in the valley. In 1925, when a Peruvian aviator, Alejandro Velasco Estete, flew a small plane over the glaciers, local shamans regarded the feat a profanation

Fig. 4.2 A view of the Urubamba Valley looking downstream. In the distance is Willkawiqui (Veronica), on which an important glacier provides water for agriculture far below. In the foreground, the shrub with yellow flowers is Spanish broom or retama (Spartium junceum), introduced from Mediterranean Europe

(Tamayo-Herrera 1978:155). After World War II, mountain climbers reached the top of several peaks in the Urubamba Cordillera. A European team led by Lionel Terray in 1956 conquered the summit of Verónica (Willkawiqui), without knowing that a neighboring peak, Padre Eterno, was 740 m higher than Veronica. On the left side of the river, the Cordillera Vilcabamba reaches its highest point in Nevado Salcantay (6271 m asl) and, although native people may well have climbed to the summit for religious rituals, a 1952 French American expedition is the first recorded ascent to the top.

Melt waters from these glaciers collect in tarns before they flow in a series of rushing streams encased in individual chasms toward, on the right side, the larger Urubamba trough. The Urubamba tributaries joining the river from the left are fewer. Only the Cusichaca River, having its source in the ice on the south side of Salcantay in the Cordillera of Vilcabamba, has an abundant flow. Where these lateral canyons join the main depression on both sides, alluvial fans form where streams lose velocity and deposit their transported materials. The rocky higher zones of the fans are unsuitable for cultivation, but farther down the slopes, finer deposited particles allow for agriculture. It was on these fans that most stone-faced terraces, pre-Inca and Inca, were built, connected to the stream by a canal diverting water to provide irrigation by gravity flow.

4.2.2 *Climate*

Temperatures in the Sacred Valley favor human settlement and agriculture. Average annual temperatures vary from 13.9 °C in the south to 14.6 °C in the north, a modest temperature spread attributable to the 500-m difference in altitude (SENAMHI 2010). The town of Urubamba has an annual temperature of 14.3 °C and daily temperatures are 5–7 °C higher than in Cusco (3400 m asl). In the dry season in Urubamba, the sky is normally cloudless and the average high temperature is 22.2 °C, whereas in the rainy season, the average is slightly less, 21.90 °C. Average minimum average temperatures vary seasonally much more: 9 °C in January and only 1.7 °C in July. Whereas in Cusco, freezes occur on most nights during the dry season, at Urubamba temperatures fall below 0 °C only on five to ten nights in each of the 3 months of June, July and August. In one year out of two, a freeze is likely to occur in the months of May and September. The microclimate on some terraces avoids freezing temperatures. Daytime weather is invariably mild and many people find it almost perfect. Climate is often understood in terms of the plants growing in a place. The pisonay (*Erythrina falcata*), transplanted from the *ceja de la montaña*, has symbolized the climate of the Sacred Valley even though it does not grow wild there. Large specimens of this crimson-flowering tree grace the plazas of Urubamba, Yucay, Huayllabamba, Calca, Pisac and in courtyards of some former haciendas. Peach trees grow, but so do avocados and highland papayas (*Carica pubescens*). Even coffee, though scantily grown, yields in protected patios.

Contributing to the valley's thermal qualities is the luminosity (Fig. 4.3). The east-southeast-west-northwest orientation of the depression allows sunlight to reach the valley floor 7 h a day, three more than if this valley were aligned in a north–south direction. Yearly total precipitation receipts are normally between 450 and 550 mm, with most of it coming between November and April. Evaporation amounts to ca. 1250–1350 mm a year, creating a water deficit of 800 mm. That ratio of rainfall to evaporation explains the importance of irrigation in the drier months. Rainfalls during the two warmest months of January and February are often diluvial as convectional currents set up thunderstorms. Hail, a major concern in the valley for its devastating effects on standing crops, accompanies some of these storms every year. Occasionally snow falls in the heights above 3900 m asl, but not on the valley floor.

4.3 The Sacred Valley in the Pre-Columbian Past

4.3.1 *Pre-Inca Presence*

Lithic materials of andesite and sandstone, as well as rock art, indicate that hunters and gatherers inhabited the valley at least as early as 6000 years ago (Covey and Griffis 2014:46–47). The transition to agricultural-based village life seems to have begun by 800 years BC. The archaeologist Alan Covey (2014a:65–73) speculated

Fig. 4.3 Orientation of the Sacred Valley, more east–west than north–south, prolongs daylight to benefit crop growth. Irrigated maize on the valley floor contrasts with rainfall-dependent wheat or barley on steeply sloping fields. This rich land was once part of Hacienda Compone. (Photograph by D.W. Gade)

that the first villages in the valley participated in an ecological complementarity of camelid husbandry above and crop growing below. Archaeological evidence elsewhere in the Andes suggests that the diffusion of maize triggered, if not sedentary living, then certainly the rise of ranked polities (Perry 2007). Social, religious and political consolidation first appeared with the Wari polity (400–1000 AD), which had distinctive architecture, town planning, textiles, ceramics and religion. The Wari civilization appears to have expanded in the Highlands in discrete clusters rather than in successive waves of territorial annexation. In the Sacred Valley, the Wari presence was most evident at its southern end (Bélisle and Covey 2010). Near San Salvador at Huanca, a Wari site lies beneath the Catholic shrine and just beyond the boundaries of the Sacred Valley as defined in this chapter lie two large archaeological sites, Pikillacta, a huge administrative center built between 600 and 900 AD, and Huaro, a residential site (Bauer 2004; McEwan 2005). Their political connections to the major Wari site at Ayacucho or to Tiwanaku on the Altiplano are still unclear, but the cultural affinities are apparent. Maize was probably the main crop during this period and chicha was made and consumed. The Incas inherited many Wari culture traits from chicha manufacture to the institutionalized practice of transferring people (mitmaq) to similar environments elsewhere and the use of entheogenic substances. Suggestions swirl that even the Inca city of Cusco was a Wari site at an earlier time. Yet, for most of the twentieth century, the standard conceit in the popular Cusco imagination was that everything pre-Conquest began with the Incas.

4.3.2 *Incas in the Valley*

During the early Inca period known ceramically as Killke (1000–1400 AD), inhabitants built their settlements on ridge tops; with the advent of the Imperial Inca (1400–1532), settlement moved from the heights to the sloping land within the confines of the depression (Covey 2006). In the 1950s, the archaeologist John Rowe (1957) identified tombs at Yucay as Killke. Since then, more than two hundred Killke sites have been found in the Sacred Valley part of the Urubama and especially its tributaries (Bauer 2004:81). Two important Killke sites were Pukara Pantillijilla, above present-day Pisac, and farther down the valley, a ridgeline settlement, Qhapaqkakhia, which had terraces, plaza, and residences. These locations suggest the importance of defense, but possibly the bottom of the Urubamba depression was too swampy and flood-prone for major construction.

During the Imperial Inca phase, monumental constructions were skillfully built in the Sacred Valley, indicating that the area played a central role in the Inca system. The Sacred Valley can be geopolitically viewed as the eastern flank of the Inca Empire's outer heartland, protecting with its fortresses the inner heartland of the Cusco (Huatanay) Valley (Farrington 1992; Covey 2006: 134, 139). One interpretation is that past battles rather than locations defined these two heartlands. The Inca story in the Sacred Valley focuses on the five rulers (Sapa Inca) who, each with his own royal lineage (*panaca*), immediately preceded the Spanish Conquest: Viracocha Inca (1408–1438); Pachacutec (1438–1471); Tupac Inca Yupanqui (1471–1493); Huayna Capac (1493–1527); and Huascar (1527–1532). More than any other rural area in the Andes, the Sacred Valley was appropriated by successive Inca rulers, suggesting the importance of its amenities. To keep his memory alive, each ruler had his *panaca* occupy one or more areas. Only the royal lands of Viracocha Inca were located in the climatically rigorous Pampa de Anta, Cusco Valley and above Calca at Huchuy Qosqo.

Pachacutec established seven estates in the Sacred Valley, not including Machu Picchu farther down the river. In Ollantaytambo and Pisac, Pachacutec built a magnificent fortress and temple dedicated to the Inti cult. Some of the labor for these projects came from conquered Colla people brought there following Pachacutec's military campaigns to the Titicaca Basin. Fortresses suggest concern for frontier security against invading forest tribes, though given their rudimentary military organization and weaponry, the idea that jungle dwellers posed a threat to the highlands seems unwarranted. Pachacutec's reign reflected more than a military preoccupation, the recasting of the valley floor. Using the labor tribute of conquered groups to reduce the possibility of flooding, Pachacutec ordered the river channel straightened for a distance of 10 km downstream from Pisac. In the same area, laborers also built agricultural terraces on both sides of the Urubamba.

North of Ollantaytambo, terraces were also built on the valley floor to take advantage of the natural gradient. Much farther down where the Cusichaca flows into the Urubamba, the Incas constructed 390 ha of terraces to add to the 226 ha of terracing built during Wari times (Kendall 1991:26). Though also stone-faced, Wari terraces differ from those of late Inca construction by their inclined, rather than flat, surface. Abundant water from the Salcantay glacier enabled food surpluses over seven decades to be sent from this zone to feed the work force constructing Machu Picchu 43 km distant. The 1000 or so people who lived at Machu Picchu after it was built continued to need food supplies from Cusichaca. The five hectares of agricultural terraces at Machu Picchu would have been insufficient to produce food to sustain that population (Wright et al. 2000). Pachacutec was followed by Tupa Inca Yupanqui who built a palace at Huayllabamba on a large property that encompassed Chinchero and the quebrada of Urquillos that leads up to Chinchero. The complex of bank terraces on the left side of the Urubamba is attributed to Tupa Inca's directives. Piscobamba near Torontoy became another of Tupac Inca's properties (Rowe 1997:282).

Huayna Capac had a profound impact on the valley beginning 7 years after he became the Inca. In 1500, he appropriated more than 10 km^2 of land down valley from the bridge at Huayllabamba, including fields around Urubamba and Yucay (Rowe 1997:282). In a grand despotic gesture, Huayna Capac forced most of the original inhabitants to resettle near Cusco and replaced them with 2000 households from other parts. Mitmaq of Canari ethnicity brought from Tomebamba (Ecuador), where Huayna Capac was born, provided a portion of the day-to-day labor for his domain. Huayna Capac also ordered brigades of workers to undertake a massive reconfiguration of the valley floor. The chronicler Juan de Betanzos (1996) estimated that 100,000 men were needed to straighten out the meanders of the Urubamba channel, level hills, install networks of canals and build terraces. Between Pisac and Chilca, more than 25 km of the river were canalized (Farrington 1983:221) (Fig. 4.4). Near Pisac, the Chongo and the Pichinmayoc near Taray were canalized for several hundred meters as they cross the floodplain to join the main river. These works increased productivity, putting more land into agriculture and reducing seasonal flooding of fields. Huayna Capac also ordered the construction of landscaped pleasure gardens and a large palace ("Quispihuanca") located near where the Rio Tullumayo (Chicón) enters the valley floor of the Urubamba (Fig. 4.5). After this Inca died in Ecuador in 1527, porters brought his mummy the 1000 km distance and placed it in that palace.

Huayna Capac's strong affinity for the Sacred Valley has regularly been reported as a probable reaction to the algid climate of Cusco. His love of a temperate climate may also explain why this Inca ordered the cultivation near his palace of coca, cotton, peanuts and chile pepper—crops that yielded well only in the warmer climate much farther down the Urubamba Valley. The perennials, coca and cotton, survived the Conquest for some years in that location, suggesting they were given protection from the elements (Villanueva Urteaga 1970:19). But they represented Huayna Capac's sentimental whim and were not part of agriculture in the zone as asserted (Morris and von Hagen 2011:127).

Fig. 4.4 Here north of Pisac the Incas canalized the Urubamba over about half of its length in the Sacred Valley. Such hydrographic engineering reduces the risk of flooding, but also ensures that the river channel does not meander, reducing the amount of agricultural land. Photograph by D.W. Gade)

Fig. 4.5 Huayna Capac, the supreme Inca between 1493 and 1527, built a palace ("Quespihauanca") near where the Chicón River emerges from its canyon and flows into the main Urubamba Valley. Huayna Capac used the Chicón gorge as a hunting preserve. The palace, now on the outskirts of the town of Urubamba, is a reconstruction project of the Instituto Nacional de la Cultura. (Map by M.K. Gade)

The last Inca who had an architectural impact on the Sacred Valley was Huascar who built a palace at Calca in what was an Inca planned town (Niles 1988:62). The Sacred Valley as a retreat for the high elite suggests the high value placed on its climate and agriculture. But the Inca perspective on landscape division may have followed a different logic than environmental unity (Kosiba and Bauer 2013). The elite zones at Chinchero and above Pisac suggests that siting on the valley floor was less important than creating areas reserved for high status people. Nevertheless, in the Inca period the valley was a zone of maize growing that produced food surpluses which helped to consolidate the political apparatus of the Empire. The Inca did not use crop rotation or fallow land on the valley floor, so the only way to maintain soil fertility and produce abundant surpluses year after year was by fertilization. By the time of the Conquest, maize monoculture was an unquestioned tradition.

4.3.3 *Valley Maize Before the Spanish Conquest*

The prodigious effort of landscape modification not only benefited the cultivation of flour maize (*Zea mays* var. *amylacea*). The land races within that amylaceous grouping have soft and mealy endosperm unlike the hard and vitreous endosperm of flint corn. In the context of Latin American maize, flour maize is relatively uncommon because it does not do well in hot weather. Two maize scientists, Major Goodman and Robert McK. Goodman and Bird (1977) analyzed 12,000 collections of maize from Latin America and organized 219 races into 14 "racial groupings." The group they called "Cusco maize" includes races of different colors and sizes of kernel that have in common a predominance of soft starch in the kernel. This flouriness, genetically expressed as a locus on a chromosome, arose as a result of spontaneous mutations and hybridization perpetuated by human selection.

The evolution of flour maize in the Andean Highlands far predates the Inca. Since the Wari civilization was especially fond of chicha, it stands to reason that they preferred to grow flour maize, high in starch content and easy to chew, for making that yeasty-tasting beverage. Manufacturing chicha required maize kernels to be chewed into a cud held together by human saliva. The enzyme diastase in that saliva changed the starch in the kernel into fermentable sugar. A postulate is that food preferences influenced maize selection. Boiled corn (*mote*) and chicha have long been the main ways to prepare this grain in the valley.

4.3.3.1 Flour Maize Requirements

Successful cultivation of flour maize requires a set of quite specific environmental conditions that relate to temperatures, growing period, day length and humidity. Its temperature tolerance limits hover between 22 and 24 °C for normal maximum temperatures. Hot conditions, as in the North American Midwest where daily summer temperatures consistently exceed 30 °C, do not allow flour maize cultivars to accumulate large amounts of starch. To grow in Peru, flour races require temperatures of 13 °C, whereas in the Corn Belt of the United States, the threshold for growth is 15 °C.[2] The coolish temperatures of the Sacred Valley, combined with the daylight period of that latitude, stretch out the growing cycle to between 8 and 9 months. That growing period is almost three times longer than that of the dent corn grown in Iowa, where the cycle from sowing to harvest is 110 days or less. Photoperiod affects flour maize production. Long cool nights are instrumental in the transloca-tion of the products of photosynthesis that go into root formation and endosperm development (Brandolini et al. 2000). The long summer days of the mid-latitudes hinder the formation of ears. Still another requirement for successful yield is dry air. Even in the rainiest months, relative humidity in the Sacred Valley is below 50 %. Humid conditions make maize with soft endosperm vulnerable to plant pests and mold. Low humidity and sunny skies are also needed for the more than a month it takes in May and June for flour maize to dry.

Water requirements are also specific. Flour maize races have more biomass than flint corn and so require abundant water. Given the long growing season and the hydric deficit caused by evapotranspiration and a long dry period, some moisture must be artificially supplied. Planting occurs during a 3-week window between late August and mid-September which coincides with the dry part of the year. Before that, the plowed field is irrigated to promote germination. In the Sacred Valley frequent rainfall cannot normally be counted on until mid-November. Thus, irrigation water becomes essential until the rains come in full force. An intriguing question is whether the periods of canal and terrace construction in the valley's prehistory corresponded to perceived decreases in rainfall. The evidence can seem contradictory. Sediment samples collected at a lake 12 km above Ollantaytambo town yielded pollen profiles interpreted as the onset of a warmer, dryer period after 1400 AD (Chepstow-Lusty et al. 2003, 2009). On the other hand, ice core data from the Quelccaya glacier has indicated arid conditions between 1160 and 1490 AD, considered to be the onset of the Little Ice Age. Glaciologist Lonnie Thompson and his team have interpreted the core from between 1250 and 1310 AD as indicating a prolonged drought (Thompson et al. 1985:973).

Flour maize has its highest agricultural setting on Sacred Valley terraces. Inca terrace construction has sometimes been interpreted as a means of erosion preven-tion. More persuasive, however, is the need for a flat surface in a very uneven topog-raphy on which to distribute water (Denevan 2001). Another argument for these platforms is concern for soil fertility to enhance yields. At the lowest depths of the terrace, a layer of gravel ensures proper drainage; above that, the soil consists of rich alluvium brought up from the main river. Moreover, terraces enjoy a microclimate, escaping the stagnant cold air which settles in the bottom of the valley.

[2] E.G. Squier brought back from the Sacred Valley to the United States maize seeds that fit the description of MBGC. In 1865 Squier distributed some of those seeds to several farmers. In one Pennsylvania experiment, the corn grew between 4 and 5 m tall, but produced no ears. To explain that failure, Squier claimed the growing season was insufficiently long; more likely the photope-riod at 41° N was too long and the temperatures too high.

4.3.3.2 Incas as Plant Scientists?

Agronomic experimentation may have been among the various functions of stone faced bench terraces, especially of the exquisite constructions below the fortresses of Pisac and Ollantaytambo. As circumscribed spaces visible from above and below, terraces lend themselves to monitoring the seasonal development of the crop. Any unusual occurrence would likely have received close attention. Harvested crops that yielded, for example, a spontaneous mutation, might have been set aside and, if attractive, planted in the next cycle. Although none of the chroniclers hinted at such a thing, the suggestion that the Inca had a specialized cadre of personnel who selected and propagated plants as a kind of laboratory of protoscientific agriculture is worth considering (Earls 1989; Plachetka and Pietsch 2009). Distinct landraces occur naturally, but they require human intervention to perpetuate them. No real insight into genetics was needed to observe that maize pollen is disseminated by the wind and that, therefore, maize varieties have to be either separated in time or in space to avoid contamination. The Incas would arguably not have possessed a diversity of maize varieties had they not figured out how to maintain pure lines by isolating landraces in either space or time. The Incas derived much of their agricultural knowledge from cultures that preceded them. Certainly the domesticated crop and animal inventory of the Incas, as well as chicha manufacture, had existed long before among the Wari. Geographer Karl Zimmerer's (1993) assessment was that the Inca state apparatus was primarily concerned with placing little emphasis on innovation.

Several chronicles indicated that of their agricultural plants, the Inca gave maize the greatest attention. Its noble appearance, taller than a man, tasseled and eared, commands attention that tuberous crops do not. Maize fields received the guano brought on the backs of llamas from the bird islands off the coast. The Inca designed the irrigation infrastructure for maize cultivation; other seed crops, kiwicha or quinoa, require much less water. Potatoes could be grown successfully in the high country above the valley without irrigation. To placate the water divinities, the Incas used priestly interventions that sacralized maize cultivation. At Urco west of Calca, an Inca ritual center tied to water is suggested by the arrangement of sculpted rocks, terraces, and canals that channeled water from streams flowing down from the glacier high above. Important too, making fermented beverage from maize gave it important social and religious significance that no other crop possessed. However, the hypothesis that maize was not a staple, but instead a ritual crop of minor dietary significance, represents an idiosyncratic reading of the chronicles (Murra 1960). Isotopic studies of human bones including those from Machu Picchu indicated that maize cultivation went far beyond ceremonial uses; it was important in the diet (Hastorf and Johannessen 1994; Burger et al. 2003).

To grow maize, the Inca modified the landscape of the Sacred Valley in ways still apparent today. Of the two kinds of man-made terraces in the valley, the low lying, broad field terraces, found parallel to and on the floodplain of the Urubamba, remain in full cultivation. Much more numerous are stone-faced bench terraces, which are among the most impressive complexes of *andenes* in the Central Andes (Donkin 1979). Approximately half of the Sacred Valley's lower slopes have these constructions.

Fig. 4.6 Ollantaytambo terraces built during the reign of Inca Pachacutec between 1438 and 1471. The high construction quality suggests they were built for Pachacutec himself. Then as now, the terraces have long been cultivated with maize, a plant of ritual importance but also one demanding irrigation water available on the terraces. (Photograph by D.W. Gade)

The lavish use of human labor required to model slopes into terraces came from directives of those in high authority. Land in the Sacred Valley left unterraced was land controlled by local groups which did not have the labor surplus necessary to build them. The exquisite stonework of certain valley terraces show a level of skill that only masons on the royal estates possessed (Fig. 4.6).

4.3.3.3 Woods: Spontaneous and Otherwise

Before the Inca, native forests in the valley receded with the advance of agriculture. Even before the Inca had consolidated into an empire ca AD 1400, it is probable that the valley floor and lower slopes below 3500 m asl had lost their tree cover to make way for crops. Trees spontaneously re-established themselves only if a field was abandoned. When unirrigated fields above the valley floor were put into fallow for several years, a spontaneous grass cover took over, but tree seedlings were consumed by grazing camelids. Wood in the Inca period came from the lateral valleys and the uplands beyond the valley rim, 3700–4500 m asl, where queñuales (*Polylepis* spp.), chachacomo (*Escallonia* spp.) and unca (*Myrcianthes oreophylla*) grew (Tupayachi 2005). In the valley bottom, the Inca practiced agroforestry on the margins of fields. Manco Inca inherited not only farms, terraces, livestock, pastures, maize lands.

Lambran (*Alnus jorullensis*), quishuar (*Buddlejia incana*) and waranway (*Sambucus peruviana*) were among the tree species planted on the edges of crop fields. Carpenters made agricultural implements and house beams from these woods.

4.4 The Valley in the Colonial Period

From the Spanish perspective, the Sacred Valley was a deliciously inviting place in which to settle or visit. Here was a zone of fertile alluvial soils, lots of water even during the dry season from glacier melt, and moderate temperatures. Absence of malignant disease that encumbered so many otherwise appealing valleys at lower elevations in the Viceroyalty of Peru made this section of the Urubamba particularly attractive. Malaria, introduced soon after the Conquest of 1532, rapidly spread wherever the parasite could survive and an anopheline host was present for its transmission. Without knowing the causative agent nor how it was transmitted, Spaniards soon learned from experience, not from books, that malaria could be avoided if hot buggy places were avoided. Low nighttime temperatures in the Sacred Valley do not favor survival of either the mosquito or the parasite. But Spaniards at the time used plants to indicate climate, noting that where sugar cane could not grow, malaria was not a danger. The landscape, not thermometers or rain gauges, indicated the thermal subtleties of this and many other valleys during the colonial period (Cobo 1956:I:76–79).

Spaniards unanimously praised the valley's charms. The Franciscan padre Diego de Córdova y Salinas (1957) wrote that the "valle de Yucay" was "…*muy ameno y fértil, en medio de montes encumbrados y nevosos*," thus succinctly presenting the dramatic landscape, high agricultural productivity, and pleasant temperatures. Garcilaso de la Vega (1960:227) described the valley as very pleasant with soft fresh air, good water, a climate without extremes of heat or cold and free of flies and mosquitoes and other noxious creatures. Early in the seventeenth century, a traveling merchant wrote that the "valle de Ucay" was "*el más abundante y fértil que se puede desear*" (Anonymous 1958:95). Vazquez de Espinosa (1948:95) extolled the valley as "el Aranjuez de sus deleytes y recreos." Aranjuez, etched into the Meseta by the Rio Tagus south of Madrid, presented a similarly benign setting (Gade 1968). Revealing their European environmental preferences, visitors to the Sacred Valley continued to express their favorable opinions of it into the late colonial period (Oricain 1906:193).

4.4.1 Land and Labor

The imperial Inca improvements to the valley floor surely fired Spanish enthusiasm for the Sacred Valley. Its usefulness for agriculture had been increased at least 50 % and its attractiveness even more. Through all the colonial period, Spaniards did nothing to further improve the valley infrastructure. Europeans did introduce an array of plants and animals, multiplying food choices, and the oxen-pulled plow, increasing the efficiency of cultivation. Spaniards kept some Inca land measurements, most notably the topo.[3] Most profoundly, they changed land tenure and settlement patterns in the valley. Less than a decade after the Conquest, the Crown rewarded important individuals with encomiendas that gave them the right to extract the fruits of the land through the use of Indian labor.

In 1539, Francisco Pizarro assigned himself an encomienda that covered much of the Sacred Valley. As the leading protagonist of the Conquest, this Pizarro chose the place he considered most desirable of all. Francisco's half-brother Hernando received an encomienda at Ollantaytambo as well as grants elsewhere (Covey and Amado Gonzalez 2008). Francisco died in 1541 and his brother Gonzalo managed his encomienda on behalf of Francisco's children. However, his control of the land changed in 1558 when the Spanish Crown declared the Sacred Valley to be royal property and distributed parcels to Spaniards living in Cusco. A decade later, proprietorship shifted again when a court ruled in favor of Sayri Tupac, the grandson of Huayna Capac. That decision provided a plausible motive for the heir's murder at the hands of Francisco Chilche, the Cañari from Ecuador brought by Huayna Capac. After the Conquest, Chilche became the manager of Francisco Pizarro's encomienda and had good reason to want the property for himself. Instead, Sayri Tupac's daughter, Beatriz Clara Coya, who married the Spaniard Martin Garcia de Loyola (nephew of St. Ignatius), filed claim for the inheritence. In the early 1600s, she, with her Spanish husband, was awarded ownership of a substantial swath of the valley core. In 1744, the Crown formally abolished that tie to the Inca legacy.

Important people and institutions in Cusco acquired land in the Sacred Valley beginning in the second half of the sixteenth century when epidemics decimated the Indian communities and their population plummeted. For example, in 1552 in Yucay the population dropped from 3000 to 700 (Wightman 1990). Spaniards in Cusco acquired most of the vacated Indian land in the valley cementing the tie between the valley and the city. Properties were consolidated and later developed into haciendas. Other individuals or institutions were awarded properties replacing the use of Indian labor represented by the encomienda. Although an indigenous workforce remained necessary, the turn from encomienda to hacienda placed the focus on land. When an encomendero died, the state abolished the encomienda; when a hacendado died, his heirs took it over.

The biggest land recipients were religious orders who supported their activities by agricultural production. For their role in the 1568 pacification of Vilcabamba, the Crown awarded the Augustinians (O.S.A.) well-watered land that became Hacienda Sillque (Glave and Remy 1983:189). This award was the beginning of the Augustinian hacienda involvement, which they expanded when they bought parcels that had belonged to Indian victims of disease. In 1588 and 1589, smallpox and measles epidemics depopulated whole communities and brought about abandonment of their land (Toribio Polo 1913). In that way, the intricately terraced Cusichaca Valley, which had produced large maize surpluses in the Inca period, became part of Hacienda Sillque. In the seventeenth century Augustinian holdings further expanded to include all of the right bank of the Urubamba River below Ollantaytambo. They

[3] A topo varied in size depending on location. In the Sacred Valley, a common-sized topo was 88 varas long and 44 varas wide. A vara was approximately one yard.

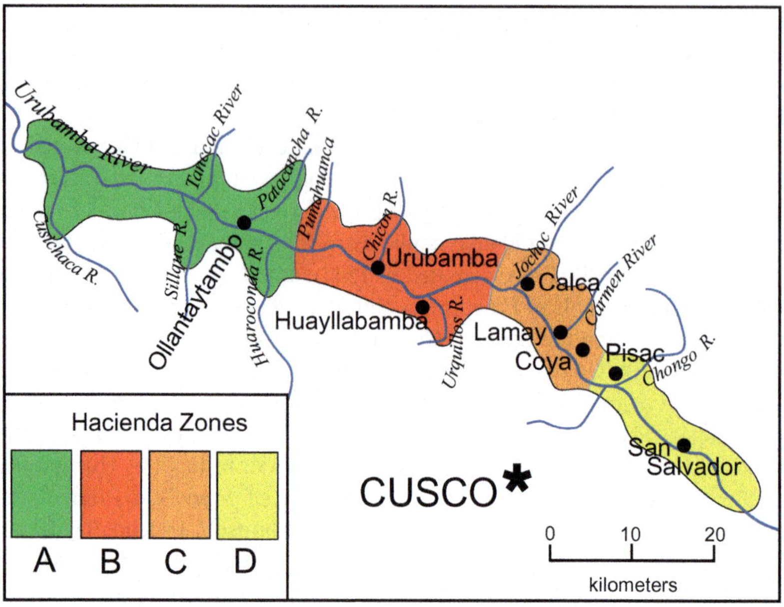

Fig. 4.7 Map of the major side streams flowing into the Urubamba and general location of the haciendas that once dominated the Sacred valley. Haciendas in Zone A (District of Ollantaytambo) were: Sillque, Chilca, Piscocucho, Taparo, Cimapuquio, Primavera, Tanccac, Phiri, Compone, Rumira, Cacchicata, Cotohuínco, Mascabamba, Pachar. Haciendas in Zone B (Districts of Urubamba, Yucay and Huayllabamba): Pumamarca, Yanahuara, Pino, Media Luna, Yaravilca, Rocafuerte, Huaypo, Recoleta, Huayoccari, Huicho, Urpihuata, Paca, Qishuarpata, Capellania, Patapatayoc, Santa Inés, Cusibamba. Haciendas in Zone C (Districts of Calca, Lamay and Coya): Huaran, Paullu Grande, Urco, Huandor Chico, Yanahuaylla, Huanco, Paullo Chico, Urqui, Unurequi, Ayaran, Paucartica. Haciendas in Zone D (Districts of Pisac and San Salvador): Sañihuasi, Chawaytire, Chongo Grande, Bella Vista, Vicho, Vilar, Uchumuca, Huanco, Sondor. Even though most estates have disappeared their names have survived as references to locations. (Map by M.K. Gade)

owned Haciendas Chilca, Phiri, and Tanccac, each of which had irrigated maize land on the valley floor and unirrigated upper slopes where cattle were raised and potatoes grown (Fig. 4.7).

A similar accretion occurred when the Cabildo of Cusco usurped agricultural land that belonged to the ayllus in Ollantaytambo and in 1557 gave it to the Convent of Santa Clara in Cusco. A tract of 58 ha of land on the fertile flats where the Huarocondo River joins the Urubamba was the original nucleus of what became Hda. Pachar (Burns 1999:49). In 1621, the convent purchased adjoining lands to make a property of more than 3000 ha. A mayordomo managed the estate for the convent until late in the colonial period when another religious order bought the property. After Peruvian Independence, the state expropriated Pachar and gave the hacienda to the Colegio Nacional de Ciencias in Cusco, a new public school founded in the Bolivarian spirit of independence.

Other religious orders also acquired land in the Sacred Valley during the colonial period. Hacienda Aguacollay, owned by the Jesuits, and located above the floor of the valley near Maras, produced wheat on rain-fed slope lands. To grind that grain into flour, the owners built several grist mills on small streams flowing into the Urubamba. At Pisac, the Jesuits from Cusco had a dual-purpose finca. On its 11 irrigated topos, they grew fruit and used the remaining space for recreation and convalescence of their members (Macera 1968: clxxi). Their largest Jesuit properties on the valley floor were south of Pisac at Haciendas Vicho and Vilcar. In 1767 the pope expelled the whole order from Spanish America and its lands in the valley were eventually auctioned off by the state. Franciscan properties in the valley at Urubamba, Yucay, and Urquillos were used as convents, places of reflection and convalescence, rather than sources of revenue from agricultural production. The Mercedarians owned three haciendas, Huanca, Sondor, and Uchumuca, in the southern part of the Sacred Valley, and proceeds from them supported their large convent in Cusco. Around Ollantaytambo they also claimed land that in the Inca period had been used to support the Sun cult. By the reasoning of religious orders, land that had been in the service of Inca religion should be reassigned to Christian religious orders. Instead, in an ultimate irony, Indian headmen (*kurakas*) in Ollantaytambo sold land to Spaniards to pay for law suits the Indians had initiated in order to retain possession of those very plots. The Seminary of San Antonio de Abad in Cusco did succeed in wresting ownership of a large parcel of sun-cult land in Ollantaytmbo. The Bethlemite order, founded in colonial Mexico, acquired Hdas. Sillque, Cacchicata, Cotohuincho, and Pachar, all on the left bank in the northern part of the Sacred Valley, to support the hospital they operated in Cusco. After 1825, Sillque passed to Mariscal Agustin Gamarra, a cusqueño who later became president of Peru, as a reward for his role in the Independence struggle. Other church property in the valley was held by parishes which rented the land to farmers. Priests sometimes owned land in their own names.

By the seventeenth century, haciendas, private or institutional, dominated the valley floor. Of the two kinds of estate workers, *originarios* were Indians who, though they never left their communities, were absorbed into the hacienda system. Originarios comprised the workforce at Hacienda Uqui near Lamay, which had six different parcels spatially separated from one another. Indians from four ayllus provided estate labor and, in return, got plots for their own use (Macera 1968). On most valley haciendas, the workers were Indians known as yanaconas who had, on their own accord, left their community to work as peons on an estate. They provided the labor for maize cultivation, firewood collection, and domestic chores. In return, yanaconas received the use of a plot belonging to the hacienda on which to grow their own food. Yanaconas had to pay tribute, but, most importantly, were free from the duty of working in the mines (*mita*). Viceroy Toledo decreed in 1574 that Indians from certain areas had to work in the mines of Huancavelica or Potosi. The mines soon acquired a reputation as notorious death traps, making peonage on a hacienda vastly preferable to working in the mines.

Fig. 4.8 On the floor of the Urubamba Valley, the village of Pisac started as a reducción in the 1570s during the administration of Viceroy Francisco Toledo. Compelling native people to live in nucleated settlements enabled the Spaniards to better control them. On the right is the plaza; an obligatory open space in all reducciónes. (Photograph by D.W. Gade)

4.4.2 *Population and Settlement*

The Indian population of 7631 in the Sacred Valley early in the colonial period had mixed origins. Natives to the valley were one component; descendants of people (*mitimaes*) brought there from distant places by the Inca were another; and still others were yanaconas who arrived somewhat later from near and far on their own accord (Garrett 2005:69). In 1572, to facilitate conversion and acculturation of indigenous people, Francisco de Toledo imposed throughout the Viceroyalty a massive resettlement scheme. In the Sacred Valley, Toledo's planners laid out the following *reducciones* in this part of the Urubamba: San Salvador de Chuquibamba, San Pedro de Pisac, Maria Magdalena de Taray, Santiago de Lamay, San Esteban de Coya, Santo Nombre de Jesús de Zamora de Calca, Santiago de Oropesa de Yucay; San Bernardo de Urubamba; San Benito de Alcántara de Huayllabamba and Pueblo de la Zarza de Ollantaytambo. In so far as the topography permitted, the reducción village had a central square with streets coming in at right angles. With the exception of San Salvador, these settlements dropped the Hispanic element of their nomenclature to become simply Taray, Pisac, Lamay, Coya, Yucay, Urubamba, Huayllabamba and Ollantaytambo. Their location on good agricultural land contrasted with Inca settlements which were placed mostly above the valley floor (Fig. 4.8). Following the plan which the Spaniards imposed on all of their New World colonies, all the reducción towns had a grid arrangement with a central plaza and a church facing it. Before the Conquest, Calca and Ollantaytambo had been Inca villages; of the two, Ollantaytambo kept much of its original pattern, whereas at Calca, the Spaniards essentially constructed on top of what had been an Inca settlement. Yucay had been an inn (*tambo*) on the Inca road system.

Five Sacred Valley reducciónes had a resident priest who, at the beginning, was the only Spaniard allowed to live in them. Later that Toledan rule loosened to permit officials and hacendados to reside in the town. Stone material for the requisite church often came from dismantling Inca constructions. Each village had geographically defined neighborhoods (*ayllus*) of 50 or so people that headmen known as kurakas used to organize community work assignments. Reducción Indians regarded themselves as members of ayllus through the colonial period. In Yucay, diverse outsiders formed an ayllu that over time became a cohesive group (Wightman 1990:89). Another Inca concept, the moiety, divided a town into an upper (*hanan*) and lower (*hurin*) half. Today ayllus and the moiety are but faint memories in the valley, though elsewhere in the Department of Cusco the tradition remained stronger (Gade and Escobar 1982).

Most of the population of valley towns was monolingual Quechua speaking through most of the colonial period. In 1689, the Urubamba area had 6199 Indians and 693 non-Indians; the Calca area had 6099 Indians and 200 non-Indians (Garrett 2005:69). Mestizos, defined as mixed blood and bilingual, increased in population as a result of intermarriage between Indians and Spaniards. The town of Urubamba became the mestizo center of the valley. About a century later, Indians still strongly dominated the demographic profile. Lamay was 96 % Indian, Pisac 95 %, Taray 94 % and Calca, somewhat less Indian at 86 % (Mörner 1977).

As the lords of the land, Spaniards determined how the valley floor was to be used and how the soil was to be prepared for cultivation. Only several decades after the Conquest did oxen become generally available, as castrating animals for plowing conflicted with the early need for bulls to inseminate cows. Eventually enough animals became available for plowing and by 1560, the oxen and plow was the farming mode among Spaniards (Garcilaso de la Vega 1960:432; Villanueva Urteaga 1970). Viceroy Toledo in the 1570s directed Indians to use the plow and oxen, a decree that could not have been readily implemented. The cost of a pair of oxen was beyond the means of the ordinary Indian farmer, but more relevant was the fact that an oxen-pulled plow could not be used on for fields on a pronounced slope, since yoked oxen cannot get the necessary traction. The chaquitaclla retained its usefulness for steep valley sides. Estates used the plow and oxen to prepare both maize and wheat fields for planting.

Plowing a maize field with oxen imposed a system of straight furrows and distance between rows, which created a geometry that maize fields did not previously have. However, maize could not be grown like wheat. It could not be sown broadcast and it required that the corn seed be placed in holes at spaced intervals. To bank the maize plant and weed the field required the use of hand tools. The only way wheat could be efficiently grown was to sow it densely as too many open spaces allowed weeds to out-compete with the crop for moisture and nutrients. To harvest maize, the whole plant was pulled up, roots and all, and carried back to the dwelling area where the ears were removed and stalks set aside to be animal fodder. Wheat harvest involved another European tool, the sickle. Later that was also used to cut maize stalks.

Wheat became important, for peninsular Spaniards had a strong cultural preference for bread. In 1553, wheat was still virtually absent in Peru and remained in short supply for more than two decades after that. Unlike wine, carrying wheat from the mother country was not practical. As with wine, however, Spaniards felt deprived without wheat bread. Maize was a poor substitute. Between 1570 and 1610, Spaniards valued wheat much higher than maize, so that land that had traditionally been in maize was sown in wheat, which continued to be in short supply and costly (Glave and Remy 1983:10; Burns 1999:55). Kurakas denounced the use of land in the Sacred Valley for wheat. During that time, the owners of Hda Sillque turned some of its 90 ha of first-class maize land into wheat fields (Glave and Remy 1983). Near streams, to make flour for either consumption or to sell, Spaniards constructed water-driven gristmills with horizontal wheels. By 1620, the rapid expansion of wheat cultivation saturated the market and prices dropped steeply. Also, Creole Spaniards had begun to accept maize and other Andean crops into their diets. Wheat, however, continued to be an estate crop grown on unirrigated slopes through the colonial period and beyond. Contrary to an assertion, irrigation was not indispensable for wheat agriculture in the Sacred Valley or elsewhere in the Andes (Butzer 1992:10).

As has happened almost everywhere with European migrants to the New World, Spaniards brought their material culture to the Sacred Valley. When examined closely, however, its content represented only a selection of what was known in Spain at that time (Foster 1960). The Mediterranean scratch plow, known in Spain as the *arado dental*, was the only kind of plow introduced into the valley. Three other kinds of plows that were known in Spain never were accepted in the Andes, suggesting the importance of Andalusians among the migrants. The scratch plow prevailed in Andaulsia to the exclusion of any other kind and arguably provided the model for the yoke used in the Central Andes. Lashed onto the horns of the oxen, it was quite unlike the collar yoke common at that time in Northern Spain. To use a horse or mule to plow required a collar yoke, and that was not available in the Andes.

Threshing involved oxen, horses and donkeys in the Sacred Valley. Neither the Spanish threshing cart nor the threshing sled (tribulum) was used in the Andes. Milling technology also reflected cultural selections. Neither the windmill nor the vertical water wheel (*aceña*) was accepted into usage to grind grain in the Andes. The horizontal water wheel, less complicated because it had no gears, became the main way to grind flour and for a long time remained in the hands of Spaniards. Other Old Word objects that did not diffuse were the noria and the animal-pulled wooden cart. The former, a device to bring water up from a well or a river, was not needed in a mountainous zone as water is more easily directed by gravity from above. Carts were not suitable on steep inclines, but their non-use also had much to do with the strong indigenous tradition of carrying heavy loads on the back. Other Hispanic elements, from the periodic market, bullfights, religious brotherhoods (*hermandades*) and the notion of certain foods as either "hot" or "cold", abounded in the valley.

In contrast to many parts of the Central Andes where poor communications and sheer distance from markets hindered the full commercialization of estate operations, the Sacred Valley had a major advantage in its relative proximity to Cusco. The grain haciendas in mid-section of the Urubamba also had a labor advantage. In addition to their own yanaconas, estates were also able to seasonally tap the indigenous communities above the valley floor for sowing and harvesting. In the colonial period, Spaniards started small farms (*quintas*) on which they cultivated fruits and vegetables of Old World origin. Fresh apples, pears, and peaches from the valley regularly reached Cusco. Dried or packed in sugar syrup, this fruit was sent by mule to distant markets in Arequipa and Potosí (Anonymous 1958:95). Quite a different valley function was as a zone of recuperation and acclimatization (Garcilaso de la Vega 1960:227; Oricain 1906:12). Spaniards from Cusco and elsewhere regularly came to escape the colder and thinner air of the Huatanay Valley, especially during the months of June, July, and August. Childless Spaniards living in Cusco also came to the Sacred Valley to conceive and pregnant women to have a safe birth.[4] The more benign conditions of the valley also turned it into a staging area for domestic animals, especially chickens, brought from the coast and ultimately destined for elevations above 3500 m asl.

4.5 Recollections of the Valley in the Early 1960s

My early impressions of this valley were that of a gorgeous and wonderous place. Dramatic vistas of the quebrada below and glacial peaks above became for me the archetypal Andean landscape. Most intriguing, the sublimity of the natural setting contrasted with the strong human imprint on the valley floor and lower slopes. Moreover, nowhere else in the Urubamba were the remains of the pre-Columbian human presence so much in evidence. In some places, especially north of Ollantaytambo, terrace abandonment reflected the lack of irrigation water, which made cultivation here too risky. Landslides had destroyed the canals that brought water from above and they were never rebuilt.

4.5.1 The Hacienda

In 1963–1964, haciendas in the Sacred Valley owned about 80 % of the valley floor between San Salvador and Chilca. Part of the other 20 % was occupied by many small properties called *quintas*, especially around Urubamba and Yucay. Some were weekend residences of people from Cusco or were small farms producing fruit, flowers, and vegetables. A few peasant communities in Calca had land on or near the valley floor, but most of their crop and pasture lands were in the lateral valleys and above the rim. Individual families cultivated them and passed on their usufruct to their offspring. Indians who were seen on the valley floor were farmers from above bringing their produce to market on the backs of burros, or occasionally llamas,

[4]The same acclimatization rationale is even better known in Bolivia, where it was said that Spaniards living in Potosí (4000 m asl) could not conceive and so went to La Plata (2850 m) to ensure successful reproduction.

in addition to carrying loads on their own backs. Members of native communities above the valley floor also descended to sow, weed and harvest maize.

In addition to dominating ownership of the land, haciendas in the Sacred Valley also controlled commerce and the social order. In Huayllabamba District, which covers territory on both sides of the Urubamba River, four haciendas owned 75 % of the land. Three of them grew giant white maize, generally known by its Spanish name of *maíz blanco gigante Cusco* ("MBGC"), for export. Hacienda Huayoccari, owned by José Orihuela Yábar, was then known as the best maize producing estate in the whole Sacred Valley.[5] Use of fertilizers and pesticides enabled high yields of up to 5000 kg of maize per hectare. Haciendas Huicho and Urpihuata also grew that maize, though the latter, when owned by Alberto Escobar, also kept a purebred dairy herd for making butter. Another notable maize hacienda, Hacienda Compone, occupied the rich alluvial soils visible from the Inca fortress northeast of the town of Ollantaytambo.[6]

From today's perspective, labor arrangements were an unjust holdover from the colonial yanacona practices. Typical of the valley was Hacienda Huarán in Calca Province, where 35 workers labored 60 days a year for the estate and, in return, got a 0.3 ha plot for their own use. Even in the 1960s when peons became more conscious of the injustices of hacienda-imposed work requirements, the Fernández family that owned Huarán demanded free labor. Such abuses harked back to the medieval analogy of feudal arrangement, yet in my observation mutual obligation best defined the relationship. The hacendado, a kind of paterfamilias or, in Quechua, *llaqta tayta*, could be demanding and tyrannical, but workers obeyed him because, in the end, he provided them with some measure of security. Hacendados frequently called peons "hijo" or servants as "hija" and in return were addressed as "papito." That fictive kin way of thinking may have emerged in the colonial period when yanaconas seeking protection flocked to estates to escape the mita. By the 1960s, peons began to seriously see their own destitution in new ways, sometimes influenced by ideas from the outside.

Relations between owner or manager and peons, while generally better in the Sacred Valley than elsewhere in the Cusco Department, varied with the property. A focus on agriculture rather than on livestock and relative proximity to the forces of law and order in Cusco may have been factors in assuring a greater degree of social peace here than in the more isolated and pastoral high provinces where jacqueries occurred and where cattle rustling (*abigeato*) was so common it dominated

---

[5] The Orihuela family, whose history in Peru goes back 300 years, represented the Cusco elite. In 1825, Don Manuel Orihuela welcomed Simon Bolivar to his hacienda. His great grandson, José Orihuela Yábar, constructed a new casa hacienda in 1950 to exhibit his extensive collection of pre-Columbian artifacts and colonial canvases. Don José died in 1979, having succeeded in keeping title to a part of his estate after agrarian reform.

[6] Before the agrarian reform, hacienda families in the Sacred Valley formed a relatively well-educated class. Hacienda Compone, which in the colonial period was known as Hacienda Huatabamba, was owned by Julio Corazao. He married Esther Giesecke Matto, the daughter of Albert A. Giesecke, North American rector of the University of Cusco.

everything (Gade 1994). Within the Urubamba itself, distinctions can be made between Canchis, for example, and the Sacred Valley. The famous novel *Aves sin nido* by Clorinda Matto de Turner (1852–1909) described ethnic and social confrontations based on Tinta where she lived as a young adult. Conditions there were different from those of the Sacred Valley where Matto had grown up on Hacienda Paullu Chico near Coya.

Call for land reform had an additional reason. In the second half of the twentieth century, the phenomenon of the hacienda renter became increasingly important in the valley. Religious and civil institutions had long leased out their properties in return for a yearly rent, but with time many absentee owners either let portions of their property to tenant farmers or rented to an individual who ran the hacienda as a business enterprise even though he did not own the land. That arrangement enabled the owner to divorce himself from day-to-day hacienda decisions and to live in Cusco pursuing other interests. One man in particular, Felipe Calderón Fuenzalida (1892–1964), became well known in the Sacred Valley as a consummate entrepreneur of rental property. Calderón first appeared in the 1940s when he bought two small quintas, but discovered that bigger profits accrued if he rented big estates. Paullu Grande and Huarán in Calca, Saniwas in Pisac, Huaypo in Urubamba and Primavera in Ollantaytambo all fell under his administration. Some kind of rental agreement covered about one-third of the land in this region of the Urubamba. Uninvolved owners left quite a few haciendas in dilapidated shape. As proprietors distanced themselves from their own properties, the traditional ties of mutual obligation between owner and worker were compromised. But the workers' primary grievances were the arrangements imposed upon them. To forestall rebellion, the landowners turned peons into salaried workers at Huarán and then at several other haciendas. The real decision was made in Lima, where the socialist government in power wanted the demise of the estate system.

4.5.2 *Agriculture in the Valley in the Early 1960s*

As the agricultural cornucopia of the department, the Sacred Valley had, aside from stony or marshy patches, an intensively cultivated valley floor. Maize occupied about 80 % of irrigated land, about the same amount as it had in the late colonial period. Though tractors were first introduced to some haciendas in 1930, generally the plow and oxen continued to be used to prepare ground for planting (Tamayo-Herrera 1978). Workers used hand tools to weed fields and bank soil around the base of individual maize plants, a procedure known as *aporque*. Banking the soil twice during the growing cycle demanded much labor. The peasants fed the weeds to their livestock or boiled edible greens (*yuyos*) for themselves (Gade 1971). Application of animal manure and hearth ashes maintained soil fertility. Several haciendas used chemical fertilizers from the then new ammonia plant at Cachimayo near Cusco. Some properties used pesticides, though not herbicides.

4.5.2.1 Sacred Valley Maize

Floury landraces of the "Cusco complex" strongly dominated maize cultivation. The valley's two main races of that group both had eight rows. Uwina sara, known to maize classifiers as "Cuzco cristalino amarillo," had its traditional market on the Altiplano where the rigors of climate allowed very little maize to grow. The second race was *paraqay sara*, which meant white maize, i.e. with no aleurone color. Other floury races were also identified by their external color: *q'osni sara* (brown), *oqhe sara* (lead colored), *chaminco* (variegated red and yellow), and *pescaruntu* (speckled) like a bird egg. The referential use of color to identify maize cultivars testifies to their phenotypic diversity. Yet color and color patterns in flour maize are superficial. In flour maize, only the top layer of cells known as the aleurone contains color, thus most of the kernel is actually white.

In the 1960s, peasant folk primarily ate flour maize as whole kernels, boiled (*mote*). Also some was made into *cocopa*, corn kernels cooked and then dried in the sun, as a product that could be stored. During the early part of the year, boiled fresh corn on the cob (*choclo*), green corn, was popular. Its cultivation required a different regime of planting than maize that was to be dried. Planted in April, heavily irrigated through the dry months, it was ready for harvest in January. Choclo from the valley was a much sought after seasonal item in Cusco, Arequipa and Lima. In the early 1960s, green corn accounted for about one fourth of valley maize production. Financial returns on growing choclo exceeded that of dried maize. Since a rainy-season harvest precluded drying, it could not be stored for sale at a later date to get a better price. Flour maize, high in soft starch content and thus easily chewed, was the raw material for chicha making. In the 1960s, chicha was still the main beverage of valley habitants. Formerly, a peasant would masticate maize into a cud-like ball and spit it into a pot to serve as the starter for chicha fermentation. By the 1960s, that old method, depending on the enzyme in saliva to trigger fermentation, had become rare in the Urubamba Valley. Instead chicha makers used sprouted maize (*jora*) to initiate fermentation.

Maize blanco *gigante Cusco* ("MBGC") is a genetic variant of paraqay sara, now recognized as a separate race of maize. It is characterized by its large kernel, 22 mm long, larger than all other maizes in the world. The plant grows up to 3 m in height, has a thick central stalk and no adventitious roots. How and when MBGC evolved is uncertain. Known as the Grobman hypothesis after the Peruvian agronomist who formulated it, the putative ancestor of MBGC may have been a proto Confite Morocho, a popcorn with a slender cob and eight rows of red or colorless kernels that diffused from the coast to the highlands before 2000 BC (Grobman et al. 1961:233). Andean farmers selected and propagated Morocho maize that spontaneously mutated toward floury endosperm. Flour maize then hybridized with a race brought from Northern Peru that the agronomists Grobman and Sevilla called Pardo. The maize scientist Paul Mangelsdorf (1974:209) favored the Mexican race Tabloncillo as the putative ancestor of MBCG. The interdisciplinary team of Bird, Browman and Dunbar essentially followed the Grobman hypothesis, though they did not agree that a separate domestication of maize occurred in the Andes (Bird et al. 1983/1984:201–202).

Plausible diffusers of this introduced landrace were Wari people who expanded southward into Cusco between 450 and 850 AD and logically brought it at the close of the Wari hegemony, when the Andes witnessed migratory movement of the first order (Bird et al. 1983/1984; Rostworowski de Diez Canseco 1999:6). When, however, Cusco flour maize hybridized with the introduced landrace remains an enigma. No archaebotanical MBGC has been recovered nor have ceramic models been found that correspond to this race (Mangelsdorf 1974; Eubanks 1999; Sevilla Panizo 1994). Notable maize producers, the Incas were less concerned with maize diversity. Although no mention is made of it in the colonial literature, Mangelsdorf (1974) suggested that this race arose in post-Columbian times. The early colonial chronicler Bernabe Cobo (1956:I:160) referred to "paracaysara," but reference to white maize is insufficient to indicate the presence of MBGC. It is possible that this outsized maize was such a minor cultivar in the Inca and/or colonial periods that it got no special attention. Even its importance during the early twentieth century is not well established. Herrera (1921), in his flora of Cusco, does not refer to it as a distinct cultivar, though he may have subsumed it under paracaysara which he wrote had kernels of different sizes. It first received attention in the literature after World War II when ethnobotanist Hugh Cutler (1946) included it in his study of Peruvian maize. Old farmers I interviewed in the 1960s remembered when yellow maize was grown in much greater quantity than MBGC. This impressive kind of maize became internationally recognized when hacendados began to sell it to exporters. The maize found its niche in the United States and later in other countries as a confection called "corn nuts." The big kernels are a convenient size to be eaten as a crispy finger food. The kernels are soaked in water, dried, deep fried in oil, and salted. The export trade made phenotypic purity important. Kernels that do not meet size and purity specifications have no market abroad. To maintain MBGC purity, the crop is either planted at a different time than other landraces or planted beyond the range of wind driven pollen from other cultivars (Fig. 4.9). In 1963 hacendados, not peasants, were largely responsible for the purity of MBGC. They also were the growers who had the fertilizers that ensure proper development of large-kernelled ears. At that time the cost of agrochemicals was the biggest barrier to peasant cultivation.

Chullpi was the main non-floury race grown in the Sacred Valley. It had rounded ears, 20 or more rows, and sugary endosperm which shrivel when dry, indicating that it had an origin different from flour maize. Low-yielding chullpi had only one or two ears per plant. It was eaten parched (*k'ancha*) or used for chicha. Its high sugar content produced a beverage of higher alcohol content than other maize. Another race, more sparingly grown, was kculli, which Padre Cobo (1956:I:160) referred to as "*cullisara*." This small-eared maize had a short vegetative cycle of only about 3 months (Fig. 4.10). The deep purple of both the kernel and the cob made it valuable for dyeing food and beverages. Yet more than color characterizes its food chemistry. Rich in anthocyanins and other phenolic compounds, kculli has now been shown, at least in mice, to have positive health benefits as an antioxidant to reduce systolic blood pressure and to prevent obesity and diabetes (Ramos Escudero et al. 2012).

Fig. 4.10 Among the maize cultivars grown in the Sacred Valley are (*left*) kculli, a small-eared purple maize; (*top center*) chullpi, a sweet corn eaten in toasted form; (*lower center*) popcorn; and (*right*) white flour maize (MBGC). (Photograph by D.W. Gade)

Fig. 4.9 The effect of crossing races of maize is apparent in two ears manifesting introgression between the white flour maize on the left and the chullpi in the back. (Photograph by D.W. Gade)

4.5.2.2 Other Crops

The valley zone around Yanahuara, Urubamba and Yucay also became known for fruits and vegetables grown under irrigation. Cabbage, tomatoes, squash, and other *hortalizas* were marketed mainly in Cusco. Though mediocre in quality, peaches, plums, pears, medlars, apples and grapes also found a ready market. Unlike most tropical produce, rosaceous fruit was highly seasonal and thus special, despite the fact that good quality fruit-bearing perennials of mid-latitude origin require a longer and deeper dormancy period than that available in the valley.

Barley, wheat, and potatoes dominated fields on unirrigated slopes. Since fertilizers were not used, that land was kept in fallow much longer than it was cultivated. Barley at the time had started to become popular due to the Cusco brewery's clever organization. Agronomists gave free seeds of two-rowed barley suitable for malting to farmers and then bought their harvest. By these means their agronomists hoped to eliminate the old six-row barley cultivars that peasants had traditionally grown. However, the brewery was the sole buyer of two-row barley and thus set the price paid peasants. Since the arrangement put cash in the hands of people who had so little of it, that monopoly had not yet become controversial.

Further up the mountain above 3700 m asl, dispersed ayllus populated by monolingual Quechua speakers had a largely subsistence livelihood. According to tradition difficult to verify, ancestors of these high country folk once had lived on the valley floor until the colonial policies forced them to move out. In their zone of refuge they farmed potatoes and raised sheep and camelids on land owned by the community. Agricultural plots in communities were transferred to descendants, but could not be sold to outsiders. Communities in the high country, necessarily more isolated, have maintained a greater cultural conservatism than has the valley. San Juan, a hamlet above Yucay, is the best known of these indigenous communities. It was recognized as a legal entity in 1920, the year the Peruvian government restored legal status to Indian-owned land. In 1969, the Velasco government, for purely semantic reasons, classified San Juan and other *comunidades indigenas* as peasant communities (*comunidades campesinas*).

4.5.3 Population, Religion & Settlement

In 1963, most valley dwellers were "medio mestizos," a category that included cholos, i.e., native people who had become acculturated to Hispanic ways in speech, dress and other practices. At that time, 21 % of the valley population wore *ojotas*—sandals made of old rubber tires—and 8 % chewed coca (Perú 1966). In the 1960s, Quechua was still the first language of most cholos. Most men then knew how to speak some Spanish, though only a minority of them could read or write it. The Sacred Valley attracted Spaniards and hacendados, officials and priests, all of whom fathered many illegitimate children by native women. In contrast, Spaniards tended to avoid the cold plateaus and there less cultural and genetic mixing occurred.

Traditional religion in the valley followed the Central Andean pattern of a focus on cosmic questions beyond human control. Peasants enlisted divine help to assure the essentials of livelihood by integrating fiestas into the agricultural cycle (Fig. 4.11). The annual feast of San Isidro Labrador (Saint Isidore the Farmer) on May 15, coincided with the harvest period in the Southern Sierra. Pisac, Lamay, Calca and several ayllus in those districts each held a San Isidro fiesta. Following

Fig. 4.11 Near Yucay, a home-made cross surmounts an ear of white flour maize as a thanksgiving symbol for a good harvest. (Photograph by D.W. Gade)

mass in the church, the image of the bearded saint, jauntily adorned with a straw hat and spikes of wheat, was removed from his niche and carried in procession. A musical band, townspeople, and garlanded oxen accompanied the *santo* to the plaza. There a yoke of young bulls hitched to a plow was ceremonially anointed with alcohol in an indigenous ritual known as *t'inkaska*. Other libations of chicha accompanied invocations to Mother Earth, the ancestors (*apus*), and the mountain divinities (*auquis*), reinforcing the cultural syncretism of this feast. The day's festivities combined the sacred and the profane with food, drink, music, and an evening bonfire and libations into it of chicha.

The mixture of indigenous and colonial European practices in this particular Andean fiesta conveyed much about how two different traditions have converged. San Isidro was clearly European: a humble farmer born near Madrid in the twelfth century. Even before his canonization in 1622, the Church in Peru had incorporated St. Isidore into the annual cycle of feast days. Perhaps because he filled the need for a harvest deity, a strong cult grew up around him in the Andes even more than in Spain itself. The San Isidro event in the valley featured the trifecta of European agricultural associations of wheat, oxen, and the plow. Since, however, maize was the main crop and was being harvested on May 15, the fiesta was much more than just a celebration of the wheat harvest. Maize had acquired some Old World elements, for by then on the valley floor, peasants used the plow and oxen, not the chaquitaclla, to prepare maize fields. Following the practice known both in Spain and the Andes, maize is sown according to the phases of the moon. One scholar inferred a layer of social meaning by opining that San Isidro Labrador was a "convenient embodiment of the values of fealty and subservience which Indians were meant to owe their Spanish masters" (Sallnow 1987:279).

In the 1960s, clustered settlement, a legacy of the sixteenth-century reducciónes, still dominated the humanized landscape of the valley (Fig. 4.12). At that time

Fig. 4.12 Young peasant woman in front of her house in Ollantaytambo, the only reducción town in the Sacred Valley whose origins predate the Spanish Conquest. Although within a village, the house compound usually has included livestock and stored crops (Photograph by D.W. Gade)

peasants walked daily from the village to their plots. Some dispersal had occurred, which was not recent. Though all towns were of colonial origin, few buildings other than the church, despite appearances, were colonial in age. Originally each village had ayllu sectors, but by the 1960s, they no longer had much meaning to inhabitants. Services in the villages were few. A poorly stocked general store or two operated in villages. As a regional market town, Pisac also had a bakery and barbershop, whereas Ollantaytambo had neither. The valley's two largest towns in 1964, Urubamba and Calca, had a notary public, lawyers and a secondary school befitting their status as provincial capitals. Colonial authorities designated Urubamba a villa, which in the Peruvian urban hierarchy gave it a prestige greater than other Sacred Valley town. Whites, mestizos, and cholos inhabited Urubamba, though some hacendados lived primarily in Cusco.

Urubamba became commercially important beyond the Sacred Valley itself in 1820 when mule trains between Cusco and La Convención began using its new stone bridge to cross the Urubamba River. Thanks to that strategic bridge the town of Urubamba became a redistribution center for coca, alcohol, coffee, and other tropical products from the lower valley. To feed the mules coming and going, alfalfa and green barley became important crops. Urubamba's commercial role was reinforced in the 1880s when hacendados of La Convención funded the construction of a mule trail (*camino de herradura*) through the canyon below the Machu Picchu ruins. That trail shortened the distance to the tropical zone and provided an alternative to the old route across the cold and high Málaga Pass. Mule trains brought manufactured goods and highland food products to La Convención, to a population almost entirely of highland origin. The demand for bread in this warm zone where wheat could not be grown made Urubamba a center of bread making. The bread was stale and hard before it reached the valley, but that seemed not to have deterred the highland consumers. When the railroad through the canyon reached Santa Teresa in 1950, Urubamba lost its role as a commercial staging area. By contrast, the town of Calca has continued to maintain its economic tie to the Yanatile jungle zone (Flores Ochoa 1985). First a mule trail, then in the 1960s a vehicle road, connected Calca, the provincial capital, with Lares and the Yanatile Valley farther down valley where tropical products, especially coca, were grown.

In 1963, the only Sacred Valley towns that had electricity were the two provincial capitals. Generated by small turbines placed in the fast streams tumbling through those towns, the undependable service was available only during the evening hours. Hacienda Huayoccari generated its own electric power that way as did Huallyabamba, for a time, beginning in 1934. Elsewhere in the valley, evening illumination consisted of candles and, exceptionally, kerosene lanterns. I recall how, when darkness descended, people would pass time telling stories sitting around the hearth or listening to their battery-powered radio receiver. Inhabitants of the Sacred Valley had little influence in directing the power grid from the recently expanded Machu Picchu hydro plant to their area. Instead, the transmission lines veered off to supply power to the ammonia plant, built in 1965, at Cachimayo near Cusco.

Roads to and within the valley in the early 1960s were either dusty or muddy according to the season. Farm trucks, which in Peru were used primarily to carry people, and ramshackle buses transported people from Cusco to the Sacred Valley via either Pisac or Urubamba. A dirt road also ran through the valley to some distance below Ollantaytambo. Highway construction started in the 1920s when President Augusto Leguia instituted the "ley de conscripción vial" during his second administration. Essentially, the state demanded free labor by requiring each male to either work on road construction or to pay someone to take his place. Under that de facto reenactment of the colonial mita, Indian peasants built the first roads into the valley and elsewhere in the Sierra.

Schools in the Sacred Valley also first came in the twentieth century. In 1907 a British Baptist missionary group founded the Inca School Society to educate indigenous children and, to that end, acquired Hacienda Urco near Calca. At Yucay, Salesian priests from Italy received a donation of land from a hacendada and started a trade school for poor children based on European ideas of vocational training. In 1962, the first contingent of Peace Corps Volunteers from the United States arrived in five valley towns with "community development" as their nebulous charge. No one has yet determined whether any Peace Corps initiative brought about a lasting change in valley behavior or practice.

4.5.4 Tourism in the Early 1960s

In the early 1960s day trippers visited the Sacred Valley in modest numbers. In 1964, only 38,839 foreign tourists came to Cusco, six times more than had come a decade earlier (Cusco 2008). Of that number, less than a third went to the valley even for a Sunday taxi excursion to see the Pisac market. Unless they were inclined to hike, most tourists who came to Pisac never saw the impressive Inca fortress above the village. The artisan and Indian parts of that market adjoined each other. What the writer Christopher Isherwood (1948:12) declared to be "too tourist-conscious to be truly attractive," fit only the contrived handicraft section. What the author apparently did not know was that its sellers had themselves come, together with their goods, from Cusco earlier in the morning just ahead of the tourists. In contrast, the peasant-oriented Sunday event in Pisac in which valley maize was exchanged for high country tubers, dates back to the colonial period. The Pisac market drew in some other communities, most notably San Salvador. Even though it had been a reducción, San Salvador never acquired a periodic market of its own. Monolingual Quechua speakers from communities at higher elevations brought potatoes and ullucus and traded them for maize grown on the valley floor. Coca leaf, brought from La Convención and the Paucartambo Valleys, was another big item. In the early 1960s, the Pisac market offered some manufactured goods, notably metal pots, aniline dyes, and radio batteries. Social interchange, lubricated by chicha, was a major function too. Many Indian market goers returned to their dwellings in a state of inebriation.

Tourists found the Indian market of Pisac of passing interest, but patronized the adjacent artisan stalls. Sundays also gave visitors the chance to observe local customs. After the morning mass in the church with its homily in Quechua, a solemn procession of officials from the ayllus in the high country part of Pisac District, wearing multicolored ponchos and knitted caps (*chullu*) and carrying staffs of authority, moved through part of the town (Fig. 4.13). To tourists, they were so much more noble-looking than the cholos from the village of Pisac, yet townspeople considered them country bumpkins and cheated them mercilessly. Sunday tourists to Pisac either returned directly to Cusco or continued down valley to visit Ollantaytambo and Chinchero before returning to their hotels. In the early 1960s, all archaeological sites in the Sacred Valley were unattended and open. Even at Ollantaytambo, with its spectacular Inca fortress and its genuine town of Inca origin, no tour guide or artisan seller interrupted the calm. Few locals even knew about a *camino* to Machu Picchu that later became famous as the Inca Trail. A government-owned "hotel de turistas" in Urubamba stood out as the sole Sacred Valley accommodation that had foreign travelers in mind. Its siting near the Urubamba River suggests the stream was then considered to be an attraction. Foreigners, generalized as "gringos," were unusual enough in most places in the valley to become centers of attention. The attitude that foreigners were useful only as sources of money had not yet taken hold. Unless they knew Spanish, communication was minimal. Though English was a subject taught in the upper grades of some schools, no one saw it as a real form of communication, one that they could ever actually learn to speak.

Fig. 4.13 Procession of native leaders from communities at higher elevations in the District of Pisac. For centuries Indians have descended to the valley floor on Sunday to participate in the market and attend Mass. (Photograph by DW Gade)

4.5.5 *Thoughts on the Environment as It Was*

In the early 1960s, environmental issues had not penetrated the valley's consciousness. Moderate pollution of the Urubamba River did not prevent boys from fishing for trout or women from washing clothes in it during the dry season. In the 1960s, trash and garbage were a minor problem. Roving pigs gobbled up all manner of organic waste, country folk picked up discarded containers to reuse, and water in plastic bottles had not yet entered commerce. Peasants accepted soil erosion as part of the natural order. Eucalyptus, introduced to the valley in 1885 at Hacienda Rocaforte near Urubamba, had become, by 1963, a conspicuous element of the valley landscape (Tamayo-Herrera 1992:II:562). It grew rapidly and the stumps of cut trees sent up many suckers, and this tree became the main source of construction wood. Used as firewood, the eucalyptus gave the valley a characteristic medicinal odor. Though some mestizo households had kerosene stoves for cooking, most folk burned firewood or dried dung (*bosta*). Charcoal was scarcely used. With no chimney, a dwelling filled with wood smoke. High particulate matter in the air contributed to the acute respiratory infections common in Highland Peru, but at the same time, the smoke from wood fires protected the dried maize hanging in the rafters from insect pests.

Few wild-growing native tree species grew in the depression, but as one moved up in elevation in the side canyons, woodlands appeared. Higher still, communally-owned grazing land, especially on west-facing slopes, had *Polylepis* woodlands. Several species in that genus dominated the woods from 3600 to 4400 m asl; above 3800 m asl, *Polylepis* occurred in almost pure stands. Cervids occupied this habitat and hacendados, since they were the people who owned guns, would shoot the taruca (*Hypoamelus antisiensil*) and white-tailed deer (*Odocoileus virginianus*) for recreation. These now well-studied woodlands are critical in conserving groundwater and maintaining flowing springs (Jameson and Ramsay 2007; Servat et al. 2002). They also hinder soil erosion and encourage the formation of mountain mists by their foliage. For those reasons, *Polylepis* forests were protected by a law promulgated already in the 1870s during the Manuel Pardo regime. If eucalyptus had not been introduced, these *Polylepis* woodlands would have disappeared, in the face of the demand for wood.

4.6 The Sacred Valley Today

4.6.1 Social Change

Much has changed in the Sacred Valley since 1963: more people, many more landowners and waves of tourists. As revealed in the four national censuses, every district in the Sacred Valley grew in population between 1963 and 2013. About 90,000 people now live on the area defined as the floor of the Sacred Valley, with the greatest growth in the zone encompassing Urubamba, Yucay and Huayllabamba. Populations of the two provincial capitals, Calca and Urubamba, each tripled since the early 1960s as a result of expanding commercial and institutional functions. District capitals also grew, though the rural population has sometimes gone down. In the case of Pisac District, the economic opportunities tied to tourism have led to more inhabitants in Pisac village, whereas considerable numbers of *comuneros* in the high country have migrated from their overpopulated farming communities to towns and cities. For a short period, the land reform of the 1970s neutralized the general trend seen all over Peru of rural departure.

Several towns are now mostly Spanish speaking, though Quechua is still spoken by older generations and most young people can understand it. The countryside has not abandoned Quechua, but it has more bilingual people than half a century ago. The long-term trend in the Sacred Valley is toward Spanish unilingualism. Educational levels have risen since the 1960s; now more than 90 % of children attend primary school. A much higher percentage than in the 1960s also attend secondary school. The 84.2 % literacy rate for Urubamba Province and 74.8 % rate for Calca Province reflect that rise in school attendance (Cusco 2008). Literacy on the valley floor is actually higher than that, since the figures include hamlets above the valley rim. Predominantly Quechua speakers with a fairly high rate of illiteracy still dominate the more traditional upper zone.

Fig. 4.14 When power came on line for the first time in 1963 from the Machu Picchu hydroelectric plant, service went to the city of Cusco and a fertilizer plant near Cusco. Although the pylon infrastructure was built mostly on the floor of the Urubamba from Machu Picchu to Ollantaytambo, the Sacred Valley was not actually connected to the power grid until years later. (Photography by DW Gade)

In religion, the trend in the valley since the 1960s has been toward more diversity. Protestants are now more numerous than they were half a century ago. Pentecostals, Methodists, Jehovah's Witnesses, Mormons, the Iglesia Evangélica Peruana and the entity called "Israelitas del Nuevo Pacto Universal" have in common their refusal to participate in fiestas honoring patron saints. Whereas Catholics have seen divine intervention as a *deus ex machina*, Protestants in the Andes have viewed religion in less cosmic and more personal terms. Women with families in particular are attracted to Protestant sects as they advocate against alcohol abuse, family violence and irresponsible spending, problems which the Catholic padres have not adequately addressed. In 2011, Protestants in Calca pressured the mayor for a more honest local government. Protestant disinclination to think hierarchically may also have influenced a decline in the notion that an administrative organization may impose its will on those below. Formerly the *gobernador* of the district capital assumed the right to demand work requirements of people living in satellite hamlets (*anexos*) in his jurisdiction. Those demands, first contested seven or eight decades ago, have now largely disappeared.

A major change since 1963 has been the arrival of electricity to the valley floor and some side canyons. In 1985, more than two decades after the completion of the Machu Picchu hydro development, the Sacred Valley finally became connected to the grid (Fig. 4.14). Between 1963 and 1985, the Machu Picchu hydroelectric plant increased its megawatt capacity five times. Electrification opened television and later the computer to valley, transforming the peasant concept of the world. Demand for electrical devices favors money economy over barter. Unlike agricultural products, manufactured goods could not be acquired by trade. The persisting lack of electrical power in indigenous hamlets in the high hills above the valley accentuates their modernization gap with the valley floor.

4.6.2 Change in Land Tenure

Land tenure in the Sacred Valley now is strikingly different from what it was in the early 1960s, a result of the agrarian reform law of 1969 that broke the back of the hacienda system that had held so much land in the hands of the few. The Sacred Valley before agrarian reform had 66 haciendas, though not all were large estates encompassing several ecological zones. Powerful owners sought to manipulate the adjudication process to their advantage. In many cases, hacendados kept ownership of the casa hacienda and choice plots. In other cases, estates lost all their property. In San Salvador District, Hacienda Huallhua had a history of illegally usurping land from indigenous communities and as retribution for those sins of the past, the agrarian reform tribunal allowed its owner to keep nothing. The agrarian reform process gutted the hacienda as an institution, sweeping away the sharp social divisions that once prevailed in the valley. The germ of a civil society was allowed to sprout.

Although it took more than a decade, workers attached to those estates eventually became the big beneficiaries. In 1973, the revolutionary government sought to impose a collectivist ideology on the peasantry. These entities, known by the acronym SAIS, incorporated into them many haciendas and peasant communities. In the Provinces of Calca and Urubamba, 15 cooperatives occupied more than 39,000 ha of good valley land. Not all CAPs (*cooperativa agraria de producción*) were on the valley floor. Yanahuaylla consisted of 32 families on 512 ha of land that ranged from 3300 to 4226 m asl. Natural pasture covered 395 ha, irrigated land 60 ha and unirrigated crop land, 40 ha. Thirty-two families farmed that cooperative, but in 1986 voted to liquidate it as a cooperative in favor of privatization (Alvarez del Castillo 2005). Elsewhere, in the early years of the collectivist movement, there were peasants who wanted to return to the hacienda system. Some CAP units survived longer than others, but eventually all succumbed to the neoliberal Zeitgeist. Most peasants, rebuking the statist ideologues who harked back to the communal land tradition in the pre-Columbian Andes, saw land ownership as their only guarantee of subsistence security. Peasants came to that conclusion when one CAP after another experienced financial mismanagement and social strife.

The demise of the cooperatives opened the door to the distribution of land to the peasant families who had once worked on a hacienda and had belonged to the cooperative formed on it. Each family got title to a parcel, ranging in size from less than 1–5 ha. In 2005, Urubamba Province had 12,714 farms on 70,499 ha and Calca had 11,972 farms on 176,474 ha, though less than half of that area lay on the valley floor (Innocente et al. 2006). The minifundia system that emerged from latifundia almost guaranteed the maintenance of palaeotechnic modes of farming. Machine technology, too expensive for most small farmers, could not compete with hiring help for the harvest. Plot quality also varied. Disparities emerged among peasants; those better off engaged people poorer than themselves to perform farm labor in return for a tiny subsistence plot, ironically mimicking the old hacienda structure of dominance and subservience.

4.6.3 Valley Agriculture

The basic distinction between the irrigated valley floor and non-irrigated slopes still holds in this region. Glacial meltwater provides the dry season life line for almost all farmers on the valley floor. In the two Provinces of Calca and Urubamba, ca. 44,000 ha above the valley floor or outside of it, are non-irrigated ("secano"), whereas about 10,000 ha, mostly on the valley and terraced lower sides, have irrigation ("riego") (Innocente et al. 2006). Elsewhere in Cusco Department that ratio is even greater. Productivity of nonirrigated land is much lower, for the plots are kept in fallow for at least 6 years to store up the necessary nutrients for the 2 or 3 years it is cultivated. Often a family that has a small plot on the *piso de valle* also cultivates a field or two on the rain-fed valley slopes. Grains long dominated the unirrigated niche. Wheat, however, has lost much of its importance as a crop. In 2010, Urubamba and Calca Provinces had about 9600 ha in wheat, less than half the amount of the 1960s (Cusco 2008). Since bakeries much prefer imported hard wheat for bread making, Peru now imports most of its wheat. That practice started when Peru first imported Chilean wheat, but gained ascendancy when the United States government sent wheat to Peru under the Food for Peace program. Unlike wheat, barley growing has greatly expanded, occupying ca. 17,000 ha of slope land above the valley. Its cultivation is in response to the large Cusco brewery's voracious demand for raw material to make malt.

Maize and horticulture still receive most of the irrigation water. Horticultural production is centered on Yucay, Urubamba, and Yanahuara, from which daily truckloads of fruits and vegetables go to Cusco. Horticulturalists enclose their parcels with adobe walls to thwart thieves and keep free-ranging domestic animals out. Fruit production includes peaches (*blanquillos*), pears, apples, plums, capuli cherries, and strawberries (*frutillas*). An equally wide array of vegetable crops broadens the risk and staggers the workload: carrots, cabbage, lettuce, squash, culinary herbs (collectively "asnapa"), and onions. Three leguminous crops once exclusively grown for their dried seeds are now grown also as green vegetables: broad beans (*Vicia faba*), peas (*Pisum sativum*) and *Phaseolus* beans. Tourist demand partly explains the shift toward fresh vegetables. The Sacred Valley also produces off-season (*maway*) potatoes, which come on the market at the time of greatest tuber scarcity and thus garner higher prices than potatoes from the regular harvest. Forage barley is grown under irrigation to feed domestic animals, among them guinea pigs (*cuys*). Once only raised in peasant dwellings on kitchen scraps, these small rodents are now also commercially produced in establishments called *cuyerías*.

4.6.3.1 Persistence of Maize

Through epic changes in land tenure and the growing influence of the city of Cusco on land use, maize has remained the most important crop in the Sacred Valley. It is also the crop most grown on prehistoric terraces. Two agronomists found that the soils on these terraces are especially rich in phosphorus derived from manure and ashes (Sandor and Eash 1995:178). Peasants now generally use pesticides. Some communities own and share machinery. Ccoya Cusco, located near the village of Coya, farms 127 ha and collectively owns a tractor and harvester (for wheat and barley). Members of the community rent these machines for use on their own plots. However, the main crop in Coya is maize and for that hand labor still dominates. Flour maize cultivation requires the banking of soil (*aporque*) around the roots with a hoe-like tool called the lampa. That operations strengthens the main stem of the plant and, in the process, eliminates weeds. Tractors have not eliminated the use of oxen to plow maize land. Bovines eat dried corn stalks and provide manure for the fields, whereas a tractor requires purchased fuel to operate and yields no fertilizer.

Maize with floury endosperm remains the valley staple. The range of cultivars grown reflects the traditional value placed on diversity, though the cultivars may be used and prepared similarly. In the 1960s boiled corn kernels (*mote*) and thick maize soup (*lawa*) were regular parts of the peasant diet and remain so today. Chicha, once the iconic beverage of peasant life, now has competition from bottled beer. The shift was apparent in the 1960s; today beer is the preferred social drink. The array of maize ears laid out on a drying floor reveal cultivar crossing, intentional or not. Crossing two different inbred lines creates "*maiz cholo*" whose advantage is higher yields. Since maize is wind pollinated, cultivar purity can be compromised by freakish weather or by a farmer's inattention in separating different landraces. Thus two different objectives can be at work in a maize field: the prospect of greater production and the possibility of creating new cultivars vs. the maintenance of recognized cultivars that are more saleable.

More than in the 1960s, MBGC now has become the signature crop of the Sacred Valley. Hacienda demise and land distribution has greatly increased the number of farmers who grow it. The largest MBGC producer in the valley has only 30 ha of land, but that is still large compared to the 2000 farmers who have only one to five ha and the 2844 who have less than 1 ha (Innocente et al. 2006). Differential use of fertilizers, pesticides, irrigation and attention to cultivation practices explains the wide variation in yields from 1500 to 6000 kg/ha. Heavy fertilization and dense planting can increase yields to 10,000 kg/ha. Total MBGC production in 2007 was 54,985 t (Innocente et al. 2006). About three-fourths of MBGC production has been for export; the top three importing countries in that period were Spain, Japan, and the United States.

4.6.3.2 Controlled Appellation

In 2005 MBGC producers from 17 communities in the valley banded together and formed a cooperative with the acronym APROMAIZ (Ravero Campos et al. n/d). This association, together with other producers and exporters in 2005, applied for and received government-approved status for the MBGC corn produced in those communities as a controlled appellation. It was placed on the international register in 2007. This legally defined geographical label, called in Peru, *denominación de origen*, guarantees the Sacred Valley as legitimate source of the product. Similar white maize produced in northern Quispicanchi Province (up to 3200 m asl) and in Paruro or Acomayo Provinces is not eligible for the appellation. Likewise, the large-kernelled *maiz blanco* produced in the Mantaro Valley of Central Peru comes from outside the legal geographical boundaries. The narrow definition for the MBGC of the controlled appellation conforms to what is believed to have been its zone of origin. However, it has yet to be demonstrated that this genetic achievement actually originated in the Sacred Valley or that the traditional knowledge of its cultivation is any different there than in any of the other mentioned zones.

Limiting the appellation to the Sacred Valley suggests protection of this maize from patent misappropriation. The Sacred Valley has a certain magic as an iconic place connecting the Peruvian present to the pre-Columbian past. In 2012, the Instituto Nacional de Cultura reinforced the status of MBGC by declaring it to be a cultural patrimony of the Peruvian nation. The assertion that Peru gave rise to domesticated maize is an honorific extension more than a plausible scenario of prehistory. An agronomist and an archaeologist have presented evidence for preceramic maize in Peru as an argument for a domestication process in Peru independent of that of Mexico (Grobman and Bonavia 1978; Bonavia 2008). No wild maize has yet been found in Peru. Phylogenetic analysis has revealed maize domestication to have been a single event and that it occurred with teosinte (*Zea mays* ssp. *parviglumis*) in Mexico (Matsuoka et al 2002).

Peru's interest in protecting agricultural plants and plant varieties motivated the appellation process. Over the past decades, much international discussion of germ plasm ownership has enhanced Peruvian awareness that experts in genetic resources regard their country as an important reservoir of agrobiodiversity. Attempts by foreign companies to acquire genetic material in Peru and then file for patent protection of commercial seeds consternated Peruvians. Plant-breeding cases brought to public light included a Peruvian sweet potato (*Ipomoea batatas*), a cultivar tested by the American fast food industry; native cultivars of cotton, a major Peruvian export crop and wild tomatoes, whose wild germ plasm is found only in the Andes (Ruiz 2001). In 1963, few individuals in the agronomic establishment evaluated Andean crops for their agrobiodiversity. Germ plasm for crop breeding was viewed as a freely exchangeable public good until the early 1980s when the genetic resource issue came to be seen as having strategic political and monetary implications (Pistorius 1997). Now peasants and not only agronomists have become aware of the issues involved under that rubric of genetic resources. Stories that circulate about foreigners appropriating seed to patent and then sell as a commercial variety for their own profit have been caused indignation in Peru.

Concern for germplasm protection explains in part the establishment, above Pisac in 2003, of the Parque de la Papa as an in situ gene bank. In 2005, about 450 cultivars were repatriated from the International Potato Center in Lima, an entity of the Food and Agricultural Organization of the United Nations. The decision to repatriate the cultivars was full of symbolism: placing native agricultural knowledge in the hands of peasants rather than scientists. Six Quechua-speaking communities in the District of Pisac grow these cultivars in a zone 500–700 m above the floor of the Urubamba. The farmers in the Potato Park are curators of a living collection of the diversity of several native Andean crops. In the Chacras, hybridization can continue to occur with weedy relatives from which new landraces and cultivars can evolve. In situ cultivation is an indispensable backup for the preservation of variation in crop plants in gene banks. Seed collections are not a failsafe way to preserve crops.

As for MBGC, legal protection was sought by applying to it the appellation concept to prevent foreign firms from commandeering the cultivar. Its association with the Sacred Valley suggests the prestige of antiquity. But the appellation has not been legally tested, nor will it prevent transgenic introgression of other kinds of maize. Over two decades, public policy in Peru on transgenics has wavered from one administration to the next, but more to the point, controlling the planting of transgenic maize seeds from the outside may be impossible. The appellation concept as defined also does not prohibit the land designated for this kind of maize from being used for another crop. With the oncoming convergence of economic forces in the valley, alternative agricultural and nonagricultural uses of the valley floor have the possibility of usurping maize.

Maize, MBGC or other cultivars not only require much water, but are also expensive in labor. The high cost of production puts MBGC at a disadvantage in competing for land in the valley. Whereas potato production takes only 50 days of work (*jornales*) per hectare, maize requires 180 *jornales*. Even when a tractor plows the field, hand labor is needed for sowing, banking and harvest. In May the maize stalk is cut and carried to the dwelling area where the ears are laid out on a drying floor (*tendal*). If left to dry on the stalk in the field, much of it would rot, be eaten by parrots or be stolen. Sun drying of the ear without the husk requires 1–2 months. Even then, moisture remains in the cob, so that the ears must be shucked and kernels stored in large sacks. Thus, growing maize is almost a full year's project which starts with plowing in June or July, sowing in August and ending with corn drying in May and June. As labor availability tightens, labor costs increase.

Climatic change looms above the future of maize in this and other valleys. Declining water supply conflicts with the enormous amounts of irrigation water needed to successfully grow this crop. Close to one third of the irrigation water in Cusco Department is distributed in the Sacred Valley and most of that is for maize (Cusco 2008). As the 35 km^2 of glaciers from Calca to Ollantaytambo continue to melt, a water shortage will surely come. In 1985, Cusco geologist Carlos Kalofatovich became aware the Chicón above Urubamba had receded 60 m in 50 years. Melting of the ice has dramatically occurred everywhere in the Andes (Vuille et al. 2008). Cores in the Quelccaya icecap, located southeast of the Sacred Valley in the Cordillera de Vilcanota, indicate recessional melt at the rate of 60 m a year (Hanshaw and Bookhagen 2013). In that cordillera as a whole, one third or 118.7 km^2 of glaciers was lost between 1988 and 2010 (Vuille et al. 2008). To the north, the Cordillera de Urubamba is now on a path to total disappearance of its glacier field before the mid-twenty-first century. Along with less water, warmer temperatures also compromise maize cultivation. One calculation is that the average temperature in the Central Andes as a whole is increasing at a rate of 0.10 °C a decade, but that climate in the Urubamba Valley is warming at the rate of 0.30° per decade (Avalos et al. 2011). When temperatures reach above 28 °C, the MBCG kernel decreases in size and the incidence of insect pests feeding on the soft endosperm increases. Smaller ears and kernels of MBCG observed on some farms may reflect warmer than usual weather.

4.6.3.3 Alternative Crops

Other crops besides maize increasingly find a place in the valley. As demand in Cusco for fruits and vegetables grows, horticulture is increasingly profitable. Since the 1990s, urban development around Cusco has pushed out many farmers that specialized in market produce. Now water and labor issues make other crops more remunerative than maize. Agricultural alternatives include two native seed crops cheaper to produce, less demanding in water and much in demand for the export trade. Quinoa (*Chenopodium quinoa*) has not historically been a commercial crop in the Sacred Valley, but growing world demand and high export prices make it attractive to grow above 2800 m asl. Quinoa cultivation does not need hand soil banking, and its growing season is shorter than that of flour maize. Another crop of considerable potential is kiwicha or cultivated amaranth (*Amaranthus caudatus*), which, like quinoa, is inherited from the pre-Columbian past. Its cultivation has been commercialized and in the 2000 decade it began to take land away from maize, especially around San Salvador. In 1963, kiwicha was rarely cultivated in the Urubamba, but by 2008, more than 100 growers produced 130 metric tons of it in the San Salvador District (Sierra Exportadora 2009). But international demand is now greater for quinoa and farmers have shifted more to the latter. Kiwicha thrives better than quinoa in the warmer temperatures of the north end of the Sacred Valley. Cultivation trials of kiwicha on the restored terraces of the Cusichaca Valley yielded well (Kendall 2005:214).

Other potential competitors with maize for valley land are crops exported to the Northern Hemisphere in its reverse season. Once the proposed Chinchero Cusco Airport is operational, high-value specialty crops may be profitable. Artichokes, though not in the local farming tradition, produce well in the cool valley climate. Among the native Andean food plants with potential export value are pepino (*Solanum muricatum*), a melon-flavored visually attractive fruit with good keeping qualities; cape gooseberry or ahuaymanto (*Physalis peruviana*), a vitamin-rich, sweet–sour fruit with a protective husk that reduces its perishability; and lucma (*Pouteria lucuma*), a nutritious fruit with an unusual texture and sui generis flavor especially good in ice-cream. The delicious and aromatic strawberry or the valley (*Fragaria chiloensis*) comes into production in November and December. Grown especially on the Inca-built terraces above Yucay, this pinkish fruit is somewhat different in flavor than the now-world-common otcopoid strawberry. The valley's commercial future could also include the cut flower trade. Near Bogotá and Quito, the combination of cool tropical highland climate, abundant sunshine, low-cost labor, and accessibility to international air freight has created a flourishing floricultural sector.

Looming over the future of the Sacred Valley is pressure from non-agricultural use that may eventually trump any kind of competition. Between 1996 and 2000 in Yucay District, almost 200 ha of land were lost to cultivation. The tourist business and second-home development increasingly target the Sacred Valley farm land. Without land use controls, these two activities could eventually gobble up this part of the Urubamba. Middle and upper-class Cusqueños, whose numbers have increased with the growth of tourism, find the area around Urubamba, Yucay and Calca attractive for weekend retreats. Tourism entrepreneurs covet the area for hotels. By 2012 more than half a dozen first-class hotels operated around Urubamba and Yucay. Choice properties are the main houses ("casa hacienda") of former estates; having colonial charm, they can be renovated to become luxury inns. Two former haciendas, Rocaforte and Yaravilca, were converted into amenity-laden hotels. Yaravilca's large buildings betray its origin as an *obraje* where Indians made cloth on Spanish looms during the colonial period. Owners of Huayoccari, which as a property was greatly reduced in size under agrarian reform, turned the main house into a gourmet restaurant. In 2010, Hacienda Urpihuata and its remaining 62 ha was put on the real estate market. The other seekers of land are middle and upper-class Cusco people whose numbers have increased with the growth of tourism. They have always found the area around Urubamba, Yucay and Calca attractive as weekend retreats.

Land speculation has affected the peasant class. Euphoric when they acquired title to land only four decades ago, the next generation of peasants now increasingly views their property in terms of prospective financial windfalls. The traditional Andean view of land being a mystical ground connecting them with Mother Earth has ceded to the idea of land as a negotiable commodity. For those members of registered peasant communities, land transfer has been hindered by the lack of freedom to sell their plots. In the two provinces of Calca and Urubamba, 87 legally recognized communities now exist, most of them on land that had once been part of haciendas. It would not be surprising if a movement emerged among those with land of potential tourist value to change their legal charter so that parcels could be sold to outsiders. Current land prices in the valley, among the highest in the Department, have contributed that change in thinking. Whereas in 1963 a one m^2 of terraced maize land in the Districts of Urubamba and Yucay commonly sold for US$0.10, in 2010 that same-sized sliver sold for US$50.00. In Calca Province, the price of comparable land ranged from US$30.00–40.00/$m^2$. In 2012, frontage on the highway through the valley sold for US$90.00/$m^2$, suggesting its desirability for tourist-oriented businesses. One choice parcel of two topos (5400 m^2) sold for an astronomical US$270,000. Property transfer further enters the mindset when the teenaged children of farmers prefer to be truck drivers or tourist operators rather than farmers. Maize cultivation, once viewed as the heart and soul of an ancient tradition, is increasingly seen by them as a life of toil with slim returns for effort expended. Minimal farming profits, the results of government price control on food imposed to favor city dwellers, feed this disillusionment.

4.6.4 *Sacred Valley Tourism*

From 1963 to the present, the number of travelers to the Department of Cusco increased steadily with the exception of a 7-year dip between 1986 and 1993. Over that period terrorist violence, a cholera epidemic elsewhere in Peru, and the Gulf War created a sense of national emergency and international foreboding. Once those issues dropped out of sight, the pent-up demand brought a sharp rise in visitors. By the end of the century, the flow had turned into a surge. In 1997, 251,544 tourists came to the Department; in 2000 that number had grown to 469,792 and in 2006, the figure more than doubled to 1,177,109 (Mincetur 2013). In that broader context, the Sacred Valley first found a tourist mode of thinking come to the fore in the 1970s. After 1995, terrorist incidents had stopped and tourism became seen as the valley's best chance for economic advancement. But with fewer than 5 % of tourists spending a night in the Sacred Valley, the challenge is to receive more travelers who do not simply pass through the valley on a day trip (Mincetur 2005:7). The city of Cusco once had a virtual monopoly on tourist lodging and still accounts for more than 80 % of the hotel space.

4.6.4.1 Kinds of Visitors

The massive flux of visitors to Cusco and Machu Picchu since the mid-1990s has overflowed into the Sacred Valley. The valley's considerable charms include dramatic scenery of the valley below towering icy peaks, peasant livelihoods and folkloric events that combine indigenous and traditional elements and, its major drawing card, the impressive evidence of Inca presence. An archaeological survey identified more than 150 sites in the 100 km^2 area from Huayacan to Yanahuara (Covey 2006). Although most of these sites do not attract tourists, they enhance the valley's pre-Columbian reputation as a place to visit.

Retrieving the pre-Columbian past is not only for tourists and those working in tourism. From the 1980s onward, peasant farmers gained when long-abandoned pre-Columbian canals were rehabilitated. In the Cusichaca Valley, the repair of the Quishuarpata canal restored to cultivation 45 ha of terraces (Kendall 2005:214). In the Patacancha Valley above Ollantaytambo, other workers reestablished the Pumamarca canal, which brought water to 43 ha of long-abandoned terraces. At Rumichaca, 7 km from Urubamba, workmen rehabilitated another Inca canal. Though the colonial heritage of the valley appeals to a smaller segment of visitors than do Inca remains, the Instituto Nacional de Cultura underwrote the cost of rehabilitating the colonial churches of Taray, Urubamba and Huayllabamba.[7] Much

[7]The Instituto Nacional de Cultura (INC), founded in 1971, has become a major bureaucracy in Cusco with the power to undertake restorations, carry out digs, and deny authorization of archaeological work by others. Once subservient to Lima, the Cusco office no longer sends to the Ministry of Education a portion of the entrance fees collected at the archaeological or historical sites.

more than tourist stops, these churches serve local populations who no longer worry about their condition.

Tourism in the Sacred Valley owes much to its easy reach from Cusco on all-weather roads. In the 1980s, the Inter-American Development Bank (BID) provided a US$70 million loan to pave the two highways from Cusco as well as the road through the valley itself. The Sacred Valley road connects to two others: the heavily traveled and now paved roads through the southern Urubamba in Quispicanchi and Canchis and from there on to Puno, and the spectacularly engineered highway from Ollantaytambo over the Málaga Pass to La Convención. Peru in 2010 had three times more kilometers of roads and four times more kilometers of paved roads than it had in the early 1960s (Webb 2013:193). The Internet and cell phones have also enormously benefited tourism in Cusco Department.

Another segment of the tourist market in the valley is adventure travelers. Since the 1950s, mountain tops in the cordilleras of Cusco Department have attracted climbers, a very select group of visitors. More numerous are those who come for rafting, kayaking, horseback riding, mountain biking, paragliding and especially trekking. The last named activity exploded in the late 1970s when hiking converged on the stone-paved Inca path that had once extended from Cusco to Machu Picchu. They were lured by guidebooks that promised "…a stunning combination of Inca ruins, unforgettable views, magnificent mountains, exotic vegetation and extraordinary ecological variety" (Murphy and Box 2001:205). Parts of the trail had been cleared by Hiram Bingham's expeditions between 1912 and 1915. The Wenner-Gren Expedition to Hispanic America in 1941 under the direction of Paul Fejos (1944:54–58) also cleared some of it. In 1968, a University of Cusco professor, Victor Angles Vargas (1980), led a group of students on a hike from Chilca at Km 88 on the railroad paralleling the Urubamba River and up into the Cusichaca Valley to the west and to Machu Picchu, 44 km from where they started. Angles called this Camino Inca; the name stuck and was translated into English as the Inca Trail. But for more than a decade, the trail's value for adventure tourism was only dimly understood.

Since the 1990s the trail has gained almost cultic status as a must-do trek for adventure travelers from around the world. That the Incas themselves used it to reach Machu Picchu places the meaning of its past in the present. Once operational, this stone-inlaid path opened a whole new employment opportunity for strong-backed young men, mostly from Ollantaytambo. Growing up as farmers carrying heavy loads, they suddenly found jobs as *porteadores*. As the trail gained popularity, littering and careless defecation occurred. Hikers were sometimes robbed and squatters set themselves up along the path to sell food to hungry trekkers. In 1981, officials enlarged the boundaries of the Sanctuary of Machu Picchu to include the Inca Trail. Within that territory, the authorities imposed a set of rules on trekkers, porters, and local inhabitants. Since 2000, regulations have required all trekkers to be accompanied by licensed porters who double as guides and who must abide by a code of ethical conduct. So many trekkers have used the trail that the major issue has since become trail deterioration. Three decades of intense use eroded the soil surface to such an extent that, in 2010, the authorities set a carrying capacity of

Fig. 4.15 Whereas in the 1960s, the small Pisac artisan market was set up on Sunday morning and dismantled late Sunday afternoon, now the market operates several days a week and has taken over most of the plaza. Looming above is a pisonay tree, a common ornamental in the Sacred Valley. (Photograph by DW Gade)

500 people a day for 11 months. Henceforth, the trail has closed for the month of February at the height of the rainy season to help it recover. High demand required that trekking the Inca Trail be by reservation only.

4.6.4.2 Uneven Tourist Flow

The Sacred Valley has very uneven tourist visitation. Whereas Pisac and Ollnataytambo on the *piso valle* and Chinchero above it have been transformed by it, San Salvador receives almost no tourists and Urubamba and Yucay see some pass through on day trips. Tourists have long visited Pisac, but now the entire village economy is clearly oriented to them. Merchants of tourist-oriented goods take over the entire plaza (Fig. 4.15). Their big day has always been Sunday, but now the market also functions two other days of the week. In 2010, Pisac had three hotels, one of which was a converted casa hacienda, five hostels, and about 20 private homes that offered lodging. An array of restaurants existed. The ruins, now connected by a road to the town, receive many more visitors than they once did. A local museum has been constructed. Seeking an economic niche in a crowded business, some travel agencies organize "mystical tours" and "shamanic journeys" to Pisac.

Fig. 4.16 On the rail line to Aguas Calientes, Ollyantaytambo holds the crucial transport link to Mach Picchu. A large tourist traffic is also generated by Ollantaytambo's location near the Inca trail and its own formidable Inca ruins. Tourism has changed this traditional community where agriculture long dominated land and life. (Photograph by DW Gade)

Local individuals learned to take on performance identities to entice tourists to give them money. Waif-like children wearing ponchos and holding lambs or puppies wait to be photographed in return for a tip. In 1963 no one would have conceived of such a ruse. Folkloric events in Pisac, once meant for local inhabitants and surrounding hamlets, now anticipate contingents of visitors sent by the Cusco agencies. Three major religious fiestas take place in Pisac during the tourist high season: "Mamacha Carmen" (Nuestra Señora del Carmen) on July 16; "Virgen Assunta" (Feast of the Assumption) on Aug. 15, and Virgen del Rosario on October 7. Carnaval, the Earth Mother (Pachamama) celebration, All Saints Day and Christmas Eve also draw visitors.

Its location on the rail line to Machu Picchu has turned Ollantaytambo into an even busier tourist town than Pisac. Of foreign tourists visiting Peru, 89 % go to Machu Picchu; 54 % go to Ollantaytambo, and 16 % see Pisac, the latter usually on a separate trip from Cusco (Mincetur 2005). Whereas Cusco was once the starting point for the noisy, smelly, diesel-powered *autovagón* that goes to Machu Picchu, now most of the train traffic begins and ends 50 km from Cusco in Ollantaytambo. Privatization of the Ferrocarril Cusco Santa Ana in 1999 led its new British operators, the Orient Express Company, to recalibrate the Machu Picchu travel experience (Fig. 4.16). By shuttling tourists from Cusco to Ollantaytambo on buses, the trip avoided the time-consuming rail switchbacks out of and into the Cusco Valley.

One train is reserved for Peruvians, who pay lower fares than foreigners, and it still begins in Cusco. Other trains begin and end at Poroy, 13 km from Cusco. Ollantaytambo has become a take-off point for most trail treks to Machu Picchu. If a road were ever built to Machu Picchu, Ollantaytambo would lose its position as a critical transfer point. Given the daunting engineering challenges for constructing such a road, such a highway is unlikely. Moreover, the narrow canyon below the ruins lacks space for vehicular traffic or storage. Given the constraints, the train is the proper conveyor of people into and out of the canyon. Moving slowly by necessity, passengers in the clean and quiet rail cars view scenery in two distinct ecological zones. For the rich in a hurry, a helicopter service started in 2006 made the trip from Cusco in 25 min. Noise and damage to monuments led the government to revoke landing rights for these aircraft in 2010.

Other than in Ollantaytambo and Pisac, tourism in 2013 in the Sacred Valley did not yet totally dominate the economy. Other places with less grand archaeological monuments or which require more effort to reach may eventually receive crowds of visitors. One possibility is Huchuy Qosqo, where the remains of Inca Viracocha's settlement perch 500 m above the valley floor between the towns of Coya and Lamay. Important ruins—Patallacta, Choquesuysuy, Quente, Ccorihuaychina and Torontoy—lie downstream from Ollantaytambo, but since none compare with Machu Picchu's setting or complexity, they receive minimal attention from visitors and the agencies. The Sacred Valley has drawing cards in addition to the ruins of a non-literate civilization. Many tourists from post-industrial countries, less drawn to remains of the past than to the folklore of the present, find country life in the Central Andes of inherent interest. More and more foreign visitors now make the yearly pilgrimage, along with native people, to Ausangate glacier.

4.7 International Development Agents in the Valley

Non-governmental organizations (NGO) also have made the international presence in the Sacred Valley many times greater than it was half a century ago. By the 1990s, NGO staff members and volunteers from North America and Europe had supplemented development programs that in the 1960s were in the hands of foreign governments. Among these entities are the following: Peruvian Hearts, sponsored by the Cleveland Clinic, seeks to improve the lives of children in Lamay; Kausay Wasi, a clinic in Coya, provides health care to peasant folk in that area; Awamkai, promotes the restoration of ancient weaving techniques in Ollantaytambo; World Vision sponsors construction of latrines in Urubamba; and Living Heart, supports children and women in education, health, and nutrition in Yucay. Some organizations target a specific improvement. WEFTA, an international organization, has had a project in Urubamba to introduce bathroom plumbing to those without, and a group known as the International Alliance for Terraced Landscapes encourages the rehabilitation of abandoned terraces.

Lost in the media attention that NGO activity in the Sacred Valley receives is the attitude they share (Shepherd 2004). Technical experts ("*promotores*") and governmental development agents share many assumptions. Both presume that "experts" have superior knowledge and abundant information. Peasants may be poor and have little formal education, but they hold stores of knowledge about local agriculture, handicrafts and a range of artisan activity. The traditional wisdom (*saberes*) that peasant folk hold in their heads and habits does not impress technical agents who see these people as being mired in thought patterns of the past. In my experience development officers are not much interested in understanding that traditional practices are rational adjustments to living on the land. From their self-reinforcing perspective, peasants cannot get out of their poverty cycle without help and for that, experts think that they must adopt outside ideas and technology that will make farming more efficient.

Not surprisingly, promoters themselves frequently define community needs and goals. The disconnection between what the NGOs think and do and what their clients think and do may come from differences in perceptions. Needs to agents are often desires to peasants, yet blanket assessments about NGOs cannot be made. A particularly successful German-funded project between 1989 and 1991 involved improving the supply of irrigation water in Urubamba (Baca 1998). During the months of highest demand, farmers had insufficient water. By funding a new system of intakes from the Chicón and Pumahuaca quebradas and paving canals to reduce infiltration, the project substantially increased production by doubling the water supply without creating new areas of irrigation. Now, however, rapid glacial melt will cause future water shortages that cannot be overcome so readily.

4.8 Environment and Landscape: Trends and Problems in the Sacred Valley

4.8.1 The River as a Sewer

From the perspective of tourism, much has gone awry in this vale of paradise. If tourism is the key to improved livelihoods, then valley folk would do well to see their valley through the eyes of the visitor. Local people have long accepted that the pollution of the Urubamba River is the normal state of affairs. Europeans and North Americans are repulsed by a visibly polluted river. The daily disposal of tons of raw wastes has turned this once limpid stream into an open sewer. Some actions have kept it from getting worse. Sanitary landfills in Calca, Urubamba and Ollantaytambo divert tons of waste materials that at an earlier time would have been thrown into the river. Plastics are recycled and sent to Lima.

The biggest contributor of the contaminants is the urbanized region of Cusco. Untreated sewage, as well as tons of garbage, are daily dumped into the Huatanay River, and from there funneled into the Urubamba near Huambutío. Although Cusco's population quadrupled over half a century and the municipal and departmental treasuries grew, waste treatment was not successfully implemented. A sewage disposal plant was built, but a design flaw made it inoperative; even if it had functioned, its processing capacity was inadequate. Streamside pig farms along the Huatanay simply dump their wastes directly into the river. The pollution of the Urubamba adversely affects the Sacred Valley more than other parts of the drainage basin.

Any discussion of Urubamba pollution must begin near its source near La Raya where wastes from alpaca farms are the most upstream additives to the innocent brook. Downstream about 10 km, the now closed Mejia textile factory near Maranganí dumped manufacturing effluents into the river for more than a century. Ten kilometers farther on, Sicuani, now a city of ca. 45,000, continues to pipe its untreated sewage directly into the river. A downstream succession of small towns like Combapata, Quiquijana, Urcos and others do the same. The mindset is that of early twentieth-century North America when rivers were used as conveyers to take discarded materials out of sight. Along the Rio Urubamba, the current pushes onward, carrying away garbage, human effluents, grease, dead animals and anything else unwanted. Maintaining the river as a source of potable water has never been important. Purer water has been available from lateral streams fed by glaciers or springs. In a rural culture that never accepted latrines, inhabitants have used the riverside for that purpose. In towns along the Urubamba, pigs and dogs eat all manner of organic waste, serving as ambulatory cleaning patrols.

Raw wastes of more than 700,000 people result in the gross pollution of the Urubamba flowing through the Sacred Valley. Counts of *Escherichia coli* of 8580 PPM in the Urubamba are higher than those of any large river in the United States (Taipe Bolaños et al. 2006). Filthy water and an unsavory river bank discourage use of the river for bathing, washing clothes and fishing. Kayaking, a sport of foreigners, has over the years been somewhat more tolerant of pollution. Together with mountain scenery, class II and III rapids on the Urubamba above Ollantaytambo and class IV rapids below that town make the river a challenge to navigate. Nevertheless, several agencies of adventure tourism now avoid the Urubamba in favor of other streams.

4.8.2 *Natural Hazards of the Sacred Valley*

This mountain environment is not free of natural hazards, always unpredictable and often deadly. Floods, tremors, landslides and debris flows have been viewed as acts of God over which inhabitants had no control. Supernatural powers were invoked for protection, more when tragedy struck than in preparation for it. The clergy could interpret the event either as the result of divine wrath or the outcome of human waywardness. Today scientific monitoring and evacuation drills mitigate the impact of these disasters. The history of floods in the Sacred Valley remains undocumented, but high water was surely an important issue. Containment walls were built by the Incas around Pisac and Ollantaytambo. During the colonial period, a dike ("tajamar") was constructed near Ollantaytambo to contain flooding (Glave and Remy 1983). In 2010, a flood on the main channel resulted from the heaviest rainfall in 15 years combined with the peak period of seasonal glacial melt. Hundreds of maize fields were inundated. Buildings were destroyed and infrastructure damaged in the towns of Taray, Pisac, Calca and Urubamba, in some cases the consequence of constructing in flood-prone zones.

In addition to overflow from the Urubamba channel, outpourings of debris from side canyons has periodically threatened the Sacred Valley. Most towns are located on alluvial fans at the mouths of these lateral streams and are thus vulnerable to this phenomenon. Steep gradients in lateral valleys increase the possibility that a surge of water, rock and mud will destroy everything in its path. In the town of Pisac in the 1930s and again in the 1990s, tons of gravel in a clay matrix surged down the Chongo canyon and into the town plaza. Calca, a town of ca. 10,000 people, located above where the Rio Jochoc meets the Urubamba floor, is even more vulnerable to debris flows. In Urubamba in 1942, 70 people died and many dwellings were destroyed when an *aluvión* roared down the Chicón Valley. Huge boulders in the middle of fields betray the presence of past debris flows in both the Chicon and neighboring Pumahuanca canyons. Above Ollantaytambo, the Patacancha River starts as a trickle at 5000 m asl and falls almost 2200 m in 24 km. The possibility of disaster downstream is ever present. Evidence of past flows litters the valley.

More recently, accelerated glacial melt increased vulnerability to massive debris flows farther down the Urubamba. In 1998, chunks of glaciers broke up from the face of Salcantay, setting up a churning mass of rock, mud and water through the Rio Aobamba Valley. The debris inundated the Machu Picchu hydroelectric plant, destroying the turbines and taking 3 years to be replaced. Five weeks before that, on January 13, 1998, saturated moraines moving down the Sacsara River, destroyed almost all the town of Santa Teresa. In April 2004 in Aguas Calientes, massive quantities of mud, water and boulders rushed down two lateral streams which form a compound alluvial fan on which the town had expanded (Carreño and Kalafatovich 2006).[8] When the debris reached the main canyon of the Urubamba, it killed 11 people and destroyed 10 homes, 300 m of rail line and two railroad bridges, and seriously damaged a large hotel. Aguas Calientes is vulnerable to more than debris flows. Its 70 % slope gradients, and heavy rains between October and May make the town a prime target for slope failure. The whole town, which should never have been constructed at that site, could be entombed in mud and rock.

Earthquakes also pose a danger in triggering devastating landslides. In 1678, a massive chunk of earth and rock, said to have been loosened in the earthquake of 1650 that demolished Cusco, fell into the valley floor 2 km upstream from the town of Urubamba. The landslide blocked the flow of the Urubamba River and flooded the town of Yucay, which lies in a narrow steep-sided valley (de Esquivel y Nava 1980:II:140). Weeks of digging were necessary for workmen to open a breach in the impromptu dam and to drain the water that partially submerged the village. Probably the 1650 earthquake was also responsible for destroying the trail that Incas had built through the Torontoy Canyon.

[8] An oral history exists that in the decade of the 1940s a landslide and flood on the same Alcamayo swept away without a trace the old sawmill called La Maquina.

4.8.3 Issues of Landscape Integrity

Beyond stream contamination and natural hazards, the Sacred Valley has flaws in its landscape. One component of that is the exotic tree, *Eucalyptus globulus*, which covers substantially more area than it did in 1963. In the valley this species is always planted; it never volunteers. Fire is necessary to release the seeds from the fruits so that germination can occur. Exotic tree plantations have expanded greatly in the valley over the last half century. During the land reform of the 1970s, fast-growing eucalyptus was the species of choice in active state-initiated forestry programs. A good deal of that expansion occurred on cultivable lands. Eucalyptus is important as a source of beams, window frames and doors in the valley and on the Cusco wood market. The two advantages of eucalyptus, fast growth and its habit of suckering from the stumps of cut trees, are countered with disadvantages. The tree extracts large quantities of water from the soil; its biochemistry has allelopathic effects on native biota, and its dominance holds the potential for monocultural disaster. The eucalyptus bark borer (*Phoracantha semipuntata*), native to Australia, and without natural enemies in the Andes, is now present. At some time in the future, this insect could destroy eucalyptus in the Urubamba.

Eucalyptus is also problematical in what it, historically and culturally, represents: an alien in a zone of rich indigenous tradition. Plants incongruent with its Inca past cover one of South America's premier archaeological areas. When juxtaposed with pre-Columbian vestiges, exotics are inauthentic elements in the valley. The main parvenus, in addition to eucalyptus, are two other planted tree species, Monterey cypress (*Cupressus macrocarpa*) and Monterey pine (*Pinus radiate*). Both conifers, introduced to the valley in the twentieth century, are native to the Central Coast of California. A shrub of Mediterranean origin, Spanish broom or retama (*Spartium junceum*), got established in the sixteenth century. This alien freely volunteers in pastures and on roadsides and riverbanks. Once a roof thatch, retama continues to be used as the ground cover for maize drying floors and its flowers have several minor medicinal uses. In Southern Europe, cloth was once made out of its fibers, but the plant was never exploited for that purpose in the Andes.

An achievable good would be to reclaim the valley for plants that the Inca knew and used. A forested landscape of native vegetation could include both planting as well as spontaneous establishment of lambran, quishuar, huaranhuay, unca, chachacomo, molle, tara, and nogal, and siwi (cedro) (Tupayachi 2005). None of these species grows nearly as fast as eucalyptus, but by comparison, none is as demanding of water. None is allelopathic to other biotic elements of the Andes. Since nine native species are involved, the threat of a monocultural disaster brought on by a massive die-off in case of disease or pests would not be an issue. Some species, nogal and siwi in particular, yield better quality wood than does eucalyptus. More than an alien species, many scientists and some farmers now recognize the price of a eucalyptus monoculture: pressure on a shrinking water supply, tainted soil, and minimal biotic diversity. Replacement with native species becomes more feasible as the demand for firewood decreases. Approximately 30 % of households in the

Sacred Valley now use liquefied gas for cooking, a proportion that is expected to increase when the gas pipeline from the lower Urubamba reaches the highlands.

Deterioration of the built landscape is another issue for the Sacred Valley. The striking quebrada setting will always be there, but the proliferation of uncontrolled sprawl and minimalist construction standards create visual disorder that repels visitors. In contrast to its tight nucleus of 1963, the town of Pisac now extends along the highway to Calca for 2 km and in a flood-vulnerable zone. In stark contrast to the graceful stone constructions of the Inca, modern adobe structures with sheet-metal roofs look untidy and are impermanent. To local inhabitants, they are simply normal houses, built by local labor from known and affordable materials. Paradoxically, awareness of other aspects of appearance seems to have increased. Street cleanliness has generally improved since the 1960s (Keisteri 1990). Pigs are no longer given free rein in the core of most villages and human waste deposition is controlled.

To meet its tourist potential as a destination, more improvements are needed in tidiness and construction standards and, most of all, land use zoning. The economic payoff of these measures comes when travelers linger rather than pass through. A system of pedestrian and bicycle paths would make exploring the countryside appealing. A cleaned-up Urubamba River would also invite use of riverine spaces. International interest has generated projects funded from outside Peru. In 2001, the National Geographic Society, in cooperation with the Peruvian government and valley municipalities, launched a project to rehabilitate the Sacred Valley along the lines of "geotourism", based on an assumption that tourism and conservation are compatible and viable options. A larger project, funded largely by the World Bank, has also been designed to impose planning order on a longer valley stretch from Pisac to Santa Teresa (Peru 2005). In 2008, a restructured World Bank project had its total funding increased from US$6,000,000 to 62,000,000.

The challenge facing the Sacred Valley is to make a space livable to inhabitants and, at the same time, appealing to visitors. The power of outside economic forces may make that goal unachievable. The biggest impending change is the planned new Cusco airport. Hemmed in by mountains and the now urbanized Cusco Valley, the Velasco Astete Airport is now programmed for replacement. Thirty kilometers from the city, the airport authority purchased 600 ha from several peasant communities near Chinchero, a large expanse of relatively flat, unbuilt land on a plateau in an otherwise topographically cut-up region. This facility, projected to cost US$460 million, is optimistically programmed for completion in 2018. The expectation is that the long runways will accommodate large aircraft flying direct to Cusco from international cities. Bypassing fog-shrouded Lima as a port of entry, air service direct to Cusco from New York, Miami, São Paulo and even Madrid would be a natural evolution in the tourist vocation of the region. An analogy can be drawn with Bali where an international airport built in 1968 made it possible to bypass Indonesia's capital. Like Jakarta, Lima does not have enough visitor appeal to compensate for the hassles of negotiating the huge sprawling metropolis.

Instead of operating only 6 h a day as has been the practice at Velasco Astete, the Chinchero facility will be designed to have a 24 h a day landing and takeoff capacity.

Albert Giesecke (1921) presciently remarked in 1921—the year that the first airplane arrived in Cusco—that tourism was the key to Cusco's economic future. In the 1950s, the trip from Lima to Cusco often took 4 days (no nights) of bus travel mostly on dirt roads. In 2013, the same bus trip, now day and night, on paved roads took 22 h of travel time. That improvement still cannot compete with the plane trip, which takes only one hour and has few of the anxieties associated with mountain road trips.

Once built, the new Cusco airport is likely to determine the destiny of the Sacred Valley for better or for worse. Although only 13 km away, the floor of the valley lies 1000 m below the proposed airport. Proximity of this large facility will funnel many visitors directly to Urubamba and Ollantaytambo. Speculators, investors and travelers will transform the zone. Both towns are convenient takeoff points to Machu Picchu and have the added benefit of lying at an elevation 600–700 m lower than Cusco. Many travelers to the Andes suffer from hypoxia in Cusco (3400 m asl) and require a full day to adjust to the altitude. Comcomitant changes will, at some point, include a four-lane paved highway to link the new airport to Cusco and Urubamba. The town of Chinchero, 1.5 km from the airport, with its strong indigenous tradition, will inevitably undergo dramatic transformation.

Proximity to the new airport will also trigger population growth in several towns. That prospect poses substantial challenges to managing the urban expansion and preserving the rural spaces around the five main towns of the Sacred Valley between Pisac and Ollantaytambo. Aguas Calientes, the town below the Machu Picchu ruins, provides an example of what happens when planning is ineffective. Subversion of the process by an array of private interest groups resulted in a chaotic landscape in what had been a hauntingly gorgeous site. High land prices in the Sacred Valley make property owners reluctant to agree to freeze land use. More likely, there will be a mad rush to cash in on the boom by land owners and commercial interests. Already in 2013, average land values between Urubamba and Ollantaytambo were six times higher than land between Pisac and Calca. Promoters of development often urge locals to accept development to relieve poverty. When poverty is viewed as a pathological condition, landscape integrity loses its argument. Yet, in the long run, a sustainable economy based on tourism requires creating a landscape that will attract, not repel, visitors.

4.9 Meanings of the Sacred Valley

In its 80 km stretch of the Urubamba, the Sacred Valley reveals a deep anthropogenic impact in which nature and culture and time and space are bound together more visibly and more powerfully than elsewhere in the Andes. Certain geographical virtues set the Sacred Valley apart from the other sections of the Urubamba. Its climate offers the human ideal of not too hot and not too cold and an environment free of most insect-borne diseases and the entomological annoyances so common at macrothermal elevations. The long flat strip of good crop land, fertile soil, and irrigation water give the valley a major agricultural advantage in this region of fractured topography, seasonal drought and frequent freezes. The Inca recognized the advantages and, with their formidable engineering skills, but also despotic power, turned this stretch into the most productive crop-growing area within a 200 km radius of Cusco.

Cultural processes starting in the early colonial period also set the Sacred Valley apart from the other parts of the Urubamba Valley. More than the Canchis and Quispicanchi sections of this depression, the Sacred Valley underwent a change from indigenous to mestizo. This shift was a measure of the saliency of the Spanish presence and how much they valued the valley floor zone for farming, convalescing and visiting. Both the Inca and colonial Spaniards appreciated the proximity of the Sacred Valley to Cusco. Changes imposed on the land by human actors usually have both positive and negative consequences. The Inca engineering of the valley floor and sides expanded spaces available for cultivation and controlled slope erosion. But these projects also displaced thousands of inhabitants. Rather than the enormous hardship and distress of those displaced, the thought is only about the outcome. Irrigation and terracing greatly increased agricultural productivity and embellished the valley's scenic value.

Other historic decisions also should be judged by complex and contradictory measures. The creation of *reducciones* in the sixteenth century had a wrenching effect on indigenous culture by forcing people to move from their scattered ayllus into tightly nucleated villages. Viewed today, these towns create a coherent settlement framework for the Sacred Valley. By contrast, the tropical part of the Urubamba north of Machu Picchu never had reducciones, with the result that a disordered, service-poor series of towns emerged in most cases only over the past 50 years on land that was once owned by haciendas. Beginning in the 1960s, the hacienda was judged to place too many of Peru's land resources in the hands of the few, and by the end of the 1960s decade, an agrarian reform program had dismembered the whole system. The minifundia pattern that eventually resulted now poses problems for agriculture in this zone. Labor costs rise, but most peasants lack the resources to mechanize their plots. Farming as a family tradition and way of life is questioned. The quickening pace of change places the valley on the cusp of transformation. Pessimism about the valley's future is not unwarranted, but there is also some reason to hope that its citizens will implement a quality land use plan for the valley's future.

References

Alvarez del Castillo J (2005) Gobernancia y gestión de la tierra: institución communal del uso y propiedad de la tierra en la comunidad campesina Vista Alegre y Yanahuaylla. Tesis maestrado. Colegio Andino. Centro Bartolome de las Casas, Cusco

Angles Vargas V (1980) Machupijchu y el camino inka. Privately printed, Cusco

Anonymous (1958) Descripción del Virreinato del Perú (ca. 1620). In: Lewin B (ed) Universidad Nacional del Litoral, Rosario

Avalos G et al (2011) Atlas climático de precipitación y temperature del aire en la Cuenca del Rio Urubamba. SENAMHI, Lima

Baca E (1998) El impacto de los proyectos de riego en la Sierra: La experiencia del plan Meriss Inka. Debate Agrario 28:56–83

Bauer BS (2004) Ancient Cuzco: heartland of the Inca. University of Texas Press, Austin

Bélisle V, Covey RA (2010) Local settlement continuity and Wari impact in middle horizon Cusco. In: Jennings J (ed) Beyond Wari walls: regional perspectives on middle horizon Peru. University of New Mexico Press, Albuquerque, pp 79–95

Betanzos J (1996) Narrative of the Inca (1557). In: Hamilton R, Buchanen D (trans and eds) University of Texas Press, Austin

Bird RM, Browman DL, Durbin ME (1983–1984) Quechua and maize: mirrors of Central Andean culture history. J Steward Anthropol Soc 15(1–2):187–240

Bonavia D (2008) El maíz: su origen, su domesticación y el rol que ha complido en el desarrollo de la cultura. Universidad de San Martin de Porres Fondo Editorial, Lima

Brandolini A, Landi P, Monfredini G, Tano F (2000) Variation among Andean races of Maize for cold tolerance during heterotrophic and early autotrophic growth. Euphytica 111:33–41

Burger R, Lee-Thorp JA, Van de Merwe NJ (2003) Rite and crop in the Inca state revisited: an isotopic perspective from Machu Picchu and beyond. In: Burger R, Salazar L (eds) The 1912 Yale Peruvian Expedition collections from Machu Picchu. pp 119–137

Burns K (1999) Colonial habitats: convents and the spiritual economy of Cuzco, Peru. Duke University Press, Durham

Butzer KW (1992) Spanish conquest society in the New World: ecological readaptation and cultural transformation. In: Wong ST (ed) Person, place and thing: interpretive and empirical essays in cultural geography. Geoscience Publications, Baton Rouge, pp 211–242

Carreño CR, Kalafatovich S (2006) The Alcamayo and Cedrobamba castastrophic debris flow (January, March, and April 2004), Machu Picchu area, Peru. Landslides 3(1):79–83

Chepstow-Lusty A, Frogley MR, Bauer BS, Bush MB, Tupayachi Herrera A (2003) A late Holocene record of arid events from the Cuzco region, Peru. J Quat Sci 18(5):491–502

Chepstow-Lusty AJ, Frogley MR, Bauer BS, Leng MJ, Boessenkool KP, Carvaillet C, Ali AA, Gioda A (2009) Putting the rise of the Inca Empire within a climatic and land management context. Clim Past 5:1–14

Cobo B (1956) Historia del Nuevo Mundo (1653), 2 vols. Ediciones Atlas, Madrid

Córdova y Salinas D (1957) Crónica franciscana de las provincias del Perú. In: Canedo LG (ed) Academy of American Franciscan History, Washington

Covey RA (2006) How the Incas built their heartland. University of Michigan Press, Ann Arbor

Covey RA (2014) Formative period settlement patterns in the Sacred Valley. In: Covey RA (ed) Regional archaeology in the Inca heartland: the Hanan Cuzco surveys. Museum of Anthropology, University of Michigan, Ann Arbor, pp 65–76

Covey RA, Amado Gonzalez D (eds) (2008) Imperial transformations in sixteenth-century Yucay, Peru. Museum of Anthropology, University of Michigan, Ann Arbor

Covey RA, Griffis N (2014) The preceramic occupation of the Janan Cuzco region. In: Covey AR (ed) Regional archaeology in the Inca heartland: the Hanan Cuzco surveys. Museum of Anthropology, University of Michigan, Ann Arbor, pp 45–52

Cusco (2008) Plan estratégico sector agrario 2009–2011. Gobierno regional Cusco. Dirección regional de Agricultura, Cusco. http://www.minag.gob.pe. Accessed 14 Oct 2014

Cutler HC (1946) Races of maize in South America. Bot Museum Leaflets Harv Univ 12:257–291

de Esquivel y Navia D (1980) Noticias cronológicas de la gran ciudad del Cuzco, 2 vols. Fundación Augusto N.Weise, Lima

Denevan WM (2001) Cultivated landscapes of Amazonia and the Andes. Oxford University Press, New York

Donkin RA (1979) Agricultural terracing in the aboriginal new world. University of Arizona Press, Tuscon

Earls J (1989) Planificación agrícola andina. Ediciones COFIDE, Lima

Eubanks MW (1999) Corn in clay: maize paleoethnobotany in pre-Columbian art. University Press of Florida, Gainesville

Farrington I (1983) Prehistoric intensive agriculture: notes on river canalization in the Sacred Valley of the Inca. In: Darch JP (ed) Drained field agriculture in Central and South America. BAR Publications, Oxford, pp 221–235

Farrington I (1992) Ritual geography, settlement patterns and the characterization of the provinces of the Inka Heartland. World Archaeol 23(3):368–385

Fejos P (1944) Archaeological explorations in the Cordillera Vilcabamba Southeastern Peru. Viking Fund, New York

Flores Ochoa JA (1985) Interaction and complementarity in three zones of Cusco. In: Masuda S, Shimada I, Morris C (eds) Andean ecology and civilization. University of Tokyo Press, Tokyo, pp 251–276

Foster G (1960) Culture and conquest: America's Spanish heritage. Viking Fund Pub. in Anthro. 27. Wenner-Gren Foundation, New York

Gade DW (1968) Aranjuez of the new world. Americas 20(3):12–19

Gade DW (1971) Setting the stage for domestication: Brassica weeds in Andean peasant ecology. Proc Assoc Am Geogr 4:38–40

Gade DW (1994) Regional isolation in the high provinces of Cuzco and Apurimac. In: Poole DA (ed) Unruly order: violence, power and identity in the high provinces of southern Peru. Westview, Boulder, pp 31–62

Gade DW, Escobar M (1982) Village settlement and the colonial legacy in southern Peru. Geogr Rev 72(4):430–449

Garcilaso de la Vega EI (1960) Comentarios reales de los Incas. Ediciones de la Universidad Nacional del Cuzco, Cusco

Garrett DT (2005) Shadows of empire: the Indian nobility of Cusco, 1750–1828. Cambridge University Press, New York

Giesecke AA (1921) El Cuzco, meca del turismo. Revista Universitaria (Cusco) 35:3–17

Glave LM, Remy, MI (1983) Estructura agraria y vida rural de una region andina: Ollantaytambo entre los siglos XVI-XIX. Central de Estudios Rurales Andinos "Bartolome de las Casas," Cusco

Goodman MM, Bird RM (1977) The races of maize IV: tentative grouping of 219 Latin American races. Econ Bot 31:204–221

Grobman A, Bonavia D (1978) Preceramic maize on the north-central coast of Peru. Nature 276(5686):386–387

Grobman A, Salhuana W, Sevilla R, Mangelsdorf PC (1961) Races of maize in Peru. National Academy of Sciences-National Research Council, Washington

Hanshaw MN, Bookhagen B (2013) Glacial areas, lake areas and shorelines from 1975 to 2012: states of the Cordillera Vilcanota, including the Quelccaya ice cap, Northern Central Andes, Peru. Cryosphere Discuss 7(2)

Hastorf C, Johannessen S (1994) Becoming corn-eaters in prehistoric America. In: Johannessen S (ed) Corn and corn culture in the prehistoric New World. Westview, Boulder, pp 427–443

Herrera FL (1921) Contribución a la flora del Departamento del Cusco. Universidad del Cusco, Cusco

Innocente O, Sumar L, Loayza A (2006) Denominación de origen maíz blanco gigante Cusco: reflexiones sobre la experiencia y recomendaciones, 89 p

Isherwood C (1948) The condor and the cow. Random House, New York

Jameson JS, Ramsay PM (2007) Changes in high-altitude Polylepis forest cover and quality in the Cordillera de Vilcanota, Peru, 1956–2005. Biol Conserv 138:38–46

Keisteri T (1990) The study of change in cultural geography. Fennia 168(1):31–115

Kendall A (1991) Los patrones de asentamiento y desarrollo rural prehispánico entre Ollantaytambo y Machu Picchu. Universidad Nacinal de San Antonio de Abadk, Cusco

Kendall A (2005) Applied archaeology: revitalizing indigenous agricultural technology within an Andean community. Public Archaeol 4:205–221

Kosiba S, Bauer AM (2013) Mapping the political landscape: toward a GIS analysis of environmental and social difference. J Archaeol Method Theory 20:61–101

Macera P (1968) Mapas colonials de haciendas cuzqueñas. Universadid Nacional Mayor de San Marcos, Lima

Mangelsdorf PC (1974) Corn: its origin, evolution, and improvement. Belknap Press of Harvard University Press, Cambridge

Matsuoka Y, Vigouroux Y, Goodman MM, Sanchez J, Buckler E, Doebley J (2002) A single domestication for maize shown by multilocus microsatellite genotyping. Proc Natl Acad Sci 99(9):6080–6084

McEwan G (ed) (2005) Pikillacta: the Wari Empire in Cusco. University of Iowa Press, Iowa City

Mincetur (2005, 2006, 2007, 2010, 2012, 2013) (statistics on tourism)

Mörner M (1977) Continuidad y cambio en una Provincia del Cuzco: Calca y Lares. Historia y Cultura (Lima) 9:79–112

Morris C, von Hagen A (2011) The Incas: lords of the four quarters. Thames and Hudson, New York

Murphy A, Box B (2001) Footprint Peru Handbook, 3rd edn. Footprint Handbooks Ltd, Bath

Murra JV (1960) Rite and crop in the Inca state. In: Diamond S (ed) Culture in History. Columbia University Press, New York, pp 393–407

Niles SA (1988) Looking for "lost" Inca palaces. Expedition 30(3):56–64

Oricain P (1906) Compendio breve....formado por Pablo J Oricain año 1790 (1790). In: Maurtua VM (ed) Exposición de la Republica del Perú...en el juicio de limites entre el Perú y Bolivia conforme al tratado de arbitraje de 30 de diciembre de 1902, 10 vol Imprenta de itenrich y comp, Barcelona, II:319–377

Perry L (2007) Starch remains, presentation biases and plant histories: an example from highland Peru. In: Denham T, Iriarte J, Luc V (eds) Rethinking agriculture: archaeological and ethnoarchaeological perspectives. Left Coast Press, Walnut Creek, pp 241–255

Perú (1966) Censo de población de 1961. Oficina Nacional de Estadistica y Census

Perú (2005) Proyecto de reordenamiento y rehabilitación del valle del Vilcanota. Ministerio de Comecio Exterior y Turismo, Lima

Pistorius R (1997) Scientists, plants and politics—a history of the plant genetic resource movement. IPGRI, Rome

Plachetka UC, Pietsch SA (2009) El centro vaviloviano en el Perú: un conjunto socio-ecológico frente a riesgos extremos. Tikpa Pachapaq 1(1):9–16

Ramos Escudero F, Muñoz AM, Alvarado Ortiz C, Alvarado A, Yáñez JA (2012) Purple Corn (*Zea mays* L.), Phenolic compounds and its assessment as an agent against oxidative stress in isolated mouse organs. J Med Food 15(2):206–215

Ravero Campos G, Riveros Serrato G, Riveros Serrato H (n/d) Estudio de caso sobre el proceso de obtención de la Denominación de Origen del Maiz Blanco Gigante Cusco-Peru y principales logros alcanzados. LIMA: FAO and IICA.0 http://www.fao.org/fileadmin/templates/olq/documents/Santiago/Documentos/Estudios%20de%20caso/Mais_Peru.pdf. Accessed 14 Feb 2014

Reinhard J, Ceruti MC (2010) Inca rituals and sacred mountains: a study of the world's highest archaeological sites. Cotsen Institute of Archaeology, University of California, Los Angeles, Los Angeles

Rostworowski de Diez Canseco M (1999) History of the Inca realm (trans: Iceland HB) Cambridge University Prress, New York

Rowe JH (1957) La arqueología del Cuzco como historia cultural. Revista del Museoe Instituto Histórico del Cuzco 10(16–17):34–48

Rowe JH (1997) Las tierras reales de los incas. In: Johannessen S, Hastorf CA (eds) Arqueologia, antropologia e historia en los Andes: Homenaje a Maria Rostworowski. Institutio de Estudios Peruanos Ediciones, Lima, pp 277–287

Ruiz M (2001) Protección de conocimientos, innovaciones y práctices indígenas en el Perú. Debate Agrario 33:73–89

Sallnow MJ (1987) Pilgrims of the Andes: regional cults in Cusco. Smithsonian Institution Press, Washington

Sandor JA, Eash NS (1995) Ancient agricultural soils in the Andes of southern Peru. Soil Sci Soc Am J 59(1):170–179

SENAMHI (2010) Servicio Nacional de Meteorologia e Hidrologia Cusco, Peru

Servat GP, Mendoza W, Ochoa JA (2002) Flora y fauna de cuatro bosques de *Polylepis* (Rosaceae) en la Cordillera del Vilcanota (Cusco, Perú). Ecologia Aplicada 1(1):25–35

Sevilla Panizo R (1994) Variation in modern Andean maize and its implications for prehistoric patterns. In: Johannessen S, Hastorf CA (eds) Corn and culture in the prehistoric new world. Westview, pp 219–244

Shepherd CJ (2004) Agricultural hybridity and the "pathology" of traditional ways: the translation of desire and need in postcolonial development. J Lat Am Anthropol 9(2):235–266

Sierra Exportadora (2009) Gaton Benza Pflucker, president. http://www.sierraexportadora.gov.pe

Squier EG (1866) The Cuzco or Peruvian corn. Am Agriculturalist 25–26(9):219–220

Taipe Bolaños M, Cabrera Carranza M, Cabrera Carranza C (2006) Identificación y evaluación de las principales fuentes de contaminación del rio Vilcanota en el sector Calca Urubamba. Revista del Instituto de Investigación de la Facultad de Ingeniería Geológica, Minera, Metalúrgica y Geográfica (Lima), 9: 1–18 revistas investigacion.unmsm.edu.peru/index.php/iigdo/article/view/2871

Tamayo-Herrera J (1978) Historia social del Cuzco republican. Privately printed, Lima

Tamayo-Herrera J (1992) Historia general del Qosqo: una historia regional desde el período lítico hasta el año 2000, 3 vols. Municipalidad del Qosqo, Cusco

Thompson LB, Mosely-Thompson E, Bolzan JF, Koci BR (1985) A 1500-year record of tropical precipitation in ice cores from the Quelccaya ice cap, Peru. Science 229:971–973

Toribio Polo J (1913) Apuntes sobre las epidemias en el Perú. Imprenta. Nacional de Frederico Banioquero, Lima

Tupayachi HA (2005) Flora de la Cordillera de Vilcanota. Arnaldoa 12(1–2):1–16

Vazquez de Espinosa A (1948) Compendio y descripcion de las Indias occidentales. Smithsonian Miscellaneous Collection, Smithsonian Institution, Washington

Villanueva Urteaga H (1970) Documentos sobre Yucay en el siglo XVI. Revista del Archivo Histórico del Cuzco 13:1–148

Vuille M, Francou B, Wagnon P, Juen I, Kaiser G, Bryan GM, Bradley RS (2008) Climate change and tropical Andean glaciers: past, present and future. Earth Sci Rev 89:79–96

Webb R (2013) Conexión y despegue rural. USMP Fondo Editorial, Lima

Wightman AM (1990) Indigenous migration and social change: the forasteros of Cuzco, 1570–1720. Duke University Press, Durham

Wright KR, Valencia Zegarra A, Wright RM, McEwan G (2000) Macho Picchu: a civil engineering marvel. American Society of Civil Engineers, Reston

Zimmerer KS (1993) Agricultural biodiversity and peasant rights to subsistence in the central Andes during Inca rule. J Hist Geogr 19(1):15–32

Chapter 5
Vilca in Andean Culture History: Psychotropic Associations in the Urubamba and Beyond

Abstract Vilca (*Anadenanthera colubrina*) is a small leguminous tree occurring as a species component of the dry tropical forest of the Urubamba and other Andean valleys. The powerful psychotropic properties of its seeds account for the long and important place of this plant in Andean culture history. Archaeological evidence from painted pots, snuff tubes, bone pipes and clyster tubes indicate its diverse modes of past use. Wari and Inca artifacts, as well as the reconstruction of Inca history from early colonial documents, suggest the role of vilca in shamanic-style religion and medicine. When understood that the tryptomines in vilca trigger a characteristic three-stage hallucinogenic experience, new interpretations emerge of several aspects of the Andean past. Vilca uses can be implicated as a feature of oracle shrines at pre-Columbian religious sites as well as the behavior of the Chanka people, enemies of the Incas. After the Conquest, vilca was the substance behind the drug-induced manifestations of the so-called Taqui Onccoy movement. Strong Spanish opposition to vilca which was viewed as a diabolical intervention of Satan, had much to do with the competition it was perceived to pose to Catholic conversion.

5.1 The Plant Called Vilca

The alternative name for the Rio Urubamba is Vilcanota, and at an earlier time it was known as the Huilcamayo or Vilcamayo. The word Vilcanota can be translated as either "sacred river" or "river of the vilca tree."[1] Likewise, a western tributary of the Vilcanota/Urubamba is called the Vilcabamba (in the old orthography, Huilcapampa). Although a plant known as vilca was an important species in the Andean past, the word also has a broader meaning. Thus a question is presented as to whether these names were originally derived from and are referential to a sacred

[1] The anarchy of Quechua and Aymara transcription has resulted in half a dozen orthographic possibilities: huilca, huillca, bilca, wil'ka or willka as well as in the doubling the substantive to become vilca vilca or willka willka. Aside from this nomenclatural cluster, Ratsch (2005:50, 54) has recorded 128 other folk names for the two *Anadenanthera* species. No name makes a distinction between the entheogenic material (mainly the seeds) and the tree from which it comes.

D.W. Gade, *Spell of the Urubamba*, DOI 10.1007/978-3-319-20849-7_5

Fig. 5.1 Dried specimen of vilca (*Anadenanthera colubrina*) showing leaves and seed pods. Herbario, Museo de Historia Natural, Lima (Photography by D.W. Gade)

quality, or to the vilca plant. However that may be, the issue in this chapter is less the referential meaning of a toponym than how vilca, the plant, fits into Andean culture history.

Vilca designates in its narrowest sense *Anadenanthera colubrinam*, but with its congener *Anadenanthera peregrina* that has similar entheogenic properties, the two can be considered a complex. Since 1964, these two species have comprised the genus *Anadenanthera* that which was formerly in the genus *Piptadenia.* In each case, they are small trees, 5–18 m, with finely pinnate compound leaves, small white-yellow flowers, and leguminous pods holding from 5 to 20 flattish reddish brown seeds (Fig. 5.1). *A. peregrina* has a bark with warts and thorns, rough (rather than smooth) pods and is a somewhat larger tree when mature than the *A. colubrina* which may or may not have thorns.

The documented history of vilca starts not in the Andes, but in the Antilles. Fr. Ramon Pané, who accompanied Christopher Columbus on his second voyage in 1494, provided an eye witness account of Taino Indians on Hispaniola inhaling through a tube into the nasal passages a substance called *cohoba*. Pané (1999:15). Pane believed, incorrectly, that cohoba was powdered tobacco. The strong psychotomimetic reaction he described was too extreme to have come from *Nicotiana*. Not until the twentieth century did W.E. Safford (1916) determine the true source of cohoba as *Piptadenia colubrina* and rightly assign its origin to South America. Carl Sauer (1966:56) concurred that prehistorically the seeds were transported northward into the Caribbean, though he corrected its binomial to *Piptadenia peregrina*. Several other biotic elements, among them manioc, guinea pig and sweet potato, had also diffused through human carriage from the continent to the Antillean islands.

Richard Evans Schultes (1967) placed *A. peregrina* as a savanna species, a designation based mostly on historic references. In 1801, Alexander von Humboldt recorded in his travel account in the Orinoco region the preparation and use of the snuff made from the seeds. He wrote that the Otomac Indians collected the leguminous pods, fermented the seeds, and from the seeds they made a paste which they took in their nostrils. They were described as "…a noisy and uninhibited people as a result not only of their various fermented beverages, but also their particular state of intoxication emanating from taking niopo (=vilca) powder, that one could almost say was demented" (von Humboldt 1970:II:620). However, Humboldt erred in stating that lime, not the powdered seed with which it was mixed, was the active agent. In 1851, the botanist Richard Spruce (1908:II:429) also traveled to the Orinoco and described how the seeds of *Piptadenia niopo* (now *Anadenanthera peregrina*) were toasted, pulverized, and mixed with equal parts of alkaline ashes of certain barks and leaves. Neither Humboldt nor Spruce described the behavioral sequence of inhaling the snuff. Ethnographically, use of *Adenananthera* snuff was centered in the Orinoco Basin where its main names were *yopo*, *niopo* and *nopo*. A different species was also used in the Andes. *A. colubrina* has fewer contemporaneous ethnographic associations, though traces of it survive in the Western Chaco of Argentina (Flury 1958).

5.2 The Psychoactive Dimension

Anadenanthera spp. contain a dozen alkaloids of which two tryptamines, DMT (*n,n*-dimethyltryptamine) and bufotenine (5-OH-DMT), have been studied most (Stromberg 1954). How these alkaloids act on the human organism was not fully defined until the psychedelic investigations of the 1960s. A composite of testimonies of the psychotropic experience identifies three phases. First, within 40 s after insufflation, mucous begins to flow profusely from the nasal passages and mouth. Arm muscles quiver, the face becomes contorted, and the whole body undergoes an extended period of uncontrolled movement. The individual prances about in a way that, though involuntary, resembles dancing. Loud shrieking accompanies these kinetic excesses. Louria (1966) described this phase as the sensation of being hurtled through time and space. Ambulatory, gesticulative, and sonic unrestraint typically last an hour, after which the individual falls into a trance-like stupor. No other entheogen in the Andean repertory has that distinctive rush, though its full manifestations depend on dosage. In a second phase, hallucinations occur, followed by the third phase described as a calm period characterized by quiet reflection and sage utterances.

Only glimmerings are available on how vilca use patterns transpired over time and through space. The sequence of effects previously described is associated most clearly with snuff taking, for entry through the nasal passage allows this fast-acting compound to move directly into the bloodstream. Vilca has also been smoked in pipes and taken as an enema. Oral intake of vilca is less immediate and physiologically the most complicated. The enzyme monoamine oxidase, found in the human gut, suppresses one of the tryptamines (DMT). But that inactivation can be biochemically blocked when another plant, one that contains a beta-carboline, is introduced to counteract the action of the enzyme. Ayahuasca (*Banistoperis caapi*) is an effective hallucinogen when ingested orally only because a plant called harmine is used to prevent demolition of the tryptophane. With vilca, however, effects may be controlled by the size of the dosage. Large quantities of vilca without a beta-carboline are said to induce hallucinations. Ott reported that his own experiment of swallowing more than 100 mg of bufotenine had an entheogenic impact lasting 2 h (Ott 1993).

Archaeological evidence for vilca depends much less on pod remains of vilca than on the objects connected with its presumed use. Containers of wood, bone, or shell that are thought to have had vilca powder are artistically decorated. Spoons were used to measure dosage of powder to control the associated risks and effects. The presence of snuff tubes and clyster tubes strongly point to the prehistoric use of vilca (Torres and Repke 2005). Pre-ceramic sites (2130 BC) have yielded bone pipes (Fernǻndez-Distal 1980). Ceramics offer suggestive evidence.

Without a retrievable archaeological trail, oral ingestion of vilca has no certain prehistory. However, early colonial chroniclers reported on chicha to which vilca had been added, suggesting a traditional mode of ingestion. Exactly what part or parts of the vilca plant was used is unclear. Polo de Ondegardo (1916) referred to its juice (="jugo") and de Gonçález Holguin (1952) mentioned its sap (="*seva*"). If a liquid is meant, the green pods or leaves may have been pressed or the exudations of nectar-like resin on the tree branches collected. Seed pods, resin, and even bark of vilca contain the tryptamine DMT. Rusby (1922) observed in early twentieth century Bolivia that "the bark of a tree called vilca, apparently a *Piptadenia*, finds an apparent use in hastening the fermentation of cane juice." But use of vilca "juice" has no reference in the ethnographic or entheogenic literatures, making it likely that the word juice was a metaphorical allusion to the powdered seed. Adding vilca powder to maize beer would have cut the alkaloid bitterness, added ritual content, and may have had a physiological function. A pleasant effect may have resulted when the alcohol in chicha, acting as a vasodilator, counteracted the vasoconstriction caused by the amino acid tryptophan in the seed. Perhaps that effect explains why vilca has been one of the few entheogens that native peoples used with alcohol. Drinking a concoction may also have been safer than nasal insufflation. Vilca snuff is caustic to the mucous membrane and an overdose can result in cardiac arrest.

5.3 Prehistoric Vilca Use in the Andes

If, as is now commonly accepted, psychoactive substances first came to human attention in the pre-agricultural era, then entheogenic use of vilca in the Andes could have originated more than 5000 years ago. Shamans, concerned with both divining hunting success and deriving transcendental meanings, determined the therapeutic uses of many plants through a long trial and error process. Passed on to subsequent generations, that knowledge eventually became part of settled agricultural societies. Some of these plants, among them vilca, were used in shamanic curing, exorcism, sorcery and divination. Among the Mochica (200–700 AD), evidence for its use is clear (Fig. 5.2). As polities formed and civilization advanced in the Andes, vilca and other psychoactive substances became spiritual vehicles for other groups in the society besides shamans. It is clear that vilca played an important role in Andean culture history (Fig. 5.3). The challenge is to interpret the prehistoric and historic records to draw out the roles this plant has played at various times and in different places. Much uncertainty remains about its distribution, place in religion and functions. References to vilca in the chronicles scantily hint at its use in the pre-Conquest Andes where its presence goes back to before the Common Era. That vilca was used as an entheogen in the Chavin horizon is an assessment based

Fig. 5.2 A stirrup pot of the Mochica culture (200–700 AD) from the Chicama Valley. The spare legend was "sleeping warrior," but when placed in a larger context, it is clear that the man was in an altered state of consciousness. Dark round seeds of vilca and a clawed feline-like beast favors that reinterpretation. (Source: W. Lehmann and H. Doering, The Art of Old Peru, 1924)

Fig. 5.3 Artistic montage of the vilca tree, seedpods, seeds, leaf and trunk and cultural associations. (By Donna Torres, reproduced with permission)

not on recovery of seed fragments, but of mortars (*vilcanas*) used to grind the seeds. Decorated with felines and raptors, mortars offer persuasive evidence that shamans used psychoactive plant substances (Burger 2011; Larco Hoyle 1946). Strong evidence for vilca use exists particularly in the Middle Horizon (500–1000 BC). The Wari polity that emerged in the Ayacucho Basin seems to have had intensive use of these seeds. Stylized representations of *Anadenanthera* plant parts in ceramic and textile designs suggest Wari fixation on vilca (Knobloch 2000). The Ayacucho area had ready access to seed material, for the vilca tree grew and still grows in the open woodland found deep in the Apurimac River drainage below 1500 m asl elevation (Weberbauer 1945; Vargas Calderon 1946). Violence and the use of oracles were consequences of the Wari use of vilca. Skeletal remains and trophy heads suggest that the Wari had a military class that established control through violent acts (Tung 2012). Wari bellicosity plausibly came about as a result of vilca insufflation, which has an intense and immediate psychological effect on the brain, triggering aggressive behavior. A clue is Wari symbolic identification with the jaguar (*Panthera onca*). Vilca transformed a person into a jaguar: ferocious, raspy-voiced, alert, and visually acute. Moreover, the Wari were known adepts of the oracle. Hallucinogenic substances, vilca included, triggered oracular response. Pachacamac on the coast near Lima, the premier oracle site in pre-Columbian Peru, rose to its greatest height during the Middle Horizon.

Wari expanded from Ayacucho, not in territorial waves, but in discrete clusters. At the pre-Conquest site of Tiwanaku on the northern altiplano, there was found evidence of a strong cultural, including religious, identity with Wari. Max Uhle uncovered inhaling tubes at Tiwanaku in 1895. Some years later, Bandelier found an exquisitely carved bone spoon used to measure dosage (Wassen 1972). Mortars and pestles, stone bowls, cups and snuff holders suggest the preparation of vilca powder from its seeds. A carved monolith holding a drinking goblet in one hand and a snuff tablet in the other places vilca squarely in a religious context (Wassen 1972; Browman 1978; Torres and Repke 2005). The discovery of fragments of tubes, spoons and spatulas near Tiwanaku at the residential site of Lukurmata persuaded the archaeologist Bermann (1994:142) that vilca had been used there. Vilca dispensed at Tiwanaku would logically have come from trees growing in semi-arid valleys extending from La Paz to Northwest Argentina (Troll 1952:135). What appears to have been an ancient pattern of exchange continued among the Chicha Indians around present-day Tarija, Bolivia. They traded vilca (there called *cebil*) seeds to the Altiplano and Pacific coast in exchange for salt and other products (Zelada and Capriles Flores 2000). The Middle Horizon site of Huaro in the Cusco area yielded implements used for insufflating vilca powder (Glowacki 2002).

Given the many accretions from the Andean past, it is reasonable to believe that aspects of the vilca tradition in the Middle Horizon carried over into the Late Intermediate Period. Centuries after the Wari, the Chanca (1400 AD), a polity comprised of different ethnic groups, arose in the Pampas Valley (Aguirre Cárdenas 2008:65–66). Later, their center corresponded to the Andahuaylas zone, but part of their home territory continued to lie in the hot valleys of the Apurimac, an area rich in wild vilca trees. The mythic origins of the culture has some associations with vilca (de Betanzos 1968:15). Of their two founding chiefs, Uscovilca and Vilcaquire, the latter was a noble warrior who left instructions to be buried next to trees whose seeds "removed bad humors." (de Santacruz Pachacuti Yampqui 1968:297). No tree but vilca would fulfill that description. El Inca Garcilaso de la Vega provided clues to the importance of vilca among the Chanca when he mentioned their boast of descending from a lion, which fits into the Middle Horizon idea of the shaman as a raspy-voiced feline of acute vision and always alert (Garcilaso de la Vega 1960:164). Garcilaso de la Vega (1960:165) also mentioned the Chanca as having ayllus in the Pampas Valley (formerly known as the Vilcas). Descriptions of Chanca behavior suggest that unusual physical courage may have induced when they were under the influence of vilca. The chronicler Pedro Cieza de Leon (1986:133) described Chanca warrior behavior during their siege of Cusco as consisting of "...whooping and uproar because the people are very noisy in their fights and we fear their shouting more than their deeds." Shrill vocalization is one of the manifestations of the excitation phase of vilca ingestion. Relevant here are two ethnographic parallels. The Otomac Indians of the Orinoco Basin used *Anadenanthera* as a stimulant before battle (von Humboldt 1970:II:620–621). Among the Yanomamo, men take vilca—known to them as *hisioma*—to gain prowess in battle (Chagnon et al. 1971). Shouting, gesticulating and copious mucus flow are part of a belligerent posture related to the excitant phase of vilca snuff. From their center in the Andahuaylas region south of Ayacucho, the Chanca were known to be courageous in military actions. Their bellicosity was the biggest obstacle to Inca expansion northward. In the sixteenth-century account of the chronicler de Santacruz Pachacuti Yamqui (1968:297), the Inca arrived with 40,000 warriors at Vilcashuaman where they came up against "seven guacas and devils" of the Chanca.[2] Those words, added to what is known about the metaphoric use of "devils", can be interpreted as indicating that their leaders were in a battle mode after taking vilca. Guaman Poma de Ayala (1980:255) related how the Chancas hanged themselves when they were "very drunk," a likely reference to one of the violent consequences of vilca-taking gone awry. The possibility of self-harm is suggested by José Gumilla who described the violence using of vilca snuff, called there yupa, among the Otomaco Indians in Venezuela: "[they]..would get into a frenzy with yupa, injuring themselves, and smeared with blood and full of range, would enter the battle like raging tigers" (Schleiffer 1973:8). Chanca behavior may have been at the source of the Quechua expression, as recorded in the dictionary of Gonzalez Holguin, of "Chanca zazapa" meaning to "walk shakily."

Seven decades after archaeologist and ethnohistorian John Rowe's (1946:II:291) assessment in the *Handbook of South American Indians* that "narcotics were unimportant in Inca culture," another archaeologist, Alan Covey (2008:827), made a more nuanced but precise judgment that "hallucinogens do not appear to have had the prominence in Inca state religion that they played in Nazca, Wari and Tiwanaku." Inca sun worship, construction of temples, and the creation of a priestly class took religious attention away from the shaman. Unlike priests who, by definition, were functionaries in the service of a particular cult, shamans were not beholden to any particular ideology. That essential difference explained why shamans, but not priests, used hallucinogens to trigger the ecstasy that enabled them to communicate with the gods on their own terms. Shamanic unpredictability and nonconformity had the potential to confound the social order. In contrast, state-level societies codified the power structure. Any deviation, such as the ability to bewitch by the use of entheogens, threatened institutionalized power (Pearson 2002:165). The competition between the local and the state may account for why, when the Inca found that a shaman did "prohibitive things," they put him to death (de Murua 1987:433). By that logic, shamans may have been marginalized as Inca religion focused on the sun became codified. How that may have fit into the Inca timeline is open to speculation. In Cusco, a stone possibly used for grinding vilca was found at the Coricancha which is associated with early Inca ("Killke") shards (Bejar Navarro 1990:174).

However, the elite in Inca culture could not control religious practice at a local level. Rites of divination, white magic, and witchcraft, for which the ritual vehicle was vilca ingestion, prevailed in the countryside. The Spanish priest Cabello Valboa (1951:287) wrote about shamans "exercising power with the force of their dangerous plants" The same chronicler made reference to the "abominable arts," which I interpret as meaning the ingestion of entheogenic substances. Shamans depended

[2]Huaca, guaca or waka also has a wide variety of meanings. Early colonial writings defined it as an idol or temple of idols. In other contexts, a huaca was a body, an animating force or a landscape object. When huaca was used to refer to any localized source of pre-Columbian sacredness, it is easy to see how that term could be applied to vilca, which induced a quasi-religious experience. References to a "huaca that speaks" denote a shaman offering comment under the influence of vilca.

upon the biochemical power of vilca to achieve an altered state of consciousness to communicate with higher forces. Festive rituals among the Canchis people south of Cusco included vilca use if one interprets Cieza de León's (1986:269) comment to mean that when he wrote "they spoke with the devil in the same way as others." In addition, vilca use of the Inca period was associated with oracles. Two traditional pilgrimage shrines south and north of Cusco had that function: Vilcanota at La Raya Pass (4310 m asl) and Vilcaconga (meaning "voice of the vilca"), located at a point at 3900 m asl where one can see the wide expanse of the Apurimac gorge. Cieza de Leon asserted that the sacred quality of this latter site came from the presence of a vilca tree and buried gold (Cieza de Leon 1986:256). The connection between oracles and vilca is circumstantial, but the nature of oracular response in various cultures is one that benefited from an induced physiological state (Vallejo Berrios 2006).

Rowe (1946:II:291) incorrectly asserted that coca was the strongest drug-containing substance used by the Inca, and that tobacco was next in importance and, finally, "perhaps wil'ka." Rowe published that remark before the entheogenic properties of vilca had been analyzed. In fact, the biochemical efficacy of vilca, especially when taken as a snuff, rapidly and dramatically induces an altered state. Rowe's (1946:II:291) remark that the "Inca sorcerers put the juice of the wil'ka into their chicha to give it more strength" implied ethylic intoxication rather than hallucinogenic inducement. Rowe based that remark directly on a phrase from the chronicler Polo de Ondegardo; another chronicler, Father José de Acosta (1940), writing seven decades after Polo, curiously appropriated those same words. The interrogator Pérez Bocanegra (1631) also wrote of making chicha "stronger" by adding not only vilca, but also toads and bones of the dead. The concoction was both swallowed and taken "the other way," that is, as an enema.

At least one entheogen besides vilca was added to chicha. The Jesuit de Arriaga (1968:205) who wrote of chicha made of sprouted maize ("zora," i.e. jora) and chewed maize, to which the powder made from a plant called espingo was added. The shaman poured this strong concoction, called *yale*, over the sacred object (huaca) and then drank it which "made him crazy" (de Arriaga 1968:207). Arriaga described espingo as a "strongly odiferous, little dried fruit, like a round almond," a description that best fits either *Nectandra floribunda* or *N. reticulata*. Those who partook of this mixture were shamans (*hechiceros*) of a particular kind known as *huacavillac* which means "he who speaks with the huaca," implying an altered state of consciousness. The huacavillac were those who expressly had dealings with the devil, made "plant offerings," and did "deceptive" things. Arriaga also implied that huacavillac were those who had epilepsy ("*mal de corazón*"), which fits into a wider shamanic configuration. As a good Jesuit, Father de Arriaga (1968:250) asked these shamans how they could pretend to be speaking to the huaca and whether their "craziness" (i.e. altered state) resulted from drinking the chicha or from having been affected by the devil, i.e. taking entheogens.

5.4 The Vilca Tree Under Inca Cultivation

The development of an Inca state religion put vilca in a complicated position. Vilca was a symbol of the "old religion," i.e. shamanism inherited from the past. In the late Inca period the Inti and other cults devoted to the celestial emerged to place emphasis on rituals and celebrations organized by the state with its priestly functionaries, who, unlike shamans, followed a pre-determined script. The shaman was too unpredictable a figure to fit into this codified belief system designed to provide a spiritual foundation for Inca expansion. Thus, two belief systems coexisted, the official state religion orchestrated from Cusco and the shamanic practices prevailing in the countryside and in traditional oracles. Vilca had a role and the tree may even have been cultivated where the climatic conditions made that possible.

The possibility of vilca gardens is better appreciated by an ecological assessment of the zone in which vilca grew naturally. Examining vilca distribution in the river valley, formerly called the Vilcamayo/Huilcamayo/Wilkamayu and today called the Vilcanota or Urubamba, gives a micro-geographic perspective of the human and non-human agencies of this plant. In this valley, the wild *Anadenanthera colubrina* tree grew in the dry forest zone between 1600 and 650 m asl. Rainfall of ca 990 mm a year is highly seasonal, which explains the presence of savanna species. Trees in the genera *Bombax*, *Ceiba*, *Chorisia*, *Jacaranda*, *Curatella*, *Anadenanthera*, *Dilodendron*, *Ochroma*, *Sapindus*, *Triplaris*, *Inga*, and *Prosopis* are mixed with grasses of the genera *Andropogon*, *Trachypogon* and *Aristida*. In addition, shrubs, particularly *Eupatorium* and *Baccharis*, *and* the cacti *Cereus vargasii* and *Opuntia Brasiliiensis*, also occur there. All these plant species have adaptations to reduce moisture loss during the dry season; vilca does so by dropping its leaves during that period. The above mentioned taxa disappear farther down the Urubamba near Koribeni (ca. 650 m asl), where greater rainfall in June and July favors tropical rainforest (Fig. 5.4).

Collections of vilca from the dry forest zone started in 1931 when the Cusco botanist Fortunato Herrera (1941:273) collected "huillca" (*Piptadenia colubrina*) at Huadquiña at an elevation of 1500 m asl. The hard, compact, and beautiful wood was valued especially for cabinet making and for carving walking sticks. In a previous article, Herrera (1941:95) mentioned its narcotic and cathartic seeds being part of the indigenous pharmacopeia without, however, alluding to their hallucinogenic potential. Now *Piptadenia colubrina*, as well as the binomials *P. macrocarpa* and *P. grata* are folded into *Anadenanthera colubrina*. Collections in the Herbario Vargas at the Universidad Nacional del Cusco show 16 accessions of the same plant under four different names from two different valleys in Apurimac Department and La Convención Province. The botanist César Vargas collected this plant on five occasions in the Urubamba at elevations between 725 and 1300 m asl. D. Gade collected the plant in 1963 under the name *P. grata*. Other herbarium records in Peru between 1942 and 2006 document *A. colubrina* and *A. colubrina* var. *cebil* in ten more places in the Urubamba.

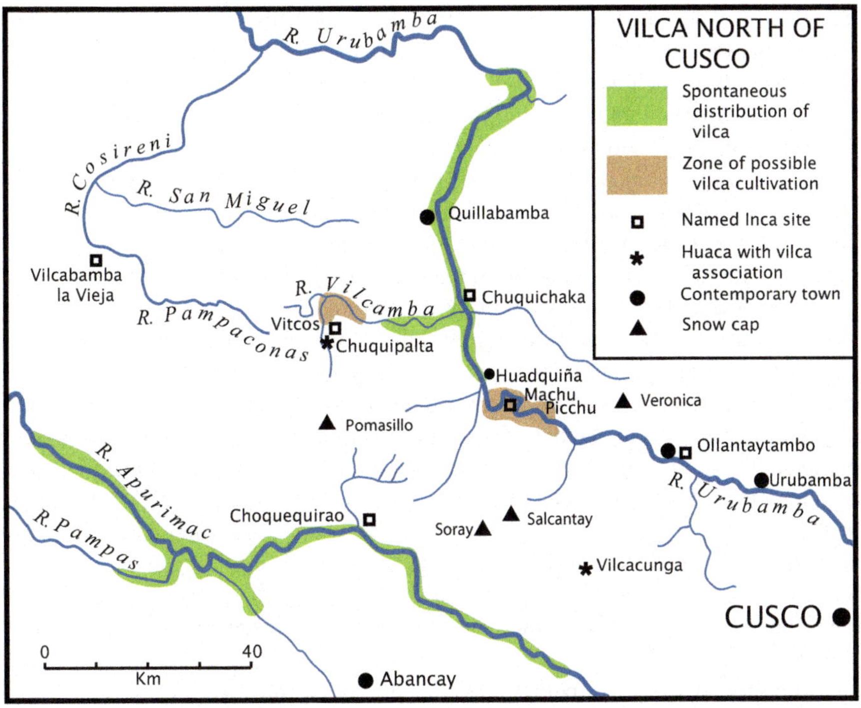

Fig. 5.4 Wild vilca distribution in the Urubamba (1600–650 m asl) and Apurimac Valleys (1500–900 m asl) as well as the zones in which the Incas probably cultivated vilca. (Map by M.K. Gade)

Vilca has been a spontaneous floristic element of the Urubamba-Vilcanota vegetation only where a wet-dry climate permits it to compete on its own, but this species has also been purposefully planted in places where otherwise it does not spontaneously grow (Fig. 5.4). This cultivation of *Anadenanthera* has been documented for the Yanomamo in the rainforest between the Orinoco and the Amazon drainages. Transplanting of the tree into gardens from seedlings gotten elsewhere started when alliance networks and warfare patterns disrupted the supply of seeds acquired by trade (Sherbondy 1988; Gade 1999). In the Urubamba, different motivations prompted the cultivation of vilca.

Vilca cultivation was part of a larger Inca agroforestry practice. For example, they planted *pisonay* (*Erythrina falcata*) using seeds or seedlings taken from the *ceja de la montaña* where the species grows spontaneously and planted it at higher elevations as a shade and decorative tree. Motivation for planting vilca came from a desire to assure a source of the raw material for hallucinatory experiences and for the symbolism that the tree incorporated. In addition, its fine wood was used for ceremonial objects such as shrines where ancestors were venerated. The mortar on which vilca seeds were ground, called a v*ilcana*, was often carved in the shape of a llama. It was commonly made from this wood, but could also be made from stone.

Its function was noted in the sixteenth century (de Albornoz 1967:22, 53). The device also served to measure vilca powder to assure the right dosage. In the Inca period and after, vilca wood was also used burned for a ceremonial fire known as a *vilcanina* (Lira 1944:153). Having a readily accessible source of seeds and wood, the Inca avoided trips to the hot country where it grew wild.

Vilca is easy to cultivate. Its dried seeds readily germinate and seedlings grow fast. When grown in partial shade, the mature tree is only about 20 m tall and does not cast dense shade. At Machu Picchu vilca was near its thermal limit so that site's microclimatic advantages minimized the risk. But the risk is not total: whereas *A.* peregrine does not tolerate any front, *A. colubrina* can withstand occasional freezing temperatures. On stone-faced terraces the stones absorbed heat in the day and reflected it at night. Stepped terraces on a steep slope permitted cold night air to drain downslope rather than to collect and damage the plants. Cultivated vilca trees might also have been planted at symbolic sites such as near rock outcrops or entrances to caves. While excavation at Machu Picchu yielded no archaeobotanical remains of vilca, the American plant scientist O.F. Cook (1916) in 1915 found the tree growing near the San Miguel Bridge (1900 m asl) suggesting its capability to successfully grow at Machu Picchu during the Inca period (Bingham 1948:114).[3] Snuff tubes of bone, one bifurcated at the distal end and the other incised with dots, were found in a burial cave at the ruins (Safford 1916; Bingham 1922:236; Burger and Salazar 2004:202).

Transverse depressions below 2300 m asl that led into the Urubamba also had potential to sustain plantings of the vilca tree. The largest such side valley was Huilcapampa, hispanicized as Vilcabamba, which Rowe (1946:II:189) translated as the "valley of the narcotic berry." Forty kilometers upstream from its confluence with the Urubamba, the Inca Pachacuti constructed an important ceremonial center later called Vitcos (2850 m) where Inca shamans performed rituals (Cobo 1956:II:79). Bingham (1922:264) made a not implausible intellectual leap when he wrote: "If one may judge by the name of the place, Uilcapampa, the wizards and sorcerers were probably aided by the powerful effects of the ancient snuff made from huilca seeds." In a later book Bingham (1948:123) again made the association of vilca the plant with Vilcabamba the valley and the fact that the shamans at the shrine called Chuquipalta were under the influence of that hallucinogen (Fig. 5.5). If Bingham (1922:249) had realized the tie between the devil and vilca, his interpretation of the hostility of Spanish monks would have had a deeper basis of understanding.

Also known as Yurac Rumi (white rock) or Nusta Ispanan (urinating virgin), Chuquipalta was a site at which four stone paths converged. An eight-meter high granite boulder with sculpted steps formed the centerpiece around which terraces, two buildings, and courtyards were arranged. Although the worked stone at

[3] Since Cook, two other records have been filed in the Herbario Vargas of vilca collected at elevations of 2400 m in the Urubamba. That elevation raises the question if they were planted or if they were volunteers from previous generations of planted trees.

Fig. 5.5 The carved stone known variously as Chuquipalta, Yurac Rumi ("white rock") or Nusta ispanan ("urinating princess") was encountered in 1569 by Augustinian friars who tried to destroy it. Spanish clerics described it as the place where "the devil speaks," a metaphor for describing the shamanic descent into an altered state of consciousness. This photograph was taken by H. Bingham 1911. (Source: Bingham 1922, facing p. 256)

Chuquipalta is clearly Inca, its status as a shrine may have long predated the Inca. The Chanka and, before them, the Wari, may both have had Chuquipalta as one of their shrines. An archaeological survey of the area documented the presence of structures of characteristic Chanka construction (Ramos Condori 1983; Bauer et al. 2010:80–89). Vilcabamba la Vieja has proven to be Wari long before it was Inca. In this valley, vilca use may long have also preceded the Inca. After the Inca assumed possession of the area, they continued the oracular use of the shrine found there. A vilca source existed 25 km down the valley where the warm climate permitted its growth as a planted species. Just as it had been for the Wari people, vilca was the sacred tree of the Incas (Marin Moreno 1961:153). Large, long-necked ceramic jars (*urpu*) with a flared rim from the Imperial Inca period have an artistic motif that suggest vilca leaves and seeds (Fig. 5.6).

5.5 Vilca After the Spanish Conquest

Vilca continued for a time after the Conquest. Decline in the use of vilca for psychoactive purposes in the Andes coincided with the imposition of Christianity. Ecclesiastical authorities could not tolerate such insidious challenges to their

Fig. 5.6 Aryballus from the Imperial Inca period with a stylized motif that can be interpreted as the round dark seeds and bi-pinnate leaves of vilca. (Source: Museo de Arqueología y Antropología, Lima, Peru)

religious hegemony and this partly explains why it is rarely directly mentioned in the chronicles. Suppression of the hallucinogenic use of vilca as a fight against the devil extended to prohibiting talk about it.

5.5.1 Chuquipalta and Its Entheogenic Foundation

Several clues strongly suggest that after the Conquest the site of Chuquipalta near Vitcos had a vilca connection. In 1569, two Augustinian monks, Marcos Garcia and Diego Ortiz, received authorization from Titu Cusi, who in 1560 was crowned emperor of the Inca, to enter Vilcabamba The two missionaries had gotten word that in this redoubt the Inca religion was still being practiced and their aim was to convert the neo-Inca to Christianity. In 1570, the Spanish monks sought to destroy the site of Chuquipalta as a symbolic gesture because of its associations with idolatry. Most of what we know came from the chronicler Antonio de Calancha who recounted the lives of these two fellow members of his order (de Calancha and de Torres 1972:I:72). Calancha noted that the devil appeared at that site, which he judged was the principal Inca shrine in that region. The devil in this case was a shaman who took vilca and provided answers. The chronicler indicated its oracular significance when he pointed out that "the devil gave answers from a white stone outcrop and was visible on various occasions." Calancha's ironic allusion to this site as the "University of Idolatry" was in reference to the question and answer invoked by the shaman (as "professor"). His reference to "professors of the abominations" was a metaphor for vilca-ingesting shamans imparting their supposed wisdom to others. In a moment of religious zeal, Garcia and Ortiz burned the site to "purge the devil from the rock," an act that led to Ortiz being killed.

5.5.2 *Vilca and the Taqui Onccoy Movement*

Vilca came starkly to the center of Spanish attention three decades after the Conquest during the beginning period of evangelization. An indigenous uprising between 1555 and 1571 against Spanish practices and values has been the subject of more than a dozen books and many articles. Known as Taqui Onccoy ("dancing disease"), the movement involved itinerant shamans who spread a message of cultural purity. From its beginnings in Ayacucho, Taqui Onccoy gained adherents as far south as Chuquiabo (La Paz, Bolivia). It may not have been a coincidence that Ayacucho had been the center of vilca use in the Andes since Wari times. The Jesuit Pablo José de Arriaga (1968:255) noted the presence of large numbers of shamans in Ayacucho ("hechicheros de Guamanga") where vilca use had an ancient tradition. The Spaniards persecuted those adherents, driving them into the mountain fastness of Vilcabamba north of Cusco. In 1536, following the trauma of the Conquest, Manco Inca had retreated to that neo-Inca refuge in his search for the wellspring of Inca identity. Since Spanish authority had not yet reached Vilcabamba by 1536, these people could maintain all their old customs and practices without interference. Rugged topography, as well as the control of the crucial bridge crossing ("Chuqichaca") held the Spaniards at bay. Suspended across the tumultuous Urubamba River at Chaullay, the span could be cut at a moment's notice, forestalling invaders moving up the Vilcabamba Valley.

Whereas in the Inca period, when only shamans normally went into drug-induced ecstasy, the Taqui Onccoy cult invited the assembled masses to ingest vilca as a communal repudiation of Spanish religion. The ceremonies, apocalyptic in tone, represented a yearning for a return to the cultural integrity of the ancestors. Yet Taqui Onccoy was not about a manifestation of Inca state religion; rather it was a reaching back to Andean folk beliefs. The colonial written record provides only hints of what actually went on. In his inquiry into the movement, the sixteenth century chronicler Fr. Cristobal de Molina wrote that "there were many Indians who trembled and rolled on the ground, others threw stones like people possessed by the Devil, grimacing and then they rested (de Molina and de Albornoz 1989). People came up to them and asked them what was happening and what they were experiencing. They responded that a sacred force—a "huaca"—had entered their bodies." This account uncannily describes the behavior that vilca can induce. The dancing, quivering, trundling and scowling described by Molina occurs in the excitation phase of vilca ingestion. Since the phase that follows involves hallucination, observers were in no position to describe that orally or in written form. Molina himself did not observe this collective frenzy; his comments about it came from talking with witnesses.

Fig. 5.7 Vilca seeds are flat and generally reddish-brown in color. Their size ranges from 2 to 4 cm. In shape, the seeds are mostly round, but sometimes oblong and even rectangular. (Source: D.W. Gade)

Since no one at the time explicitly connected vilca ingestion and extravagant behavior, most scholars of Andean ethnohistory failed to recognize this linkage. However, Cristóbal de Albornoz (1967:22), the major Spanish contemporary investigator of this issue, did seem to understand when he wrote that vilca was "a kind of poisonous fruit" found in the hot lands, "the size of a copper coin"—actually a description of the seed not the fruit—with which they purged and cured themselves (Fig. 5.7). Without directly stating why, Albornoz focused on the need to find and destroy the seeds, as well as the carved wood and stone objects (*vilcana*) on which they were ground. He commented that the sorcerers "invite the devil before they begin to cure and that is the truth," but did not explain that ingestion of the seeds brought on an altered state of consciousness. In 1657 Albornoz had recently arrived in Peru from Europe, bringing his Eurocentric perspective about demonic forces. That background, plus his strong desire to move up the ecclesiastical hierarchy, accounts for Albornoz's inability or unwillingness to empathize with indigenous customs or values (Ramos 1992).

None of the scholars of Andean ethnohistory who have examined the Taqui Onccoy phenomenon connected the performance described by Molina with vilca

(Brundage 1967; Castro-Klarén 1993; Cornejo Bouroncle 1959:73–194; Gareis 1990; Heilman 2002; Wachtel 1971). Several of them described the state of mind that the participants would get into and a few of them implicated possible exterior influences. But none specifically linked the particular behavior with the effects that vilca can produce. Ethnohistorian Classen (1993:133–135), who took an organic perspective on Inca cosmology as it related to the human body, failed to link Taqui Onccoy with vilca. Wachtel's (1971:275) elegant remark that in this movement, "la divinité se trouve interiorisée" (divinity became internalized) nevertheless failed to see the entheogenic connection. Another scholar wrote that the "possessed Indians would tremble, shake, and dance insanely" without explaining just how they could have collectively gotten themselves into that state (Heilman 2002:123).

Two more recent ethnohistorians, Roy (2010:9–58) and Brosseder (2014:49–52), missed the vilca connection entirely. However, some of the Taqui Onccoy literature considers the possibility of an outside element at work. Marco Curatola (1990) hypothesized that pellagra was at the root of the noisy demonstrations associated with the movement. He argued that the extravagant outbursts were a patterned form of exorcism of the disease, one that struck most regions of Tawantinsuyo at the end of the dry season. Peruvian ethnographer Luis Millones (1990:15) noted that the followers of Taqui Onccoy "offered their own bodies as vessels through which the huacas could speak." Peter Gose (2008:97) came even closer when he noted that dancing, oracular possession, strange sounds and drug use all converged. French ethnohistorian Thierry Saignes (1989:100) was more specific when he hypothesized that San Pedro cactus (*achuma*), jimson weed (*chamizo*) and vilca were sources of possession. However, he did not distinguish between the effects of the different substances. The first two hallucinogens mentioned do not cause an excitation rush as the chronicler Molina described. Likewise, the ethnohistorian Redden mentioned ayahuasca and San Pedro cactus, but not vilca, which is the critical substance that causes the "dance" (Redden 2008).

Millones (1990:11–18) identified the fundamental contradiction of Taqui Onccoy as having been inspired by the memory of the Inca past, but with an aim of reviving local huacas and using shamans in those cults. The movement did not attempt to recall the Inca state religion and its worship of the sun, moon, Viracocha, and Illapa (lightning). The key point of the Taqui Onccoy public demonstrations, missed by many scholars, is that such an intense and coordinated reaction of a whole group of people would have required a psychotomimetic substance. If just one individual had gone into a state of emotional possession, a range of factors, from mental illness to epileptic seizure, could have triggered the display. But a collective ecstasy required a force that acted on all participants in a predictable way.

Simultaneous ingestion of vilca created the tumultuous scene that made a Taqui Onccoy demonstration so impressionable. The huge drama of the moment was short-lived. In 1572, this, the first public threat to Spanish cultural and religious authority, ceased. Under orders of Viceroy Toledo, the Spaniards initiated an audacious military incursion into the Vilcabamba redoubt. They captured Tupac Amaru, the last Inca ruler and son of Manco Inca, and took him to Cusco where they beheaded him on Huacaypata, now the Plaza de Armas. Vilcabamba, where Indians had practiced all their native customs including drug-induced hallucinations, fell under Spanish colonial control. From then on for about a century, vilca use in shaman-led rituals increasingly went underground.

5.5.3 Psychoactive Agency and the Devil in Colonial Peru

In the early seventeenth century, the setbacks to evangelizing Andean peoples came to a head. Viceroy Toledo's *reducción* plan made Spanish priests live in much closer contact with the indigenous population than they had before, and they became increasingly aware of the strong hold that shamanic practice continued to have on the native mind. In the confessional booth and from headmen (*kurakas*) reporting on their own people, priests learned of heterodox religious practices. Spanish religious authorities interpreted shamanic activity as idolatry of the devil. In 1609, the Church in Peru instituted a campaign to snuff it out. For the next 28 years, the destruction of drug-using shamans became a clerical obsession.

Though the concept of a devil did not exist expressly in native Andean religion, Spaniards interpreted shamanic states of ecstasy in demonic terms (Griffiths 1996:117). Extirpation of idolatry mostly involved suppressing shamanic states of ecstasy, which were interpreted as the work of the devil (Fig. 5.8). That explanation

Fig. 5.8 This early seventeenth-century panel bespeaks of an association of the devil with idolatry. As described, "all wizards and witches speak first of all with the devils of hell in order to know what is what and what is happening in the world" (Source: Guaman Poma de Ayala 1980:251)

seemed reasonable if the shamanic experience that initiated an altered state of consciousness was not understood. Unable to conceive of psychotomimetic qualities being inherent in a plant, Spaniards viewed the devil as the source of what they considered idolatrous inebriation, which they called "borrachera" (de Matienzo 1967:80). As a drug-induced trance, borrachera involved more extreme behavior than that caused by the stupefying excess of alcohol. Among the colonial sources conflating borrachera and idolatry was a seventeenth-century account made at Cajatambo in the north central Peruvian Highlands (García 1996:II:26). The sixteenth-century chronicler Polo de Ondegardo (1916:30), who clearly described what he claimed was an invocation of the devil, described use of vilca as drunken behavior. Most scholars have interpreted the word "borrachera" to mean excessive alcohol. For example, Morales (2012:65) rendered one of Viceroy Toledo's *ordenanzas* that included the phrase "todas las idolatrías son borracheras" as "all idolatries they practice are drinking bouts." Toledo's phrase is best translated as "all altered states of consciousness are invested with idolatry."

The seventeenth century Jesuit Bernabé Cobo's (1956:II:230) description makes clear that the Spanish use of the word 'devil' meant more than it seemed to. He noted that a diabolical consultation ("*consulta con el demonio*") involved a sequence in which the shaman entered a closed room, got intoxicated, lost consciousness and later regained his senses to provide lucid answers to what had been asked. Even the mestizo Guaman Poma de Ayala (1980:253) expressed the Spanish view when he discussed various kinds of shamans (*hechiceros*) including those who "speak with the devil." The inference drawn from that remark, six decades after the Spaniards had taken control of Vilcabamba in 1572, is that drug-induced ceremonies used vilca as the instrument of idolatry. The chronicler de Vargas Machuca (2008:38) was one of the few who defined vilca as "…a seed that the Indian chews to speak to the Devil." The chronicler Cabello Valboa (1951:288) would seem to have had vilca in mind when he made the connection between "the power of the herbs and the power of the Devil."

Contemporary scholarly accounts of the colonial campaign against idolatry lack discernment of the entheogenic linkage in colonial devil discourse. Historian Sabine MacCormack's (1991:183) comment that the Andean "devils" possessed the "indwelling of gods through human beings" makes little sense unless hallucinogens are taken into account as the source of that "indwelling." This same historian of Andean religion wrote that "most missionaries…convinced themselves that the demons held an almost invincible command over the Indians, a command that could ultimately be broken only by religious coercion." The coercion is what the missionaries did, but MacCormack (1985:452) did not describe it. The demonic in colonial Peru did not simply spring out of the mind. Without grasping the metaphorical nature of the devil, the abundant literature on the campaign has omitted a discussion of hallucinogenic involvement.

The idea that the devil is lord of hell and adversary of God did not emerge until the twelfth century (Deane 2011:194). During the trials of the Inquisition, ecclesiastical concern about demon worship became an obsession. Manuals printed during the early years of the Inquisition described the evil. *Malleus Maleficarum*, written by a Dominican inquisitor was one influential tract (Maxwell-Suart 2007). "Speaking with the devil" was tantamount to demon worship and called for severe punishment. Penalties were harsh because the chaotic, rebellious and possessive energies of the devil idea threatened to disrupt the whole medieval cosmology. The Protestant Reformation exacerbated concern about the devil.

In the New World, ecclesiastical scrutiny of putative devil worship took its cue from European experience. There the Church had previously beaten into submission the witch idea, which involved anything potentially capable of moving people into an altered state of consciousness. Ecclesiastical pressure explains, for example, why the drum, which had the capacity to put people into an altered state, disappeared in Europe. In the Americas, the Spanish clergy conflated the use of entheogens with a pact with the devil. In early colonial Mexico a Catholic priest described the psilocybin mushroom, called *teunonacat*l (translated from the Nahuatl as the "flesh of god") as "the devil whom they adore" (de Motolina 1973:20). In 1620, the Inquisition in Mexico declared peyote to be a heresy. In the same time period in the Andes, vilca, a sacred inebriant, was being condemned. The use of these substances represented a threat to Spanish authority. Not only did native use of drugs have a holy quality, mimetically inverting the Catholic idea of sacrament, vilca also had a power greatly exceeding anything in the Christian repertoire of divinatory performance (Cervantes 1994:25). Neither the dogma of transubstantiation nor exorcism by incantation could compare to the pharmacological power of hallucinogens to physiologically, not just emotionally, envelop the participant. Bystanders too were affected by the dramatic sequence of behavior brought on by vilca.

Such was the Church's power and influence in the seventeenth century that it effectively eliminated public use of hallucinogens as well as its use in private séances. The experience of police-like tactics during the Inquisition effectively stamped out the psychoactive use of vilca. Religious orders developed their own theologies of the devil. One scholar wrote that the Franciscans "imagined the whole of the Indian past as a diabolical hallucination" (Cervantes 1994:27). But it was the Jesuits who most focused on flushing vilca out of indigenous religious practice by attributing it to the devil. More than any other expositors of the Church's magisterium, the Society of Jesus led the charge. In Peru, José Arriaga S.J. organized the extirpation campaigns. He had plenty of help: in 1610, only four decades after they first arrived, the Jesuits had 341 members in Peru (Vargs Ugarte 1959:II:235). The order's obsession with the idea of the devil can be attributed to the ideas of its canonized founder, Ignatius of Loyola, who depicted Satan in his *Spiritual Exercises* as laying siege to the individual.

de Arriaga (1968:246) had a bevy of ammunition to enforce this pronouncement that "ministers of idolatry" were prohibited from curing. Inquisitorial methods and the threat of execution drove entheogenic use of vilca deep underground. In 1639, Ramírez del Aguila (1978), a priest in Charcas (present-day Sucre, Bolivia), refrained from referring to vilca though he wrote of the hallucinatory effects of other plants. Ramirez del Aguila noted use of achuma or San Pedro cactus (*Trichocereus puchanoi* or *Echinopsis pachanoi*) whose entheogenically active element is mescaline. It continued to be used in public through the colonial period

(Glass-Coffin 2010:69–71). By contrast, vilca, though readily available from trees growing nearby in the dry canyons, was banned. Even the mention of vilca was anathema.

Decades of relentless repression of vilca severed the intergenerational transmission of indigenous knowledge about it. de Murua (1987:432) proclaimed that the shamans were on the decline in their battle with the Holy Gospel, which had rooted out the "bad seed" of ancient times. That seemingly metaphorical reference may, in fact, have been a specific allusion to vilca. By 1650, the psychoactive use of vilca had receded from public knowledge so that the great majority of people no longer knew of its visionary power. At some point in the mid-eighteenth century, its entheogenic use essentially disappeared. At the same time, however, the mania about diabolical heresy as a theological transgression also faded away. That irony suggests that the struggle was never really about idolatry or superstition, but the power of vilca. Moreover, the many other unorthodox practices that continued in native ritual the Church dismissed as inconsequential, for unlike vilca, they did not challenge the central mystery of transubstantiation. Another paradigmatic shift occurred in which the devil lost its force. The idea of the devil no longer terrorized the population as it once might have. At the same time, two centuries of Christian practices made Andean people lose trust in their native deities.

5.5.4 Vilca as a Purgative

Abandonment of vilca's entheogenic application in the colonial period did not wipe out other uses of the plant. Just like ayahuasca, dosage determines whether vilca inebriates or purges. When ingested in low doses of two to six seeds, both vomiting and diarrhea occur, but not hallucination. The dual effect of emetic and vermifuge made vilca valuable in Andean folk medicine as an effective cleanser of the body. Native ideas about purgation went beyond expelling intestinal parasites to a heath ritual that was believed to extend life. No less than five chroniclers of the early colonial period alluded to vilca as a purgative, with no mention of its entheogenic qualities (Vazquez de Espinosa 1948:608–610; Jiménez de la Espada 1965:I:190; de Albornoz 1967:21; de Molina and de Albornoz 1989:172; de Gonçález Holguin 1952:352). It is not clear if that reflected a taboo of the period. The colonial chronicler Vazquez de Espinosa (1948:608) knew about this plant, where it grew and its medicinal value. Two other chroniclers described purge recipes. Guaman Poma de Ayala (1980:608) noted the use of vilca combined with the root of maca, though the identity of the latter plant is uncertain. Maca is the common name for a high altitude cultivated plant (*Lepidium meyenii*), but Varón Gabai (1990:348), using Fortunato Herrera as his authority, described it as a plant in the genus *Oxalis* and that it had narcotic properties when taken together with chicha. Half of that compound mentioned by Guaman Poma de Ayala was swallowed and the other half taken rectally with a clyster tube appropriately called a v*ilcachina*.

Padre Cobo (1956:I:158) noted a recipe mixing two or three ground vilca seeds with the root of a *Polypodium* fern as a combination that yielded results "without distress." He also listed other medicinal applications for this plant. Swallowed with honey to mask the bitterness, vilca was used to treat congestion, kidney malfunction, female infertility, malaria ("mal del valle") and melancholy (Cobo 1956:I:272). Another early colonial source recommended vilca as a cure for cold and as treatment for epilepsy (Basto Girón 1977:69). One historian interpreted colonial sources as recommending vilca against poison and "evil spells" (Brosseder 2014:223). Folk medicine sources from the eighteenth century repeated some of these medicinal uses, but were silent about its ancient use as a hallucinogen (de Losa Avila y Palomares 1983:125). Another treatise on Andean folk medicine in that period discussed the entheogenic use of achuma (San Pedro cactus), but not vilca (de Losa Avila y Palomares 1983:26).

By the mid-nineteenth century, vilca had lost its once lofty place in the Andean pharmacopeia. Purgation for full body cleansings fell out of favor. Alternative medicinals, native as well as European, replaced other uses of vilca. Absence of this plant in encyclopedic works of folk medicine in the Peruvian Andes has reflects its general abandonment (Moscoso Castilla 1963; Bastien 1987; Franquemont 1990; Bussmann and Sharon 2007). The ethnographic scholar José Lira (1944:162), who gave the name "willka willka" to the plant, referred to it only in the context of a magical use following the doctrine of signatures in which the shape of the useful plant part had therapeutic meaning. He described vilca seeds as fancifully resembling a human or animal body part and used as good luck charms when buried in the ground. Uses to which vilca had been put have gone from psychotropic to medicinally therapeutic to magical.

5.6 Legacy of Vilca in the Andes

So many archaic survivals exist in the Andes that any generalization about the obliteration of the entheogenic use of vilca risks being wrong. A rare contemporary mention of vilca comes from American anthropologist Billie Jean Isbell (1989) who, in the late 1960s, participated in a local ceremony in the village, Chuschi, Department of Ayacucho, in which vilca seeds were added to chicha. Whether those who drank the potion experienced an altered state beyond ethylic inebriation, was not reported. Remarkably, many Andean peasants today with ready access to vilca are oblivious to the awesome psychoactive power of its seeds.

Abundance of the word vilca in Andean place names must be interpreted with caution. Either as a prefix or suffix, vilca, villca, huilca, or vilcas occur as names of towns, features or fields. Among the long list are Huilcaralca, Huilcatira, Vilcano, Vilcalla, Vilcacancha, Vilcapata, Vilcapuquio, Vilcaqui, Vilcarayra, Vilcacancha, Vilcahuara, Vilcalma, Vilcamarca, Villcaloma, Vilcataure, Vilcayacu, Vilcayoc, Vilcacancha, Vilcaconga, Vilcacoto, Vilcaran, Vilcahuara and Vilcashuaman. These places with the name of vilca do not necessarily contain any historic connection with the plant called vilca. The polysemic nature of the word, meaning also sacred, makes it unproductive to establish any link. An instructive case of etymological reductionism occurred when the French traveler Paul Marcoy (1874:133) tried to puzzle out the name "Huilccamayo" which is used for the stretch of the river from its source at La Raya Pass (4340 m asl) to Sicuani (3500 m asl). Marcoy reasoned that such a cold climate could not have sustained that tree, though he muddled the argument by confusing the taxonomic identity of huilca with algarrobo (*Prosopis* spp.), both genera of which have leguminous seedpods and compound leaflets.

Yet there may have been a connection with the plant in another way. The Inca shrine (wakahuaca) called Vilcanota ("house of the vilca" in Aymara) at La Raya Pass above 4000 m asl was an oracle site (Bauer 1998; Reinhard 1995). As such, the substance that shamans there used to go into an altered state would have been *Anadenanthera colubrina*. Two other sites with a vilca toponym may have had associations with vilca the substance. Vilcashuaman (3600 m asl) was a Wari site, then Chanka and finally an Inca settlement high above the Pampas (formerly the Vilcas) River 2000 m asl. A Jesuit missionary in 1601 documented that in the region of Vilcashuaman people adored a vilca tree and made offerings at the site (Polia Meconi 1999:42). Another suggestive name is Wari (or Guari) Vilca, a ceremonial site at 3000 m asl (de Yupangui 2005:120). At both Vilcashuaman and Wari Vilca, vilca seeds needed for ritual purposes could have been found at lower elevations in nearby canyons where the tree grew spontaneously.

Patronyms and matronyms with the name Vilca abound in the Central Andes among people with some or all indigenous ancestry. As with toponyms, Vilca/Huillca as a family name exists with various prefixes and suffixes. Descent from a shaman or warrior is a plausible origin of the Vilca family name. For example, in the Paucartambo region of Cusco Department a recent peasant leader named Saturnino Huilca was also a "high-ranking *misayoq* (shaman) in the clandestine priesthood of the Andean Indians of the Cusco region" (Mayer 2009:254). In its broader onomastic meaning, Vilca refers to the lineage of the grandfather or great-grandfather.

5.7 Conclusions

Vilca use in the Andes is a reminder that when culture history is informed by natural history, new insights can emerge. When inserted as a naturalistic element into historical accounts, vilca the plant provides an explanation for what so many have viewed in mentalist terms. Some of the enigmas of the Andean past—the bravery of the Chanca people, the impact of pre-Inca and Inca oracles, the spontaneous Taqui Onccoy demonstrations, and the ferocity of the Church's idolatry campaign—become comprehensible when the use of vilca is factored into them. While much has been learned since John Rowe wrote his article in the *Handbook of South American Indians*, the culture history of vilca is still shrouded in uncertainties. Discovery of additional colonial documents and new archaeological finds can place vilca's past on firmer empirical ground. My interpretation of vilca's importance postulates that its use reached its zenith in the Wari period and declined in the Inca period with the rise of state religion. In the early colonial period vilca use came under full assault. Failure of most colonial writings to identify vilca is especially puzzling until one realizes that its very mention may have been proscribed. A lack of intellectual curiosity to track down the evil substance to its source seems also to have been a factor. Contemporary scholars who have studied the early colonial chronicles have also failed to get to the heart of the Taqui Oncoy eruptions.

The long-hidden importance of vilca in the Andes is part of the human quest documented in so many other cultures to go beyond ordinary reality. Reconstruction has revealed its importance in Andean culture history in a way unthought-of even 50 years ago. Some of this long silence can be traced to what the French philosopher Henri Bergson (1975:41) referred to when he noted that "the eye sees only what the mind is prepared to comprehend." The progression of knowledge about vilca also validates the remark of the folklorist James Frazer (1953:826) that the "dreams of magic may one day be the waking realities of science."

References

Acosta J (1940) Historia natural y oral de las Indias (1590). Fondo de Cultura Económica, Mexico City

Aguirre Cárdenas M (2008) Ayacucho: Vilcashuaman y Cangallo (Gloria y ocaso de una heroica región andina). Servicios Multiples El Sur EIRL. San Juan de Miraflores, Peru

Bastien JW (1987) Healers of the Andes: Kallawaya herbalists and their medicinal plants. University of Utah Press, Salt Lake City

Basto Girón LJ (1977) Salud y enfermedad en el campesino peruano del siglo XVII. Universidad Nacional Mayor de San. Marcos, Lima

Bauer BS (1998) The sacred landscape of the Inca: the Cusco ceque system. University of Texas Press, Austin

Bauer BS, Kellett LC, Silva MA (2010) The Chanka: archaeological research in Andahuaylas (Apurimac) Peru. Cotsen Institute of Archaeology Press, Los Angeles

Bejar Navarro R (1990) Arquetectura Inka, el templo del sol o Qorikancha. CONYTEC, Cusco

Bergson H (1975) Creative evolution (1911) (trans: Mitchell A). Greenwood Press, Westport

Bermann M (1994) Lukurmata: household archaeology in prehispanic Bolivia. Princeton University Press, Princeton

Bingham H (1922) Inca land: explorations in the highlands of Peru. Houghton Mifflin, Boston

Bingham H (1948) Lost city of the Incas. Atheneum, New York

Brosseder C (2014) The power of huacas: change and resistance in the Andean world of colonial Peru. University of Texas Press, Austin

Browman DL (1978) Toward the development of the Tiahuanaco (Tiwanaku) state. In: Browman DL (ed) Advances in Andean archaeology. Mouton Publishers, The Hague, pp 327–350

Brundage BC (1967) Lords of Cuzco. University of Oklahoma Press, Norman

Burger RL (2011) What kind of hallucinogenic snuff was used at Chavin de Huantar Nawpa Pacha. J Andean Archaeol 31(2):123–141

Burger RL, Salazar LC (eds) (2004) Machu Picchu: unveiling the mystery of the Incas. Yale University Press, New Haven

Bussmann RW, Sharon D (2007) Plants of the four winds: the magic and medicinal flora of Peru. Editorial Graficart, Trujillo

Cabello Valboa M (1951) Miscelánea Antártica. Instituto de Etnologia, Universidad Nacional de San Marcos, Lima

Castro-Klarén S (1993) Dancing and the sacred in the Andes: from the Taqui Oncoy to Raso-Niti. In: Greenblatt S (ed) New world encounters. University of California Press, Berkeley, pp 159–176

Cervantes F (1994) The devil in the new world. Yale University Press, New Haven

Chagnon NA, Le Quesne P, Cook JM (1971) Yanomamo hallucinogens: anthropological, botanical and chemical findings. Curr Anthropol 12(1):72–74

Cieza de Leon P (1986) Cronica del Peru seguna parte (1553). In: Franklin Pease GY (ed) 2nd edn. Fondo Editorial Pontificia Universidad Católica del Perú, Lima

Classen C (1993) Inca cosmology and the human body. University of Utah Press, Salt Lake City

Cobo B (1956) Historia del Nuevo Mundo (1653), 2 vols. Ediciones Atlas, Madrid

Cook OF (1916) Staircase farming of the ancients. Natl Geogr Mag 29(5):474–534

Cornejo Bouroncle J (1959) Tres ensayos: las comunidades indígenas, la idolatría en el Antiguo Perú, socio-economía de la región del Cuzco. Ediciones Inka, Cuzco

Covey RA (2008) The Inca Empire. In: Silverman H, Isbell WH (eds) Handbook of South American Archaeology. Springer, New York, pp 809–830

Curatola M (1990) “Mal de canto” y “mal del maiz” Etnologia de un movimiento milenarista. Antropológica 8:121–144

de Albornoz C (1967) Instrucción para descubrir todas las guacas del Peru y sus camayos y haziendas. In: Duviols P (ed) Journal de la Société des Américanistes 66 (1)

de Arriaga PJ (1968) Extirpación de la idolatría del Piru (1608). In: Barba E (ed) Crónicas peruanas de interés indígena. Ediciones Atlas, Madrid, pp 191–277

de Betanzos J (1968) Suma y narracion de los Incas (1551). In: Barba FE (ed) Crónicas peruanas de interés indígena. Ediciones Atlas, Madrid, pp 1–56

de Calancha A, de Torres B (1972). In: Merio M (ed) Crónicas agustinianas del Perú, 2 vols. CSIC, Madrid

de Gonçález Holguin D (1952) Vocabulario de la lengua general de todo el Peru llamada quichua o del Inca (1608), 3rd edn. Universidad Nacional Mayor de San Marcos, Lima

de Losa Avila y Palomares G (1983) De los áboles; frutos; plantas; aves y de otras cosas medicinales.tiene este reyno (1780). In: Loza-Balsa G (ed) Sociedad Geográfica de La Paz, La Paz

de Matienzo J (1967) Gobierno del Perú (1567). In: Guillermo Lohmann Villena G (ed) Institut français d'Études andines, Paris-Lima

de Molina C (1959) Ritos y fábulas de los Incas. Editorial Futuro, Buenos Aires

de Molina C, de Albornoz C (1989) Fabulas y mitos de los Incas. In: Urbano H, Duviols P (eds) Historia 16, Madrid

de Motolina T (1973) Historia de los indios de la Nueva España (1568). Porrua, Mexico City

de Murua M (1987) Historia general del Perú. In: Ballesteros M (ed) Historia. Madrid, p 16

de Santacruz Pachacuti Yampqui J (1968) Antiguedades deste reyno del Peru. In: Barbara FE (ed) Ediciones Atlas, Madrid, pp 279–319

de Vargas Machuca B (2008) Indian militia and description of the Indies (1599). Duke University Press, Durham

de Yupangui DC (2005) Titu Cusi, a 16th century account of the conquest (trans: Della Legnani N). Harvard University/David Rockefeller Center for Latin American Studies, Cambridge

Deane JK (2011) A history of Medieval Heresy and inquisition. Rowan and Littlefield, Lanham

Fernández-Distal AA (1980) Hallazgos de pipas en complejos precerámicos del borde la a Puna jujeña (Republica Argentina) y el empleo de alucinógenos por parte de las mismas culturas. Estudios Arqueologicos (Antofagasta) 5:55–74

Flury L (1958) El caá-pí y el hataj: dos poderosos ilusiógenos indígenas. América Indígena 18(4):293–298

Franquemont C (1990) Ethnobotany of Chinchero, an Andean Community in Southern Peru. Field Museum of Natural History, Chicago

Frazer JG (1953) The golden bough: a study in magic and religion. Macmillan Co, New York

Gade DW (1999) Nature and culture in the Andes. University of Wisconsin Press, Madison

García JC (1996) ¿Por qué mintieron los indios de Cajatambo? La extirpación de idolatrías en Hacas entre 1656–1665. Revista Andina 14(1):7–52

Garcilaso de la Vega EI (1960) Comentarios reales de los Incas. Ediciones de la Universidad Nacional del Cuzco, Cusco

Gareis I (1990) La "idolatría" andina y sus fuentes históricas: reflexiones en torno a Cultura Andina y represión de Pierre Duviols. Revista de Indias 50:607–626

Glass-Coffin B (2010) Shamanism and San Pedro through time: some notes on the archaeology, history, and continued use of an entheogen in northern Peru. Anthropol Conscious 21(1):58–82

Glowacki M (2002) The huaro archaeological complex. In: Isbell WH, Silverman H (eds) Andean archaelogy: variations in sociopolitical organization. Kluwer Academic/Plenum, New York, pp 267–285

Gose P (2008) Invaders as ancestors: on the intercultural making and unmaking of Spanish colonialism in the Andes. University of Toronto Press, Toronto

Griffiths N (1996) The cross and the serpent: religious repression and resurgence in colonial Peru. University of Oklahoma Press, Norman

Guaman Poma de Ayala F (1980) El primer nueva coronica y buen gobierno. In: Murra JV, Adorno R (eds) 3 vols. Siglo Veintiuno, Mexico City

Heilman J (2002) A movement misconstrued? A response to Gabriela Ramos' interpretation of Taki Onqoy. Colon Lat Am Rev 11(1):123–145

Herrera FL (1941) Sinopsis de la flora del Cuzco. Privately printed, Lima

Isbell BJ (1989) To defend ourselves: ecology and ritual in an Andean village. University of Texas Press, Austin

Jiménez de la Espada M (ed) (1965) Relaciones geográficas de Indias, 3 vols. Ediciones Atlas, Madrid

Knobloch PJ (2000) Wari ritual power at Conchopata: an interpretation of *Anadenanthera colubrina* iconography. Lat Am Antiquity 11(4):387–402

Larco Hoyle R (1946) A cultural sequence for the north coast of Peru. In: Stewart JH (ed) Handbook of South American Indians. Smithsonian Institution, Washington, pp 149–175

Lira, JA (1944) Diccionario kkechuwa español. Universidad Nacional de Tucumán, Tucumán, Argentina

Louria DB (1966) Nightmare drugs. Pocket Books, New York

MacCormack S (1985) The heart has its reasons: predicaments of missionary Christianity in early colonial Peru. Hisp Am Hist Rev 65(3):443–466

MacCormack S (1991) Religion in the Andes. Princeton University Press, Princeton

Marcoy P (1874) Journey across South America, from the Pacific Ocean to the Atlantic Ocean (trans: Rich E), 2 vols. Blackie, London

Marin Moreno F (1961) Panorama fitogeográfico del Peru. Revista Universitaria (Cuzco) 50(120):9–66

Maxwell-Suart PG (2007) The malleus maleficarum. Manchester University Press, New York

Mayer E (2009) Ugly stories of the Peruvian agrarian reform. Duke University Press, Durham

Millones L (1990) Introducción. In: Millones L (ed) El retorno de las huacas: estudio y documentos del siglo XVI. Instituto de Estudios Peruanos, Lima, pp 11–22

Morales MP (2012) Reading inebriation in early colonial Peru. Ashgate, Burlington

Moscoso Castilla M (1963) Secretos medicinales de la flora peruana y guia de la maternidad, 3rd edn. Tipografia Americana, Cuzco

Ott J (1993) Pharmacotheon: entheogenic drugs, their plant sources and history. Natural Products Co, Kennewick

Pané R (1999) An account of the antiquities of the Indians (1498). In: Arrom JJ (ed) (trans: Griswold SC). Duke University Press, Durham

Pearson JL (2002) Shamanism and the ancient mind: a cognitive approach to archaeology. Altamira Press, Walnut Creek

Pérez Bocanegra J (1631) Ritual formulario e instruccóon de curas para administrar a los naturales de este reino. Gernomico de Contreras, Lima

Polia Meconi M (1999) La cosmovisión religiosa andina en los documentos inéditos del archivo romano de la compañia de Jesús, 1581–1752. Fondo Editorial Pontificia Universidad Católica del Perú, Lima

Polo de Ondegardo J (1916) Informaciones acerca de la religion y gobierno de los Incas (ca 1545). Coleccion de libros y documentos referentes a la historia del Perú, vol 3. Sanmarti y Cia, Lima

Ramírez del Aguila P (1978) Noticias políticas de Indias (1639). Imprenta Universitaria, Sucre

Ramos G (1992) Politica eclesiastica y extirpacion de la idolatria: discursos y silencias en torno al Taqui Onqoyu. Revista Andina 10:147–169

Ramos Condori F (1983) K'intuna: occupación aldeana pre-inca en Vilcabamba. In: Arminda M, Oviedo M (eds) Vilcabamba: Arqueología Andina. Instituto Nacional de Cultura Departamental, Cusco, pp 23–28

Ratsch C (2005) Encyclopedia of Psychoactive Plants (trans: Baker JR). Inner Traditions, Rochester

Redden A (2008) Diabolism in colonial Peru, 1560–1750. Pickering & Chatto, London

Reinhard J (1995) House of the sun: the Inka temple of Vilcanota. Lat Am Antiquity 6(4):140–149

Rowe JH (1946) Inca culture at the time of the Spanish conquest. In: Stewart JH (ed) Handbook of South Americans, vol 2. Smithsonian Institution, Washington, pp 183–330

Roy H (2010) En torno del Taqui Oncoy: texto y contexto. Revista Andina 50:9–58

Rusby H (1922) Report of work on the Mulford biological exploration of 1921–1922. J N Y Bot Garden 23(272):101–111

Safford WE (1916) Identity of cohoba, the narcotic snuff of ancient Haiti. J Wash Acad Sci 6:547–562

Saignes T (1989) Borracheras andinas: ¿por que los indios ebrios hablan en espanol? Revista Andina 7(1):83–126

Sauer CO (1966) The early Spanish Main. University of California Press, Berkeley

Schleiffer H (comp: 1973?) Sacred narcotic plants of the new world Indians. Hafner Press, New York

Schultes RE (1967) The botanical origins of South American snuffs In: Efron DH (ed) Ethnopharmacologic search for psychoactive drugs, Publication No. 1645. U.S. Public Health Service, Washington, pp 29–306

Sherbondy JE (1988) Ancestros y árboles en los Andes. In: Mendieta RM (ed) Sociedad andina: pasado y presente. FOMCIENCIAS, Lima, pp 101–136

Spruce R (1908) Notes of a botanist of the Amazon and Andes (1864), 2 vols. Macmillan, London

Stromberg VL (1954) The isolation of bufotenine from *Piptadenia peregrina*. J Am Chem Soc 76:1707

Torres CM, Repke DB (2005) *Anadenanthera*: visionary plant of ancient South America. Haworth Herbal Press, New York

Troll C (1952) Die lokalwinde der tropengebirge auf niederschlage und vegetation. In: Studien zur Klima-und Vegetationskunde der Tropen. (Bonner Geograpische Abhandlungen Heft 9), Verlag F. Dúmmler, Bonn, pp 124–182

Tung TA (2012) Violence, ritual and the Wari empire. University Press of Florida, Gainesville

Vallejo Berrio F (2006) Alucinogenos y practices chamanicas en los ritos asociados al oraculo de Pachacamac. ConTextos I Estudios e investigaciones en arqueologia peruana 1:1–38

Vargas Calderon C (1946) Diez años al servicio de la botánica en la Universidad del Cuzco. Universidad Nacional de Cusco, Cusco

Vargs Ugarte R (1959) Historia de la iglesia en el Perú. Aldecoa, Burgos

Varón Gabai R (1990) El Taki Onqoy: las raíces andnas de un Instituto de Estudios Andinos, Lima, pp 331–406

Vazquez de Espinosa A (1948) Compendio y descripció de las Indias occidentales (1628–1629) Smithsonian miscellaneous collections, vol 108. Smithsonian Institution, Washington

von Humboldt A (1970) Relation historique du voyage aux régions équinoxiales du Nouveau Continent (1819), 3 vols. Brockhaus, Stuttgart

Wachtel N (1971) La vision des vaincus: les Indiens du Pérou devant la conquête espagnole 1530–1570. Gallimard, Paris

Wassen SH (1972) A medicine man's implements and plants in a Tiahuanacoid tomb in highland Bolivia. Etnologiska Studier 32, Goteborg

Weberbauer A (1945) El mundo vegetal de los Andes peruanos. Ministerio de Agricultura, Lima

Zelada DA, Capriles Flores JM (2000) La importancia de las plantas psicotrópicas para la economía de intercambio y relaciones de Interacción en el Altiplano sur Andino. Anales de la XII Reunión Annual de Etnología, 2 vols. MUSEF, La Paz, Bolivia, pp 97–110

Chapter 6
Mysterious Ucumari: The Andean Bear in Nature and Culture

Abstract The only ursid native to South America, the spectacled bear (*Tremarctos ornatus*) still occurs in parts of the Urubamba Valley and its tributaries where human impact has been low. Small size (90–150 kg), distinctive facial markings, and a strong tree-climbing habit set this species, called ucumari in Quechua, apart from bears elsewhere in the world. Never abundant, the bear's chief habitat has been in the upper montane zone from Western Venezuela to Southern Bolivia. Although its present geographic spread approximates the past, its distribution has contracted into non-contiguous "island" zones. Human associations with the spectacled bear give it a minor role in folk medicine and peasant nourishment, but a striking place in Andean folklore. Certain anthropic resemblances between man and beast reinforced a belief that male bears sexually paired with women. Far from being an indigenous notion, evidence indicates that Spaniards transferred this to the Andes from a similar folklore involving the European bear species. Shy and elusive, the ucumari is better known from inferential evidence of it, such as claw marks, than from encounters with the animal itself. To reverse its declining numbers, hunting and habitat must be addressed. When spectacled bears eat crops (especially maize) or attack livestock, peasants target them. Recently, agents have begun buying bear parts that are valued in Asian medicine. The future of this bear, a wild animal of the Andes, also depends upon maintaining the pristine character of the upper montane forest where human impact has been low, but is now increasing.

6.1 The Power of Evanescence

My fascination with the spectacled bear (*Tremarctos ornatus*) dates from 1963 when I glimpsed one as I climbed the hillside near the archaeological site of Choquesuysuy above the Urubamba canyon about four km south of Machu Picchu. The sighting of this endearing animal with a black and white face marked the beginning of my five decade-long pursuit of knowledge about it. Never again, except in zoos, would I see this elusive creature. The spectacled bear became for me the emblematic denizen of the Andean forest for which the progressive removal of the latter decisively impacts the former. Two dimensions collide here: the scientific and the mythopoetic aspects

D.W. Gade, *Spell of the Urubamba*, DOI 10.1007/978-3-319-20849-7_6

of this animal. Academic disciplines rarely allow themselves this holistic approach, that of joining the rational with the non-rational.

Hiram Bingham (1930:27) spotted a bear at Machu Picchu in 1912 as it was slowly making its way along the mountain side, feeding on clumps of "wild pineapple." The occasion was the first time anyone had photographed this species in its native habitat even though, at the time, *Tremarctos* was not uncommon around Machu Picchu. As tourism expanded especially in the 1950s and 1960s, the bears retreated. At that time, inaccessible forested tracts between ca 2500 and 1000 m above sea level (asl) in the Urubamba Valley still provided cover and food for a small bear population. Antonio Raimondi (1898:189) had seen one in 1865 near the Yanatile River. In most places these animals were so rare that even few Quechua-speaking folk actually saw them. In 1964, a farmer in the lower valley reported glimpsing a bear near his property on a high hillside. Several others related second-hand accounts of friends or neighbors having encountered one. Bear sightings were most likely in the less populated lateral valleys of the Urubamba. The Lares Valley had a concentration of bears although they had disappeared from the main valley nearby (Chávez Chaparro 1950:287). In the lower Paucartambo (also known as the Mapacho or Yavero) Valley, a landowner from around Lacco (1200–1800 m asl) mentioned to me that bears were a yearly pest in his maize fields. The scarcity of bear sightings below 1000 m asl on the Urubamba suggests that the selva lies outside the bear's ecological habitat. Perhaps at the lower elevations too many other animals compete for the same food, or perhaps the species is simply better adapted to the edible plants which are found at higher elevations.

In 1982, the mammalogist Bernard Peyton conducting a formal study of the bear focused on the Historical Sanctuary of Machu Picchu and the three adjacent valleys of Santa Teresa, Lucumayo and Phiri north of Ollantaytambo (Peyton 1987). He collected information from 87 sites in which the bear's presence was inferred. Peyton found bear signs such as black hair and scratch marks on trees in six of the nine habitat types between 2010 and 4170 m asl, which indicated its presence in the *ceja de la montaña* as well as in the grassy high puna. In the forests below 2700 m bears found their paired requirements of food and cover. Evidence of feeding was most often found on the bromeliad *Puya* spp., the bulblets of orchids, the fruits of laurel (*Nectandra* sp.) and the succulent internodal tissue of bamboo (*Arthrustylidium* spp.). Bears also fed on maize, moving up in elevation as the crop matured, preferring those fields closest to forests. Peyton talked to farmers who used Parathion to poison bears that entered their *maizales*, although during Peyton's 7 month stay, he saw no bears. Suzanne Paisley, whose Ph.D. fieldwork in the 1990s was in the Apolobamba region of Bolivia, saw a bear only once and even that was momentary.

From all evidence, T*remarctos* had disappeared even before the Conquest above Ollantaytambo at 2700 m asl in the Urubamba Valley. The native forest cover had already been removed in this zone, and although bears can feed outside a forest, they breed in a treed environment. No one I talked with in the upper valley had ever

Fig. 6.1 Spectacled bear (*Tremarctos ornatus*) at forest edge in Ecuador. (By James Clare, used with permission)

seen one. Here, as in much of the Andes, the bear, known by the Quechua name of *ucumari*, has a quasi-mythical status.[1] It is better known in regional folklore than as an element of the regional fauna.

6.2 Description

The only ursid in South America, *Tremarctos* is also the world's smallest living bear. Since 1825 when Cuvier scientifically identified this species, certain basic facts about it have become established. The full-grown male averages only 150 kg and the female at ca. 90 kg is substantially smaller. Count Buffon (1833) interpreted the small size of this bear as evidence of how "degenerate" New World mammals were compared to members of the same family in Europe. Whitish markings on the head, especially on the face, contrast with the thick black fur covering most of its body (Fig. 6.1). Most striking is a circular pattern of white fur around each eye, the feature that gave rise to its vernacular name in several European languages: *oso de anteojos* in Spanish, *ours à lunettes* in French, *Brillenbär* in German, *brilbeer* in Dutch, and spectacled bear in English. However, the facial markings suggesting the eyeglass analogy are not invariable with the species.

[1] Various names for the spectacled bear in the Andes exist. In Aymara it is *jucumari, an* obvious variant of the Quechua term found in Peru. Among highland people in Ecuador, the Quichua name seems to have been lost in favor of *usu*, borrowed from the Spanish *oso*. In Southern Peru, the word ukuku also refers to a bear, though more commonly a person disguised as a bear. In Colombia and Venezuela, native names are *manoba, manaba, mashirano* or *masira*. The two most common Spanish names in these countries are *oso de anteojos* and *oso frontino*; others are *oso enmascardo, oso careto* and puma (a word that elsewhere applies to the mountain lion).

The bear's present distribution grew out of the palaeogeographic drama of the Western Hemisphere when the rise of the Panama Isthmus during the Pleistocene created an overland connection from the north. That new land passageway enabled four genera in the subfamily Tremarctinae to migrate to South America from North America. The genus *Tremarctos* had evolved in North America during the Upper Pliocene ca two million years ago. Less certain is whether *Tremarctos ornatus* evolved in South America from *T. floridans*. Fossil evidence of *T. floridans* indicates a much larger bear than *T. ornatus* (Kurtén 1966:91). Until more fossils of T. *ornatus* are found, its evolutionary lineage remains conjectural. One recently discovered skeleton of *T. ornatus*, found in a cave at 2950 m asl in the Department of Amazonas, Peru, had anatomical features that closely resembled the living species (Stucchi et al. 2009).

6.3 Historical Knowledge of Tremarctos

Chroniclers of the early colonial period made only brief mention of this New World bear, suggesting its marginal place in an environment already heavily transformed through human action. Polo de Ondegardo (1916:190) mentioned bears that people worshipped. José de Acosta (1940: 200), who confused the Quechua name of the jaguar ("otoronco") with that of the spectacled bear, further jumbled the issue by asserting that the bear of the Viceroyalty of Peru was of the same species as the European and that they both ate ants.

This mammalian confusion was part of a 250-year misconception about the relationship between two very different animals: one a member of the bear family (Ursidae) and the other, the giant anteater or, in Spanish, *oso hormiguero* (*Myrmecophagia jubata*). The giant anteater is not only of a different family (Myrmecophagidae) but of a different order (Pilosa) as well. Acosta regarded the *oso* of the Indies to be uncommon, an assertion also mentioned by Garcilaso de la Vega (1960:385). The latter reasoned correctly that bear numbers greatly diminished when farm land replaced highland forest. Augustin de Zarate (1947:46) and López de Velasco (1971:222, 224, 240) in the same century, referred to the presence of "*osos negros*," a novelty to Europeans who knew only brown bears. The *Relaciones geográficas* of the sixteenth century alluded to bears occasionally in reference to Guamanga (Ayacucho), Loja (Ecuador) and Quito (Jimenez de la Espada 1965:I:192; II:213, 296). Europeans who wrote of them may not have ever seen one. Vespuccio (1951:304), writing in 1563, mentioned bears in the Paria Peninsula of Venezuela, though it is doubtful that he actually saw them there. Polo de Ondegardo (1916:195) placed bears in the forested lands together with "lions, tigers, and snakes." Vargas Machuca (2008:195), writing at the end of the sixteenth century, asserted that "there are very large bears living in hot and cold lands, they are black and are not harmful." In that sentence he had conflated the giant anteater with the bear, just as other colonial observers, including Polo de Ondegardo above, did.

Gonzalo Fernández de Oviedo's (1959:II:35) remarks about "osos" suggest that same confusion of names as well as of information about them. He mentioned "bears like those of Spain" in Nueva Granada (Colombia) and said that they were very bold toward dogs and hunters and that it was necessary to kill them. But the Andean bear is less than half the size of the European brown bear, of a different color, and could not in the least be described as bold. His subsequent remarks, that the *oso hormiguero*, that is, the giant anteater, was found in the highlands around Bogotá could only have referred to *Tremarctos*. Gonzalo Fernández de Oviedo (1959:II:47) then wrote that the "other" kind of *oso hormiguero* was black and had a tail. Vázquez de Espinosa (1948:311) distinguished between "osos negros como los de España" and "osos hormigueros que sacan la lengua larga," and stated that both were in Colombia. Although anteaters were restricted to the hot country east of the Andes, they entered the ken of travelers much more than *Tremarctos*.

In the early seventeenth century, Padre Cobo (1956), who reported on natural history of the Andes and of Mexico where he spent years, had much more to say about the North American bear species than about the Andean bear. Also writing in the seventeenth century, Ordóñez de Ceballos's (1614:414) reference to "osos muy grandes" in Peru suggests that he knew about them only in the abstract. Likewise de Lizárraga (1987:201) mentioned "very large bears" around Pocona in Charcas, in the hot country beyond what is now Cochabamba, Bolivia. Ramírez del Aguila (1978:47), another priest in colonial Charcas, made passing reference to the bear as an animal of the hotter, lower zone, which was then—and still is—largely forested. Given the nomenclatural confusion of the era, it is not certain if any of these authors were referring to the spectacled bear or to the giant anteater. Likewise in the late seventeenth century, the Quiteño priest Juan de Velasco (1989:181) in Ecuador wrote that the *ucumari* was a "very bold" bear that lived only in cold climates; the other kind, the *iznachi*, taller than the European bear, was all black and lived in hot climates. In fact, the *Tremarctos ornatus* is not bold nor does it live only in cold climates. Like de Oviedo, de Velasco seems to have been confused, for his description plausibly referred to the giant anteater. These two animals were also described in the late eighteenth century for the Piura region of Northern Peru (Lequanda 1793:178–179). In the same century, Llano Zapata's (2005:517) description of the "oso" was of the giant anteater, not the spectacled bear. Two nineteenth-century Europeans with scientific training distinguished between the two kinds of "bears" in their travels around Peru. Johann von Tschudi (1849) did not describe them, whereas Antonio Raimondi (1862) wrote that one kind of bear fed on meat and the other ate plant matter especially fruit. *Tremarctos* will eat meat, though its intake is primarily plant matter; the giant anteater is quite specialized on insects. Much information disseminated in the nineteenth century about bears in the Andes was not based on first-hand observation.

Scientific understanding of this animal began to emerge when living specimens were first taken to Europe beginning in 1825 (Festa 1904–1905:104) Cuvier (1831) in France named it *Ursus ornatus* and named Chile as its source. Since no other record of this species is known from Chile, the live specimen brought to Europe most plausibly came from northern Peru. Inland from Trujillo, a bear population had adapted to life in a dry forest. This desert population of *Tremarctos ornatus* exists up to the present; how or if it morphologically differs from bears elsewhere in the Andes is still not clear. Zoologists once defined races for the species based on geographical distributions, but these distinctions have since been discarded. Individual variations occur, but they do not follow spatial groupings. Bears in the northern Andes are often larger than those of Peru and Bolivia, but enough exceptions exist to foreclose the establishment of subspecies. Likewise more variation in body markings occurs among bears within one region than between bears of different regions.

6.4 Dietary Habits

The bear family is classified as omnivorous. More than other ursids, the diet of the spectacled bear is heavily weighted toward plants. Its dentition and jaw musculature are designed to grind vegetal matter and, unlike other species in the order Carnivora, its premolars are quite small and weak (Sicher 1944). Several species of the Bromeliaceae, especially *Puya* spp., provide an important part of its intake. Beyond the carbohydrate content of *Puya*, the stems harbor protein-rich insect larvae. Schulenberg and Awbrey (1997:32) have suggested a close connection between bears and these bromeliads in Bolivia; likewise Suarez (1988) in Ecuador confirmed a strong dietary emphasis on *Puya* on the subalpine páramo (3550–4500 m asl) on the Antisana volcano in Ecuador. Abundance of *Puya* in the treeless páramo and puna of western South America accounts for the forays many bears make into exposed landscapes. Within the *ceja* forest, epiphytic bromeliads of the genus *Tillandsia* are common bear food. In Venezuela, Goldstein (2004) documented bears feeding on *Tillandsia fendleri* by climbing into trees on which it grows. Bamboo in the genus *Chusquea* is also consumed. Unlike the giant panda of China, the digestive tract of the spectacled bear is not so specialized that the species feeds only on a single plant. In highland Colombia and Ecuador, bears eat the fruits and petioles of the wax palm (*Ceratoxylon*). That and various other fruit-bearing trees offer a seasonal feast for bears. To feed in peace, they construct platforms made of branches.

In semi-arid environments, other plants provide nutrition. Early in the twentieth century, Osgood (1914:171–172) found bears in the desert of northern Peru feeding on chapote, a shrub bearing a pear-shaped, heavily-seeded fruit. There and elsewhere, *Tremarctos* also eats the pads and fruit of *Opuntia* and other cacti (Morrison 1972:53). Cacti, which abound on the dry floor of the upper Apurimac canyon between 3000 and 2000 m asl, support an indeterminate bear population. Fruit-bearing trees of any kind are likely to attract this bear. In Chuquisaca, Bolivia, the bear eats the fruit of *janchicoco* (*Parajubaea torallyi*), a palm growing in forests up to 3000 m asl and yielding a greenish fruit which humans also consume (Ribera and

Lieberman 2006). Bear presence in semi-open areas suggests that a dense forest is not a requirement for its survival. Protection from human persecution may be as important as food availability.

6.5 Distribution

All five Andean countries fully within the tropical latitudes—Venezuela, Colombia, Ecuador, Peru and Bolivia—have spectacled bear populations in a discontinuous distribution pattern (Fig. 6.2). Within each of these countries, *Tremarctos* has contracted into blocks, about half of which correspond to protected areas. The bears now occupy areas that cover less than half the extent of their former territory (Kattan et al. 2004). Presence of the bear on the northern and southern fringes of its range is not well known. Helms (1998) and Hershkovitz (1957) provided evidence that the bear lived in Panama within historic times. In northwestern Argentina, evidence of the bear is circumstantial. Historically the bear may have lived in Argentina, although many natural history observers, including Argentine mammologist Cabrera, assumed its absence (Cabrera and Yepes 1960:I:148; Del Moral 2005; Cipolletti 1983; Rumiz et al. 2012; Brown and Rumiz 1989).

Clusters of bear populations explained by an appropriate land cover, availability of suitable food resources, and minimal human presence in its breeding area, follow fairly specific regional patterns. About 80 % of the ursid population in the Andes dwells in the upper and middle parts of the cloud forests (*ceja de la montaña*) of the eastern slope. The human population there has been low and the vegetation provides maximum cover. In Venezuela, bears breed in forest cover between 2900 and 3200 m asl in the mountainous states of Mérida, Táchira, Trujillo and Lara. Above that zone, bears roam into the treeless landscape of the páramo to feed. Rarely have they been found below 1000 m asl. In Colombia most of the remaining bear population occurs on the western slope of the Cordillera Occidental and the eastern slope of the Cordillera Oriental. In both regions, abundant rainfall favors dense forest vegetation of which sizeable tracts remain. Elsewhere, bears survive on the eastern side of the Cordillera Central, the Sierra de Perijá and the Sierra Nevada de Santa Marta. Small populations can also be found in the southern Colombian department of Pasto where the topography jumbles to form a massif. In Ecuador, *Tremarctos* occurs mainly on the eastern slope of the eastern range from above 900 m to the high páramo above 4000 m asl. The bear population in surviving forest tracts on the western slope of the western range is believed to be considerably smaller than in Colombia.

Southward in the Peruvian Andes, the humid elfin forest of the eastern slope overlooking the Amazon Basin holds the largest number of bears. In the west, on the arid foothills from 300 to 1500 m asl back from the north coast, an anomalous bear population was first described in the early twentieth century (Osgood 1914:171–172).

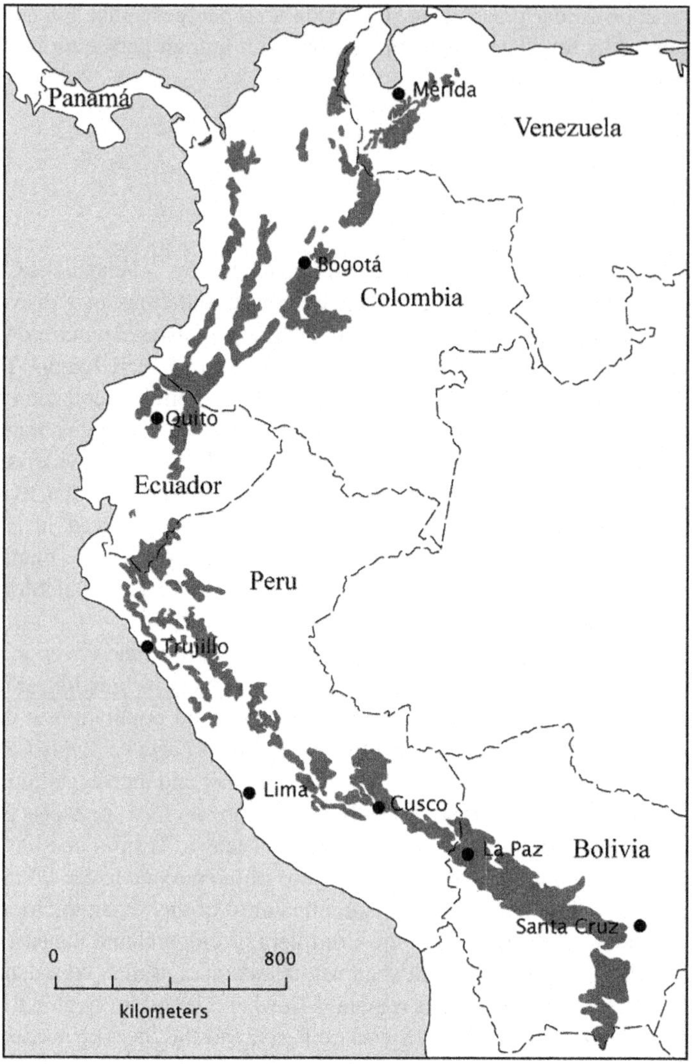

Fig. 6.2 Distribution of the spectacled bear in the Andes. Although its spread covers more than 1500 km from north to south, its disjunction into blocks of forest hinder its movement for breeding purposes. (Modified from Stolzenburg 1997)

Since then, they have also been observed living on xerophytic vegetation in the sparsely populated intermontane valleys of the highlands. In Bolivia, bear populations have occurred in the well-watered band of forest along the eastern Andean front between 500 and 3200 m asl from the Yungas of Northern La Paz Department to Cochabamba, western Santa Cruz, Chuquisaca, and Tarija Departments (Salazar et al. 1990; Schulenberg and Awbrey 1997). In 2001 and 2002, circumstantial evidence of the bear indicated their presence in the southern half of Tarija Department close to the Argentine border (Varga and Azurduy 2006).

6.6 Population Numbers

As a creature that flees at the sight or smell of people, determining population size in the Andes is one of the challenges of bear research. Alexander von Humboldt spent several months in the Northern Andean Highlands, but did not mention this creature. Perhaps he did not see one, for if this exceptionally curious scientist had caught sight of an animal new to him, he would have certainly recorded it in his travel account. Jean Dorst (1967:187), a French biologist, recounted his experience of searching long days for the ucumari without success in the eastern foothills of Peru between 2530 and 3050 m asl. Were there no bears or was Dorst simply not sufficiently patient? A team seeking to observe the bear in a remote ceja zone of Amazonas Department of Northern Peru at an elevation of 2300 m asl, spotted the bear only once in a 2-month period (Butchart et al. 1995). How much of this pattern is attributable to bears avoiding humans or, alternatively, to the Andean bear being less inquisitive than its black bear cousin of North America. *Ursus americanus*, not nearly as shy as the spectacled bear, approaches and manipulates objects. That practice has not been observed in *Tremarctos*.

The North American mammalogist, Bernard Peyton (1980), for several decades the leading *Tremarctos* specialist, accumulated a corpus of information about this animal over a 2 year period in South America. Yet Peyton described only eight sightings, some of them for not more than a few minutes. Understanding this bear's habits thus required observing bear signs and making extrapolations. A recent study in Bolivia reported no direct observations of the bear. Instead it tallied the frequency of different kinds of bear signs—hairs, scratch marks, scats, tree nests, beds, paths, and feeding sites—in different habitats to conclude that the upper ceja zone had the highest bear density (Rios-Uzeda et al. 2006). In another study on the eastern cordilleran slope in Bolivia near Pelechuco, the mammalogists Paisley and Garshelis (2005) captured two males and monitored their movements through radio telemetry. Researchers in Ecuador have also used radio telemetry to track bears and were able to film bear behavior in corn fields. They used that information to generate a data base for a compensation program (Castellanos 2004:27).

Shyness in showing itself to humans accounts for apparent miscalculations of bear numbers. In the late 1990s, 60 observers conducted a bear survey over a 4-year period for the whole of the Andes that led to major readjustments of the estimated population. Rather than the total of 2000–2400 individuals that for years conventional wisdom had estimated, the survey established a threshold population of 18,000 with the possibility that the actual number of surviving bears could be as high as 65,000 (Peyton 1999:160). Such a wide divergence in population estimates reflects the difficulty in counting such an elusive creature. The zoologist Grimwood, considering only Peru for the 1960s, described the bear distribution as that of small, widely separated pockets. By using a multiplier effect, Grimwood (1968) estimated the total population in Peru of between 800 and 2000 individuals, whereas Peyton (1999) some three decades later arrived at an estimate of 5700 individuals. Beside human impact or lack thereof, local densities are tied to food supply. A zone from Loja in southern Ecuador to Huancabamba in northern Peru consistently reports

Fig. 6.3 Spectacled bear in a tree in the high montane forest in Ecuador. Of the world's seven living bear species, this is the most arboreal in its habit. (Photo by James Clare, used with permission)

high bear numbers. Woodlands of wild fruit-bearing trees in the Annonaceae provide a rich food supply for bears.[2] Likewise hearsay of country people in the Zudáñez area of Chuquisaca Department (Bolivia) suggests that concentrations of bears are related to the presence of a native fruit-bearing palm. The arboreal habit of this bear facilitates its food-getting and food storage (Fig. 6.3).

---

[2]That the spectacled bear may have been an agent in the domestication of the cherimoya (*Annona cherimolia*) is suggested by an analogy with the Eurasian brown bear (*Ursus arctos arctos*) which may have played a part in the domestication of the apple (*Malus pumila*) in Eurasia. Both bear species find ripe fruit delicious and cull those that are sweeter and larger, but their jaws are inefficient in crushing seeds, permitting the seeds to pass through digestive tracts and remain viable in the fecal mass. The result is that the individual trees which sprout have a large fruit rich in sugar. Humans value these fruit characteristics and propagated them.

6.7 Human-Bear Associations

An ethnozoology of this South American bear includes interpretive evidence from Andean culture history. Bear representations suggest a totemic affinity that characterizes bears elsewhere in the world (Hallowell 1926). A carved stone ("El Lanzon") in Chavin culture (ca 1000 BC) suggests a bear (Paisley and Saunders 2010). A zoomorphic pot from the Mochica culture of the north coast of Peru could not be confused with another animal (Lavallée 1970). In Colombia, a carved figure from the San Agustin culture plausibly represents a spectacled bear (Helms 1998). Although the stocky body and eyes outlined are suggestive of this bear, its dentition is feline (Fig. 6.4).

6.7.1 Human Use of Bear Parts

Using bear parts for human nutrition and folk medicine may have pre-Columbian roots. At least in Peru, meat and also the pelt of the bear historically was perceived to have value in curing syphilis, a New World disease (Lequanda 1793:179). More generally, bear flesh, fat and blood were thought to be fortifiers (Festa 1904–1905; Mondolfi 1989:539; Butchart et al. 1995; Faust 1996:178). That may explain why, after the umbilical cord was cut, bear fat was rubbed on a new-born baby. The fat is

Fig. 6.4 The upright posture and eyes outlined in circles point to the Andean bear as the inspiration for this stone carving in San Agustin, Colombia. (From Helms 1998, used with permission)

also applied to heal bone fractures and cure rheumatism (Salazar et al. 1990:9). An advantage of bear fat in folk medicine is that it is does not harden. Bear fat has been used in cooking, presumably a post-Columbian practice, since Spaniards introduced the concept of frying food. Bear parts had use in sympathetic magic in the same way as a rabbit's foot. Two Peruvian researchers, Figueroa and Stucchi (2005), found part of the front leg of a bear without the thumb (sold separately) for sale in the Pisac market. The paw is most valuable as an amulet. An Andean tradition assimilates the "manito misteriosa" into the power to capture the bear's perceived qualities of strength and courage (Bastien 1982:353).

Human encounters with bears in the Andes most often occur when these animals trespass into agricultural fields. Their desire to avoid people is facilitated by the bear's ability to scent a human a kilometer upwind and to hear human voices at 300 m. When hunger exceeds fear, bears enter maize fields aided by their uncanny ability to sense when the maturing ears are at their peak of succulence. Peyton (1981) reported that the animal has been known to climb a tree above a corn field to scout out where best to enter and when to feed. Farmers also have noted that bears enter *maizales* during thunderstorms when they can feed on maize ears without fear of harassment or retribution. A sow in a corn field with cubs teaches her offspring the fine art of maize stealing. At lower elevations, bears are known to enter sugar cane fields to consume the sweet stalks. Unlike maize fields, a cane field is rarely guarded.

Depredations by wild animals prompt Andean peasants to protect their standing crops. Maize receives special vigilance, for one plant's entire food value is concentrated in just one or two conspicuous and easily removed ears. Not only do bears devastate corn fields; so do foxes. Normally a carnivore, in the Andes foxes find corn kernels in the milk stage irresistible. In the Inca and colonial periods, communities designated a guardian (*arariwa*) to protect crops in fields from thieves, both animal and human (Gade 1970; Guaman Poma de Ayala 1980) When that communal practice disappeared, individual farmers became responsible for keeping crop robbers away. When standing maize reached the green corn (*choclo*) stage, the family mobilized protection. Typically, a boy, armed with a slingshot and accompanied by a dog, slept in a lean-to. Only in more remote zones or those close to a ceja forest were bears a threat. Bears attacking livestock has not been common. Reports of bears eating meat usually involve domestic animals gotten as carrion or by predation. Carnivorous behavior by this species of bears appears to be learned rather than instinctual. In some places but not in others, calves, goats, heifers and sheep left unguarded are vulnerable to bear attacks. On the Venezuelan páramo, the mammalogist Goldstein (1991:233) found active bear predation, always by males, on small herds of young cattle. Remains of bovine vertebrae found in tree nests indicated that a bear had dragged a dead calf from the open páramo into a cloud forest, and stashed it where it could be eaten without disturbance. Predation on large livestock required a different strategy. In one case Peyton (1981:233) conjectured that a bear had pursued a full-grown cow near cliffs, forced it off and then descended to retrieve the carcass.

Bear attacks in the Andes on humans have been rare. A warning dating from the late colonial period to be careful when walking through the forest lest one become a "victim of the fury" of a bear reflects a Spaniard's misinterpretation (Descripción 1793:179). The ucumari of the Andes was in no way comparable to the much larger and more aggressive European species that was still quite common in mountainous areas of Spain in the eighteenth century. Johann von Tschudi's (1849:293) bizarre assertion that this animal was known to assail solitary travelers seems plausible only if a sow was protecting her cubs. The zoologist Chávez Chaparro (1950:256) asserted that ucumari, especially if wounded, will sometimes attack people. Aggressive encounters, though exceedingly rare, are within the realm of possibility.

6.7.2 Libidinal Dimension of Ursine Folklore

An anthropomorphic analogy has given *Tremarctos* a place in Andean folklore well beyond what one might expect from its rarity and elusiveness. Humans can feel kinship with an animal that in certain ways resembles them. Bears resemble humans in that they can stand on their posterior feet. With back support, they can also sit upright. They use forefeet like hands. Small ears, the near lack of a tail, and nipples on the chest rather than on the stomach are other anatomical features that this and other ursids share with humans. Since their lips are, like humans, not attached to their gums, bears, like humans, can change facial expressions. Both too are plantigrade, which highlights the extent to which the five-digit footprint of a bear resembles that of a barefooted man. Omnivory characterizes the food intake of both humans and bears. Fish, honey and berries are among the foods keenly sought by both bears and humans. Bear scats have some resemblance to human turds. Several other anthropomorphic features of the spectacled bear cross the human-animal divide. Its gestation period lasts eight and one half months and the cubs require a long period of maternal care before they become independent.

Similarities and analogies, real or perceived, have encouraged the invention of a libidinal dimension to the human-bear relationship. Records of that connection go back to the Middle Ages in Europe (Pastoureau 2007). Multiple analogies have unconsciously fostered apocryphal notions of ursine lubricity, adeptness at ventral-to-ventral sexual intercourse. That humans and Andean bears could be cross-species sexual partners was accepted as a truth in the colonial period. Cieza de Leon (1986:265) repeated stories of a man in the jungle mating with a female monkey and an Indian woman having intercourse with a dog. The product of such a mating was a monster, the divine punishment for such an evil pairing. Andean tales of interspecific coupling vary in their details. Most commonly, a male bear abducts a shepherd girl. The Spanish chronicler Cabello Valboa (1951), writing ca. 1580 about an area corresponding to central Ecuador, provides the earliest known account. In it the bear kidnaps the girl, takes her to a cave, and feeds her raw meat. The sexual encounters that follow result in offspring who display a combination of human and animal attributes. Cabello Valboa presented this as a factual account with roots in the pre-Columbian Andes. Over the centuries, a standard story line emerged involving a shepherdess and a bear. In one account from the highland village of Salas near Piura in northern Peru, a bear took a shepherd girl to a cave. For 4 months, the bear kept her there, having sex and bringing food. A priest found her and brought her back to the village where she died 3 days later (Lequanda 1793:179).

Unlike this narrative, bear stories usually featured a birth event. The late sixteenth-century Spaniard Martín de Murua (1987:456) recounted the birth in an Indian village near La Plata (Sucre, Bolivia) of a baby girl born with hair covering her forehead. He wrote that, to his disbelief, the mother had said that she had seen a bear above her bed. Rather, he postulated, she had copulated with a bear in a canyon on the way to her chacra. As evidence, de Murua noted the baby's features as a cross between a human and an animal. Folklorist Morote-Best (1957) and ethnographer Catherine Allen (1983) each collected in Southern Peru accounts in which the offspring of the bear-human coupling was endowed with superhuman strength. Innumerable twists on the theme are known and only some of them are published (Lara 1973:370–374; Payne 1984: 50–60; Taylor 1997; de Uña 1992; Weber 1987; Vokral and Masson 1993). Allen (2011:74–77) made the point that rural Andean people like to tell bear-human stories because they are metaphors for the kind of domestic tyranny that wives often receive from their husbands. Gender reversal in this bestial folklore is much rarer. Folklore of human males mating with female bears centers on priests who are depicted as manifesting unseemly concupiscence and so are abducted by a she-bear and taken to a cave (Robin 1996). Documented cases of sexual assaults by human males on female bears indicate that this variant of bestiality is also part of the historical record (Beirne 2002).

Cabello Valboa reported the "oso legend" as part of a larger Spanish notion that indigenous people practiced bestiality. Ironically, it was Spaniards who transferred the anthropomorphized and sexualized stories about the bear in Spain to the Andes and applied them to *Tremarctos ornatus*. The Old World to New World transfer in the Andes has its parallel in Mexico where stories about the bear-human transgression involved the black bear (*Ursus americanus*). Two European scholars, Vokral and Masson (1993:168–169), are emphatic in their belief that the son-of-the-bear tales in the Andes and Mexico are simply New World echoes of an Old World storytelling tradition. Decades earlier, José Maria Arguedas (1960/1961) made similar assertions about its transfer from the Old to New Worlds. Bears as Andean symbols of male chauvinism and sexual desire are likewise Old World in origin. The tales of the carnal pairing of bear with human extend well beyond Iberia. Much of Eurasia also has this folkloric element. In Siberia, a Yakut tale tells of a male bear that fell in love with a beautiful young woman whom it abducted and carried deep into the woods. She bore a son who as an adult became a great warrior (Corso 1955; Bychkova Jordan and Jordan-Bychkov 2001). In most of Europe, the Catholic Church combated these symbolisms through long campaigns of depreciating the animal ("*ursus est diabolus*") (Pastoureau 2007; Brunner 2007).

6.7.3 The Bear in Ritual

The bear theme in Andean dance festivities may also be the adaptation of an Old World practice. A Spanish tradition of feigning to be an animal was part of certain Carnival celebrations (Foster 1960:175–176). In Ecuador, several fiestas included dancers and other participants disguised as bears (de Carvalho-Neto 1964:324). In the northern Peruvian Andes, Taylor (1997) described a dance at a local festival that similarly featured humans in bear costumes. Still today in Southern Peru, dance troops (*comparsas*) of eight to ten adolescents in bear disguise participate in many fiestas. Each member wears a wool mask in which white circles are woven around the eyes suggesting the face of the spectacled bear. In that masquerade, they beat drums, blow whistles, or play flutes (*quenas*). An unruly aspect takes over when they start to crack whips, tell obscene jokes, and chase girls. Orgiastic indulgence might follow (Urton 1985:272). Clearly assimilated in this transgressive behavior is the legend of Juan de Oso whose libido targeted winsome country girls. That connection explains an extended meaning of 'ukuku' in southern Peru to refer to a particularly lubricious man.

At the indigenous ritual festival in the Department of Cuzco called Qoyllorriti ("Lord of the snowy star"), held every year at the Ausangate glacier, the behavior of ukuku dancers has a theriomorphic symbolism. The festival's origins may go back to a miracle reported to have occurred in the late eighteenth century. The event draws thousands of indigenous people from a wide region, two troops of ukukus (men dressed as bears) traditionally perform, one from Paucartambo and the other from Quispichanchi. That duality followed the Andean principle of moiety. Aside from the dance routines, several hundred people climb by moonlight to the glacier where they break off blocks of ice and carry them down the mountain to offer them to the Christ image in the chapel below. Water from the melted ice has magical healing properties. Traversing the glacier can be dangerous; ukukus have fallen into crevasses and disappeared. The symbolic meaning of the night-time climb is open to different interpretations. One is that ukukus on the ice protect fellow pilgrims from the "condenados," dead persons from hell who rise from their graves and start walking. Alternatively, the bear dancers are said to transform themselves into the malevolent spirits of the restless dead roaming the glaciers (Sallnow 1987:218). The belief is that condenados gain entrance to heaven only when they offer the apu, the spirit of the mountain, blocks of ice. In either case, condenados can be viewed as mediators between the upper world of the gods and the lower world of humans.

Among the broader interpretations of the ukuku is that he is a powerful yet awkward creature who faces the same contradictions faced by Indian men in a society dominated by Spanish speakers (Allen 1983). Another view is that the ukuku is a liminal figure between order and disorder: an animal found neither in the hot jungle nor in the cold highlands, a creature caught between father bear representing nature and human mother representing culture (Molinié 2005). Another idea about the ukuku is that the figure is a shamanic survival from a pre-agricultural Andean past: a human turned into a beast to guarantee hunting success (Larsen 1990:159). Thus, the bear-human relationship has involved a cognitive fluidity that combines humans as animals and animals as people. The spectacled bear has been a charismatic animal, precisely because it is a creature manifesting archetypal powers. That status hinders its survival as a species as much as it helps.

6.8 Conservation of the Andean Bear

Maintenance of the *ceja de la montaña* is the key for ucumari survival in the Andes. The higher parts of this zone of small trees rich in epiphytes long survived the deforestation that other forest types underwent. Its wood has little commercial value, but even more importantly, its environment of high rainfall, coolish temperatures, and steep slopes do not favor agricultural use. On the other hand, the lower ceja between 1000 and 2000 m asl is the prime habitat for alkaloid-rich coca. Profits from the illegal drug trade have led to its greatly expanded cultivation in several Andean countries since 1980. More than 500 km^2 of mostly ceja forest in Colombia, Peru and Bolivia—the countries that produce most of South America's coca crop—are removed every year mainly for that purpose. Human/bear encounters occur when farmers move into ceja forests to clear land for agriculture (Jorgenson and Sandoval 2005). As the forests in the lower ceja mountain zone are cut, bear populations confine themselves to the remaining wooded areas. Fear of annihilation if it crosses areas settled by humans keeps bears from migrating. In Colombia, bear populations in the three north–south parallel mountain ranges are cut off from each other and are without valley corridors through which they could safely move.

Male bears have a home range of more than a 100 km^2 so that many habitats are now not large enough to satisfy their roaming instincts. Fragmentation of forests into islands causes in-breeding that reduces the genetic diversity of the population and decreases its ability to adapt to environmental change. In his DNA sampling of 82 bears in three northern Andean countries, Ruiz-Garcia (2003:90) concluded that the spectacled bear population is more genetically impoverished in Ecuador than in Venezuela or Colombia. Since the species deployed from north to south, it is reasonable to assume that the bear is least diverse in Bolivia. Yerena (1994:565) estimated that a pool of at least 500 individuals is needed to maintain healthy genetic variability of the Andean bear population. Herein lies the logic of having corridors along which bears could move to find mates from a larger gene pool. Connecting separated wild lands facilitates movement and enhances the possibility of gene flow. Radio-collar tracking indicates that bears change their altitudinal zone and habitat use through the year (Rechberger et al. 2001).

Hunting is another major threat to the bear's survival. After 1973, when the IUCN Red List of threatened species classified *Tremarctos* as vulnerable to extinction, all five Andean countries banned hunting it, yet those who know the Andes recognize that declaring the bear to be legally protected does not ensure its survival. If peasants believe that a bear is stalking their livestock or raiding their fields, they react. Researchers believe that ranchers and farmers exaggerate ursid predation on livestock as a pretext to kill bears (Goldstein et al. 2006). More seriously, protected areas are not immune to bear poaching. Most parks have no guards, but even in those that do, corrupt officials have allowed illegal activities in return for payment. Poachers apprehended with dead bears have often received verbal warnings from judges rather than heavy fines. Such leniency emboldens bear hunters who have financial incentives to sell bear parts. Though Andean folk medicine involves some

Fig. 6.5 Eighteenth century painting of a bear hunt in Northern Peru in a work attributed to Jaime Baltasar Martinez Compañon (1737–1795)

bear products, the lucrative market is with Korean and Chinese traders. In Asian folk medicine, ancient superstitions include false ideas about the therapeutic benefits of a bear's gall bladder. It matters not which bear; all seven species around the world are vulnerable to trade in gall bladders. Poor countries are more compliant sources of this organ than are those where wildlife laws are generally respected. Historically a certain amount of hunting occurred simply for sporting purposes (Fig. 6.5). Trophy hunting is much less a factor today.

In the early part of the 2000 decade, spectacled bears in North American and European zoos numbered about 200. Another 100 or so bears, many of them captured as orphaned cubs, were kept in privately-owned enclosures in western South America (Rosenthal 2003). In Peru the conditions in such enclosures are generally abysmal.

The largest group of captive bears, about ten, are at the Parque de las Leyendas in Lima (Figueroa and Stucchi 2005). Bear cubs tamed by Andean peasants to be pets is uncommon but not unknown (Fountain 1902:274). Andean bears under various kinds of human control represent a potential breeding reservoir. With that in mind, a research group at the San Marcos Faculty of Veterinary Medicine in Lima collected ucumari semen at the local zoo (Enciso et al. 2006). Captivity also permits closer observation of spectacled bear behavior than is possible in the wild.

The hundreds of thousands of tourists who come every year to Machu Picchu have turned the 10,800 ha of the Sanctuary into the most salient place for bear observation (Figueroa and Stucchi 2005). Although the bear is rarely seen, the ceja zone which dominates the Sanctuary is the ideal habitat for this animal. The bear is protected as the emblematic species of the Sanctuary, for its rarity as a megaspecies with which the public can relate. Forest fires set by farmers, tree cutting, cattle raising and contamination of streams compromise the Sanctuary's role in the conservation of this and other species. The growing human impact of hikers outside the main cluster of the ruins impinges on the habitat seclusion that bears seek.

6.9 Conclusion: Toward a Holistic View

My curiosity about this animal transcends the science of the bear and its place in Andean culture. Of the wild mammals of the world, the secretiveness of this species and its rarity have conspired to keep *Tremarctos ornatus* poorly known. Understanding the bear's habits, distribution, population numbers and subspecific diversity rests largely on the "science of appearances." More than for most species, fragments and factoids provide the generalizations that are made about the bear. Our appreciation of the ucumari is enhanced when we learn why a researcher would go into remote areas for long periods to pursue an animal that he or she may never even see. Questions are posed about how the spectacled bear perceives its environment. How does it prioritize its quest for food? What sense of vulnerability accompanies the bear's forays on the paramo or puna? How much of ucumari shyness results from long human persecution or is it a trait genetically selected before humans arrived on the scene?

Learning about the bear goes beyond mammalogy to ethnozoological and mythopoetic dimensions. Until recently, human uses of bear parts have not threatened the species. The insanity of destroying a species to meet the false medical benefits of a culture 10,000 km away is surely one of the great distortions of globalization. Particularly interesting is how the Andean bear has been pulled into the orbit of human consciousness because of certain human-like characteristics of body and behavior. The shamanic practice of theriomorphism is still today ritually enacted in the Andes. Symbolic transformation of humans into animals suggests that vestiges of a thought pattern have survived in the Andean memory millennia after agriculture displaced hunting and gathering. Grafted onto that New World configuration was the transfer to the Andes via Spain in the sixteenth century of the Old World folkloric complex relating to bears. Combining reconstruction of the human-bear association of the distant past with the animal's present zoogeography, ecology and conservation serve to help save the bear for the Andean future.

References

Allen CJ (1983) Of bear-men and he-men: bear metaphors and male self-perception in a Peruvian community. Lat Am Indian Lit 7(1):38–51

Allen CJ (2011) Foxboy: intimacy and aesthetics in Andean stories. University of Texas Press, Austin

Arguedas JM (1960/1961) Cuentos mágico-religiosos quechuas de Lucanamarca. Folklore Americano 8/9: 142–216

Bastien JW (1982) Good luck fetishes: Andean amulets. In: Browne RB (ed) Objects of special devotion: fetishes and fetishism in popular culture. Bowling Green University Popular Press, Bowling Green, pp 352–361

Beirne P (2002) On the sexual assault of animals: a sociological view. In: Creager ANH, Jordan WC (eds) The animal-human boundary: historical perspectives. University of Rochester Press, Rochester, pp 193–227

Bingham H (1930) Machu Picchu: a citadel of the Incas. Yale University Press, New Haven

Brown AD, Rumiz DI (1989) Habitat and distribution of the spectacled bear (*Tremarctos ornatus*) in the southern limit of its range. In: Rosenthal M, Haggerty C (eds) Proceedings of the first international symposium on the spectacled bear. Lincoln Park Zoological Garden, Chicago, pp 92–103

Brunner B (2007) Bears: a brief history (trans: Lantz L). Yale University Press, New Haven

Buffon GLL (1833) Buffon's natural history of the globe and of man: beasts, birds, fishes, reptiles and insects. In: Wright J (ed) T. T. Keltie, London

Butchart SHM et al (1995) Threatened mammals of the Cordillera de Colán, Peru. Oryx 29(4): 267–274

Bychkova Jordan B, Jordan-Bychkov TG (2001) Siberian village: land and life in the Sakha Republic. University of Minnesota Press, Minneapolis

Cabello Valboa M (1951) Miscelanea antártica. una historia del Perú antiguo. Instituto de Etnología. Universidad Nacional Mayor de San Marcos, Lima

Cabrera A, Yepes J (1960) Mamíferos sudamericanos, 2 vol, 2nd edn. Ediar, Buenos Aires

Castellanos A (2004) American bear research project in the Intag region, Ecuador. International Bear News, Quarterly Newsletter of the International Association for Bear Research and Management (IBA) and the IUCN/SSC Bear Specialist Group 13(2):25–26

Chávez Chaparro J (1950) Distribución geográfica de los principales mamíferos peruanos. Revista Universitaria 99:233–321

Cipolletti MS (1983) En torno a un relato andino el ukumari. Allpanchis 19(22):145–160

Cobo B (1956) Historia del Nuevo Mundo, 2 vol. Ediciones Atlas, Madrid

Corso R (1955) L'orso della Candelora. (FF Communications No 153). Suomalainen Tiedeakatemia, Helsinki

Cuvier G (1831) Animal kingdom arranged in conformity with its organization, 4 vol (trans: McMurtie H). Carvill, New York

de Acosta J (1940) Historia natural y moral de las Indias. Fondo deCultura Económica, Mexico City

de Carvalho-Neto P (1964) Diccionario del folklore ecuatoriano. Editorial Casa de la Cultura. Ecuatoriana, Quito

de Cieza de León P (1986) Crónica del Perú primera parte (1550). P.U.C.P. Fondo Editorial, Lima

de Lizárraga R (1987) Descripción del Perú, Tucumán, Rio de la Plata y Chile. Historia 16, Madrid

de Murua M (1987) Historia general del Perú (1600). In: Ballesteros M (ed) Historia 16, Madrid

de Oviedo GF (1959) Historia general y natural de las Indias, 3 vol. Ediciones Atlas, Madrid

Descripción geográfica del Partido de Piura (1793) Mercurio Peruano 264: 176–182 de Uña JM (1992) Función Alfa (situación inicial) y doble significación morfológica de Omega (matrimonio). Approximación a la morfología de una versión de "Juan el Oso" (II parte). Revista de Folklore 139:16–24

de Velasco J (1989) Historia del reino de Quito en la América meridional, 2 vol. Corporación de Estudios y Publicaciones, Quito

de Zarate A (1947) Historia del descubrimiento y conquista de la provincia del Perú. In: de Vedia E (ed) Historiadores primitivos de Indias. Madrid Ediciones Atlas, Madrid, pp 459–574

Del Moral JF (2005) Evidence of Andean bear in northwest Argentina. Int Bear News 14(4):30–32

Dorst J (1967) South America and Central America: a natural history. Random House, New York

Enciso M et al (2006) Estudio preliminar de colección de semen en oso de anteojos (*Tremarctos ornatus*). Revista de investigaciones veterinárias del Perú 17(1)

Faust FX (1996) Apuntes a la etnozoología del Macizo Colombiano. Montalbán (Caracas) 29:173–186

Festa E (1904–1905) Osservazioni intorno agli orsi dell'Ecuador. Atti della Reale Accademia delle Scienze di Torino. 40(2–3): 186–195

Figueroa J, Stucchi M (2005) Registro del oso andino en cautiverio en el Perú y algunos alcances para su mantenimiento. Asociación Ucumari, Lima. www.alberdi.de/oso_andino.pdf. Accessed 12 Jan 2012

Foster GM (1960) Culture and Conquest: America's Spanish Heritage. Quadrangle Books, Chicago

Fountain P (1902) The great mountains and forests of South America. Longmans, Green & Co, London

Gade DW (1970) Ecologia del robo agricola en las tierras altas de los Andes centrales. América Indigena 30(1):3–14

Garcilaso de la Vega EI (1960) Comentarios reales de los Inca*s*. Universidad Nacional del Cuzco, Cusco

Goldstein IR (1991) Spectacled bear predation and feeding behavior on livestock in Venezuela. Stud Neotropical Fauna Environ 26(4):231–235

Goldstein IR (2004) Andean bear use of the epiphytic bromeliad *Tillandsia fendleri* at Quebrada el Molino, Venezuela. Ursus 15(1):54–56

Goldstein IR et al (2006) Andean bear-livestock conflicts: a review. Ursus 17(1): 8–15

Grimwood IR (1968) Notes on the distribution and status of some Peruvian mammals. American Committee International Wildlife Protection, Special Publication No 21: New York Zoological Society, New York, pp 1–86

Guaman Poma de Ayala F (1980) El primer nueva coronica y buen gobierno. In: Murra JV, Adorno R (eds) Siglo Veintiuno, Mexico City

Hallowell AI (1926) Bear ceremonialism in the northern hemisphere. Am Anthropol 28(1):1–175

Helms MW (1998) The spectacled bear in iconic imagery of ancient Panama and Colombia. In: Oyuela-Caycedo A, Raymond JS (eds) Recent advances in the archaeology of the northern Andes*:* in memory of Gerardo Reichel-Dolmatoff. UCLA Institute of Archaeology, Los Angeles, pp 129–141

Hershkovitz P (1957) On the possible occurrence of the spectacled bear, *Tremarctos ornatus* (F. Cuvier 1825) in Panama. Saugetierkunde Mitteilungen 5(3):122–123

Jiménez de la Espada M (ed) (1965) Relaciones geográficas de Indias, 3 vol. Ediciones Atlas, Madrid

Jorgenson JP, Sandoval SA (2005) Andean bear management needs and interactions with humans in Colombia. Ursus 16(1):108–116

Kattan G et al (2004) Range fragmentation of the spectacled bear *Tremartos ornatus* in the northern Andes. Oryx 38(2):1–10

Kurtén B (1966) Pleistocene bears of North America 1. Genus *Tremarcto*s: spectacled bears. Acta Zoologica Fennica 115:1–120

Lara J (1973) Mitos, leyendas y cuentos de los Quechuas. Editorial "Los Amigos del Libro," La Paz-Cochabamba

Larsen S (1990) The mythic imagination. Bantam Books, New York

Lavallée D (1970) Les représentations animales dans la céramique Mochica. Institut d'Ethnologie, Musée de l'Homme, Paris

Lequanda JI (1793) Descripción geográfica del Partido de Piura. Mercurio Peruano 264:175–182

Llano Zapata JE (2005) Memorias histórico-físicas, crítico-apologéticas de la América meridional. In: Casteñada RR (ed) Instiut Français d'Etudes andines, Lima

López de Velasco J (1971) Geografía y descripción de las Indias (1574). In: de la Espada MJ (ed) Ediciones Atlas, Madrid

Martinez Compañon JB Trujillo del Peru. Biblioteca virtual Miguel de Cerrrantes (www.bib.cerrantesvirtual.com)

Molinié A (2005) La transfiguración eucarística de una construcción andina del Corpus. In: Molinié A (ed) Etnografías del Cusco. Centro de Estudios Regionales Andinas "Bartolomé de las Casas," Cusco, pp 69–83

Mondolfi E (1989) Notes on the distribution, habitat, food habits, status and conservation of the spectacled bear (*Tremarctos ornatus* Cuvier) in Venezuela. Mammalia 53(4):525–564

Morote-Best E (1957) El oso raptor. Archivos Venezolanos de Folklore 5:157–178

Morrison T (1972) Land above the clouds: wildlife of the Andes. Universe Books, New York

Ordóñez de Ceballos P (1614) Viaje del mundo Luis Sanchez, Madrid

Osgood W (1914) Mammals of northern Peru. Field Museum Nat Hist Zool 10:143–185

Paisley S, Garshelis DI (2005) Activity patterns and time budgets of Andean bears (*Tremarctos ornatus*) in the Apolobamba region of Bolivia. J Zool 265:25–34

Paisley S, Saunders NJ (2010) A God forsaken: the sacred bear in Andean iconography and cosmology. World Archaeol 42(2):245–260

Pastoureau M (2007) L'ours: Histoire d'un roi déchu. Éditions du Seuil, Paris

Payne J (1984) Cuentos cusqueños. Centro de Estudios Rurales Andinos "Bartolome de las Casas," Cusco

Peyton B (1980) Ecology, distribution and food habits of spectacled bears, *Tremarctos ornatus*, in Peru. J Mammal 61(4):639–652

Peyton B (1981) Spectacled bears in Peru. Oryx 16:48–56

Peyton B (1987) Habitat components of the spectacled bear in Machu Picchu, Peru. Bears Their Biol Manag 7:127–133

Peyton B (1999) Spectacled bear conservation action plan. In: Servheen C, Herrero S, Peyton B (eds) Bears: Status survey and conservation action plan. International Union for the Conservation of Nature, Gland/Cambridge, pp 157–170

Polo de Ondegardo J (1916) Informaciones acerca de la religion y gobierno de los Incas (ca 1545). Colección de libros y documentos referentes a la historia del Perú, vol 3. Sanmarti y Cia, Lima

Raimondi A (1862) Apuntes sobre la provincia litoral de Loreto. Tipografía nacional por Manuel D. Cortes, Lima

Raimondi A (1898) Itinerario de los viajes de Raimondi en el Perú: valle de Lares, Santa Ana y regreso por Mollepata y Limatambo (1865). Boletin de la Sociedad Geográfica de Lima 8:1–61

Ramírez del Aguila P (1978) Noticias políticas de Indias [1639]. In: Urioste J (ed) Extension Universitaria, Sucre

Rechberger J, Wallace RB, Ticona H (2001) Un movimiento de larga distancia de un oso andino (*Tremactos ornatus*) en el norte del Departamento de La Paz, Bolivia. Ecologia en Bolivia 36:73–74

Ribera MO, Lieberman M (2006) El uso de la tierra y los recursos de la biodiversidad en las areas protegidas de Boliva. Servicio Nacional de Areas Protegidas, LaPaz

Rios-Uzeda B, Gomez H, Wallace RB (2006) Habitat preference of the Andean bear (*Tremarctos ornatus*) in the Bolviian Andes. J Zool 268:271–278

Robin V (1996) El cura y sus hijos osos o el recorrido civilizador de los hijos de un cura y una osa. Bulletin de l'Institut français d'Etudes andines 26(3):369–420

Rosenthal M (2003) International studbook for the Andean bear (*Tremarctos ornatus*) *2001–2002*. Lincoln Park Zoo, Chicago

Ruiz-Garcia M (2003) Molecular population and genetic analysis of the spectacled bear (*Tremarctos ornatus*) in the northern Andean area. Hereditas 138:81–93

Rumiz DI et al (2012) El ucumari (*Tremarctos orantus*), mito y realidad de su presencia en Argentina. Mastozoologiea Neotropical (Mendoza) 19(2):1–9

Salazar B, Anderson J, Anderson S (1990) Informe sobre el estado actual del conocimiento del oso andino en Bolivia. Ecología en Bolivia 15:3–23
Sallnow MJ (1987) Pilgrims of the Andes: regional cults in Cusco. Smithsonian Institution Press, Washington
Schulenberg TS, Awbrey K (eds) (1997) A rapid assessment of the humid forests of south central Chuquisaca, Bolivia. Conservation International, Washington
Sicher H (1944) Masticatory apparatus in the giant panda and the bears. Zool Ser Field Museum Nat Hist 29(4):61–73
Stolzenburg W (1997) Andean ambassador. Nat Conservancy 47(4):10–15
Stucchi S, Salas-Giamenbi R, Baby P, Guyot JL, Shockey BJ (2009) A 6,000+ year-old specimen of a spectacled bear from an Andean cave in Peru. Ursus 20(1):65–68
Suarez L (1988) Seasonal distribution and food habits of spectacled bears: *Tremarctos ornatus* in the highlands of Ecuador. Stud Neotropical Fauna Environ 23(3):133–136
Taylor G (1997) Juan Puma, el hijo del oso: Cuento quechua de La Jalca, Chachapoyas. Bulletin de l'Institut français d'Etudes andines 26(3):347–368
Urton G (1985) Animal metaphors and the life cycle in an Andean community In: Urton G (ed) Animal myths and metaphors in South America. University of Utah Press, Salt Lake City, pp 251–284
Varga RR, Azurduy C (2006) Nuevos registros de distribución del oso andino (*Tremarctos ornatus*) en el Departamento de Tarija, el registro más austral en Bolivia. Mastozoología Neotropical 13(1):137–142
Vargas Machuca B (2008) The Indian Militia and Description of the Indies (1599). In: Lane K (ed) (trans: Johnson TF). Duke University Press, Durham, NC
Vázquez de Espinosa A (1948) Compendio y descripción de las Indias occidentales. Smithsonian Miscellaneous Collections, vol 108. Smithsonian Institution, Washington
Vespuccio A (1951) El Nuevo Mundo: cartas relativas a sus viages y descubrimientos. Editorial Nueva, Buenos Aires
Vokral EV, Masson P (1993) Erinnerungen an den Bärensohn: elemente und fragmente schwindender Erzahltraditionen im zentralen Andenhochland von Ecuador. In: Saravia JM (ed) Die schwierige Modernität Lateinamerikas. Vervuert Verlag, Frankfurt am Main, pp 159–195
von Tschudi JJ (1849) Travels in Peru (trans: Ross T). Putnam, New York
Weber DJ (ed) (1987) Juan del Oso. Yarinacocha Instituto Linguístico de Verano, Pucallpa
Yerena E (1994) Corredores ecológicos en los Andes de Venezuela. Librería Alemana, Caracas

Chapter 7
Urubamba Ramble: Hiram Bingham (1875–1956) and His Artful Encounter with Machu Picchu

Abstract An aesthetically harmonious human construction in a spectacular natural setting above the Urubamba River, Machu Picchu is the primary tourist destination in Peru. The claim of the American Hiram Bingham presents its classic narrative of discovery. In 1911 he assembled a well-organized expedition that reached the site by a combination of fortuitous circumstances: the rail line from the coast to Cusco had just been opened; a fairly new mule trail became available through the Torontoy Canyon, and he found out where to go and with whom to speak. An American living in Cusco, Albert A. Giesecke, was the conduit that enabled Bingham to conceive his "discovery." By keeping other expedition members at bay, he arranged to be at center stage and attained the fame and status that he desired, a desire shared by many explorers. In countering Bingham's account, this chapter also contests his assertion that Machu Picchu had been unknown before his visit. Bingham's biography and the letters he exchanged provide insights into the centrality of his ego and his unwillingness to both acknowledge the crucial help he received and to correct erroneous conclusions he made about this Inca site.

7.1 Introduction

Re-evaluation of the canon of geographical discovery can be prompted not only by additional facts. Critically reading the standard narratives, giving close attention to the motives of protagonists, and emphasizing the time/space matrix may also lead to new interpretations of the classic accounts. Discovery events are primarily about claims of precedence, for unless the event is the first, there is no achievement. In the eternal human quest to position the self in the realm of discovery of a place, geography meets psychology. *Geltung*, a word meaning the cultivation of appearances and concern for reputation. It refers to how the ego maneuvers as an outcome of the evolution of human behavior (Wagner 1996). It represents a universal beyond culture. However, cultures vary in the degree to which they encourage or discourage its manifestation in individuals. Within any one culture, Geltung is sought and is tolerated to different degrees as shown by the self-effacing behavior of a monk compared to the publicity-seeking behavior of a film star. People engaged in the exploratory quest are prone to expressions of Geltung, for without claims to precedence, their activities receive little attention from journalists or underwriters.

D.W. Gade, *Spell of the Urubamba*, DOI 10.1007/978-3-319-20849-7_7

Fig. 7.1 Machu Picchu, now one of the most photographed places in South America, is now quite different in appearance (except for the clouds) than when Hiram Bingham saw it in 1911 but more like when the Inca Pachacutec lived there in 1513

The finding on July 24, 1911 of Machu Picchu, now one of the world's most iconic and visited archaeological sites, provides a notable example of unbridled Geltung on the part of its main protagonist. To a certain extent that date was also the biggest day in the twentieth-century history of the Urubamba Valley. On July 24, the town the Incas apparently called Patallacta, but which is today known as Machu Picchu, came to thc attention of someone who for the first time understood its larger meaning. Hiram Bingham (1875–1956) was the sole author of the documentary record of its unveiling, one that became memorable partly as a result of the literary skill and photographic imagery with which it was communicated. The site remains an iconic find in the annals of exploration. When the motivations of Machu Picchu's putative discoverer are dissected and the discovery event contextualized in its regional and national frameworks, this long-accepted narrative must be corrected. Unraveling the sequence of events in the regional framework, and paying close attention to the mind of the protagonist offers insight into how knowledge about a place becomes intertwined with certain individuals and not with others.

No one in 1911 could have imagined that this tight cluster of late Inca construction, 130 km northwest of the then-remote town of Cusco, would become the most internationally visited tourist destination in South America (Fig. 7.1). Visitors were few until the 1930s when the railroad reached that part of the Urubamba Valley. Many more people arrived after 1948 when the new vehicle road from the canyon floor to the ruins replaced the trip on mule back with a speedy 15-min ride on a bus. Machu Picchu has since become the destination for close to a million people a year, many of whom have no particular interest in the Inca past nor any other past, but for whom Machu Picchu presents an unforgettable monument in a spectacular setting.

7.2 Genius Loci of an Andean Jewel

Visual, ecological and historical elements contribute to the irresistible appeal and special mystique of Machu Picchu. The landscape of the Machu Picchu Sanctuary zone focuses on the tightly arranged settlement; one planned and built as a set piece rather than as a series of accretions over time. Nestled in a saddle between towering perpendicular peaks above and the deep canyon of the Urubamba below, the ensemble offers the ineffable aura of a self-contained unit, although, in fact, it was dependent on interchange with the outside.

Heavily faulted granitic intrusions provided the raw material for the noble architecture of exquisitely carved ashlars and sculptures fashioned from the living rock. Spring water flowed by gravity in stone-inland channels to 14 fountains. Stone-faced bank terraces offered flat strips of land on which to grow crops in this location of thermal moderation that the Incas called *chaupiyunga*. Maize, potato and quinoa grew side by side with sweet potato, arracacha, capsicum pepper, and coca. Many visitors wandering among the silent stones in the enveloping mist have fallen under the spell of the site. Attracted to more than its aesthetics, visitors have generally found the site to be magical and mysterious.

In addition to its visual harmony and appearance of ecological sustainability, the story of Machu Picchu's putative discovery is arresting. How, when, and why the site entered modern human consciousness is an evocative tale, partly because that account is recorded, whereas the story of the site's construction is not. Bypassed by the Spanish colonizers at least to the extent that they provided no descriptions of it, its first known recorded mention was made by Hiram Bingham in the early in the twentieth century. Since then, the basic information package about Machu Picchu stars this tall, blue-eyed gringo. Not the Inca emperor Pachacutec who had this place built in the fifteenth century and spent time there, but Bingham stars in this Machu Picchu narrative.

Early in the twentieth century, articles carried in popular magazines brought Machu Picchu to the attention of the masses. Two articles appeared within 2 years of the 1911 "discovery" and three others followed in 1915–1916. Of the two articles published in 1913, Bingham's piece in *Harper's Monthly Magazine* did not match the detail or illustrative splendor of the article that appeared in the *National Geographic Magazine* (Bingham 1913a, b). By publishing about Machu Picchu in two widely disseminated monthly magazines rather than in scholarly journals which have long publication lag times, Bingham gained more recognition as an explorer than as a scholar. His articles in the *National Geographic Magazine* had a major impact. This popular monthly had a quite well-educated readership and it featured photographs, a relatively new medium for relating knowledge about places. His long piece in the 1913 National Geographic Magazine had 238 illustrations. Of these, 141 were of Machu Picchu and its surroundings. Better than words, photography verified what Bingham said about where he had gone and what he had found. He (Bingham 1913b) himself admitted that without the Kodak images, no one would have believed him. The article included some of the 31 photos he took in July 1911

Fig. 7.2 Hiram Bingham (1875–1956), a 36-year-old in his dashing explorer mode (Photography by Herman Tucker, 1911)

and the many more he took on his return visit in 1912. Some of these illustrations also appeared in his subsequent books and, since they are no longer restricted by copyright, they also have been used by many other authors.

A decade later Bingham (1922) published a semi-popular book on Peru that contained a chapter on the site. By the time his scholarly monograph on the site's archaeology came out in 1930, he was well-known as its discoverer. A retrospective book, alluringly titled *Lost City of the Inca*, introduced a new generation to the story (Bingham 1948). Bingham's publications on Peru projected an image of himself as an intrepid man of action who managed harrowing episodes with aplomb. Accounts of inching across crumbling bridges, sidestepping paths clinging to sheer drop-offs and thwarting vipers ready strike, appealed especially to readers more entranced with the adventure of overcoming adversity in exotic places than with Inca culture history. Bingham personified the romantic explorer and showed how a personal call for adventure is answered (Fig. 7.2). Although he had a Ph.D. in history and taught Latin American history at Yale, Bingham's explorations gave him more credence as a geographer than as an historian. In 1914, the Association of American Geographers elected him a member.

7.3 Cusco Department in the Early Twentieth Century

Hiram Bingham went to Peru focusing on its distant past when the eyes of the country were on the future. The rubber boom in the lower Urubamba Valley north of Cusco created an economic optimism that has not been seen since in the region.

Between 1880 and 1912, a large influx of highland folk (*serranos*) and some foreigners descended into the uncharted forested lands of the Amazonian drainage east of the Andes to exploit the wild-growing latex-bearing trees. Before that, only the tropical lands nearest the mountain front had received settlement from the highlands. In the sixteenth century, within three decades of the Conquest, Spanish individuals had gained title to land, and the two earliest estates, Hda. Ccolpani and Hda. Huadquiña, still produced coca and rum more than 300 years later. But by then it was the prospect of riches from the rubber found 150–300 km farther down the valley that was on the mind of every entrepreneur in Cusco.[1] Investors proposed two railway projects in southern Peru to tap into that boom. The Ferrocarril del Sur (incorporated as "Southern Railway"), started in 1870 under a contract with British capitalists, and originated on the coast at Mollendo. After almost four decades of construction, the track finally reached Cusco, its highland terminus.

The other proposed railroad, less ambitious but longer in the works, was for a narrow-gauge rail connection from Cusco through the Urubamba canyon to the jungle piedmont (Robledo 1899). After years of debate, departmental authorities decided that only a pack trail (*camino de herradura*) was economically feasible. Taxes imposed on agricultural products carried by mule back largely paid for its construction, which required dynamiting stretches where a sheer precipice descended to the river. Completed in 1895, the canyon route superseded the longer and much steeper trail that followed what has been known variously as the Huayopata, Lucumayo or Amaybamba Valley over the Pantiacolla (now Málaga) Pass to Ollantaytambo. Higher than 4200 m above sea level (asl), the pass is hypoxic and frigid. In 1903, 30 people froze to death there (Cruz Ccorimaya 2009:113). The mule trail broke the long inaccessibility of the Urubamba gorge, which, in turn, gave impetus for constructing the railway that had been earlier proposed through those narrows. However the rail tracks that eventually took the place of the mule trail were not laid until more than two decades after Hiram Bingham's first visit.

7.4 The National Geographic Yale University Expedition of 1911

Hiram Bingham's 1911 expedition grew out his previous experience in South America. In a 1908 trip around the continent, it was the visible presence of the pre-Columbian in Peru's cultural landscape that most captured his imagination. In that year, the exhilaration of having reached Choquequirao, an Inca site totally off the beaten track perched high above the Apurimac gorge, was an unforgettable adventure. Hiram reveled in the V.I.P. reception he got enroute from officials and hacendados. Realizing the usefulness of historical documents for tracking down

[1] The tropical Urubamba valley consisted of two parts: a settled zone from 1600 to 700 m asl comprised of haciendas started largely during the colonial period by people of highland origin and a selva zone below 700 m asl occupied mainly by forest Indians.

such sites, in April 1911, Bingham queried the Lima historian Carlos Romero in Lima about where to find ruins "no less important than Choquequirao." Romero (1909) had published an article on that site before he met Bingham. Romero wrote back that the Inca town to look for was Vilcabamba and that became Hiram's big objective. The possibility that he could be the first to find other Inca ruins took him back to Peru. Bingham calculated that institutional support for the trip would not be forthcoming unless conceived as having a scientifically respectable objective. He decided that collecting biota, making maps, and determining mountain elevations were suitable expeditionary activities and selected his team members for their skills in those fields. Bingham earlier had learned the rudiments of organizing an expedition when he accompanied the explorer Hamilton Rice to Venezuela in 1905–1907. Rice was to envy Bingham's Peruvian success later when he, Rice, explored the Brazilian Amazon looking for a lost city at the same time as Percy Fawcett (Gram 2005). He knew about careful trip planning, enlisting cooperation of rural folk, and the importance of getting good pack animals and to avoid the wet season.

Sponsorship from the National Geographic Society and Yale University gave Bingham scientific respectability. Yale lent its name, but the financial contributions came from people with Yale ties who Bingham cultivated. To a large extent, the expedition was a front for his esoteric endeavor to find long-forgotten Inca sites. With that as his hidden motive, Bingham did not follow advice from W.C. Farabee, a Harvard anthropologist who had worked in the lower Urubamba, to enlist a trained archaeologist as expedition member (Letter WF to HB 4-21-1911). A professionally trained archaeologist would have overshadowed Bingham in his quest to be a discoverer. Bingham's correspondence while preparing for their trip conveyed extraordinary self-confidence, and also an exaggerated sense of self-importance. He wrote the President of the United States asking him to send a skilled government topographer as an expedition member. With similar chutzpah, he wrote to the President of Peru, asking him to notify the customs officials at Mollendo to not charge duties on his supplies nor delay the expedition. The first request was turned down; the second apparently was filled.

7.4.1 Bingham in Country

In July 1911, Bingham and his seven fellow expeditionaries arrived in Cusco by train from Mollendo. Most travelers to Peru went first to Lima, but Bingham's obsession controlled the expedition's organization and the proposed itinerary of the group. By landing at the port of Mollendo, the train could be taken directly to Cusco, which was precisely where the director wanted to start his expedition. The company on that trip included Isaiah Bowman, a geographer who taught at Yale; Harry Foote, a professor of chemistry at Yale, avid outdoorsman and good friend of Bingham's; William G. Erving, Bingham's Yale classmate and a practicing medical doctor in Washington, D.C.; Herman L. Tucker, a young man with mountain climbing experience who could be counted on to be an obedient factotum; Paul Baxter Lanius, who had lived in Peru, knew Spanish, and had been a student of Bingham's at Yale; Kai Hendricken, a young Danish surveyor who had come to the United States to study with the U.S. Coast and Geodetic Survey; and Casimir Watkins, an English naturalist, who knew something about mountain climbing. His companions seem not to have known that the "chief" had selected Cusco as the expedition's take-off point in order to facilitate Hiram's attempt to find Vilcabamba. That was his priority and the real reason he wanted to assemble an expedition. Other items on the trip's agenda would come later.

When they got to Cusco, Bingham knew whom to see: César (aka Cesare). A congenial and knowledgeable import–export merchant of Italian origin, Lomellini was Cusco's leading entrepreneur and a man of regional influence. Two years previously he had provided storage space in his warehouse for expeditionary equipment. On this trip, Lomellini provided mules, drafted letters of introduction in Spanish, and dispensed advice about local conditions. Hiram also went to Albert Giesecke's house facing the Plaza San Francisco, to talk with the 25 year-old North American rector of the University about Inca ruins. The two men had not met before, but they had exchanged correspondence, initiated by Giesecke, about Choquerquirao. Giesecke had tried but failed to reach that site. After a fortnight in and around Cusco, the seven-man expedition set out by mule for the Urubamba Valley. Bingham chose to go down that valley after Carlos Romero had commented to him in a letter that ruins would be found in the Urubamba drainage north of Cusco. Bingham had learned about the mule trail through the Urubamba canyon from W. Farabee of Harvard who in 1908 took that trail from Ollantaytambo to Santa Ana (Letter WF to HB 4-21-1911). The trail was a camino de herradura made with mule traffic in mind, and it provided the fastest and easiest way to get to the valleys of La Convención. When the French explorer Charles Wiener (2010:322) made the rugged trip over the Malaga Pass to get to La Convencion in 1877, he wondered why the "hacendados of Santa Ana Valley, all very rich, had not established a road along the banks of the Urubamba."

Leaving Cusco, the expeditionaries stopped first in the town of Urubamba. There Bingham pumped people about ruins, and though their information was vague, it encouraged him. In Ollantaytambo, Bingham decided that Isaiah Bowman, Paul Baxter Lanius and William Erving would go farther down the Urubamba. Bowman and Lanius had a mapping project and Dr. Erving wanted to make inquiries about tropical disease. Bingham, together with Harry Foote and Kai Hendricksen, proceed down valley as a separate traveling group. Unrevealed to his companions, the chief had his own, private agenda. The plan was for the six expedition members to later rendezvous at Hda Santa Ana to take advantage of the hospitality promised by its owner Pedro Duque. With regard to ruins, Bingham (1913a:709) incorrectly asserted that "Weiner decided to go down the valley and look for them, but owing to one reason or another, he failed to find them". However, Weiner did not go down the Urubamba Valley; when he traveled in the 1870s, a trail through the canon had not yet been constructed. Instead, Wiener entered La Convención by going over the pass above Ollantaytambo and then down the Lucumayo Valley. The route entirely bypassed Machu Picchu.

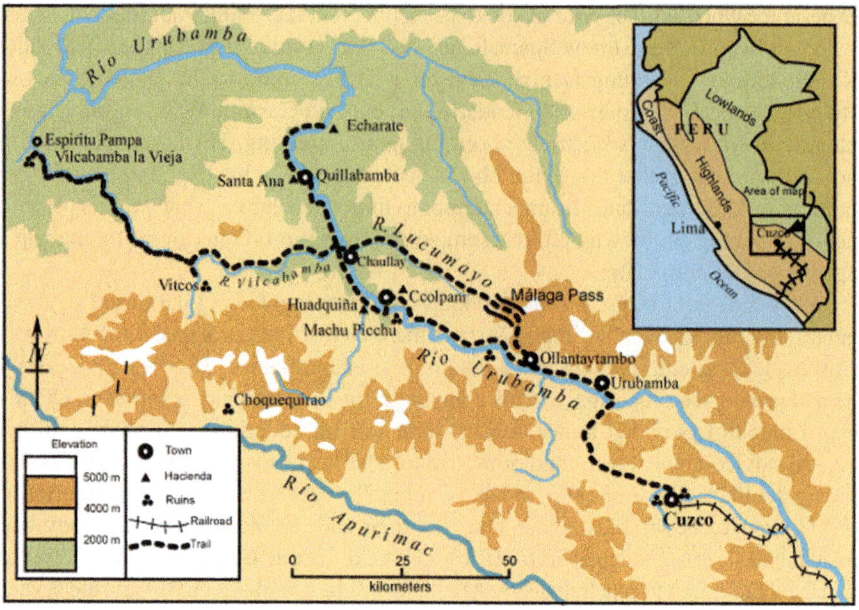

Fig. 7.3 The trail that Bingham and his team members took in 1911 down the Urubamba Valley was used by muleteers to bring tropical products from La Convención to Cusco. This map does not include the transportation adjustments there today (Map by M.K. Gade)

The Bingham party, which also included Sargent Carrasco, a Peruvian auxiliary, headed down the valley with their mules (Fig. 7.3) Thirty kilometers from Ollantaytambo and 400 m lower in elevation, the wide valley became a narrow steep-sided gorge known as the canyon of Torontoy (Fig. 7.4). The vegetation at that point dramatically changed from dried-up ichu grasses and low shrubs covering the hillsides to an elfin forest abundantly watered by frequent drizzle. Starting at 2350 m asl, the red-blossomed pisonay tree (E*rythrina falcata*), festooned with epiphytes, indicated also a sharp increase in precipitation. When the group arrived at a place called Mandor Pampa, Bingham gave instructions to set up camp. That decision must have puzzled his companions, for they could have reached a welcoming hacienda by nightfall (Fig. 7.5). According to Bingham's account, an Indian, Melchor Arteaga, who lived in a hut nearby so he could sell fodder to passing pack trains, offered to show Bingham where ruins could be found. The next morning Bingham, Arteaga and Sargento Carrasco hiked about a kilometer up the valley, crossed the roiling river on a fragile makeshift bridge, and clambered about 400 m above the valley floor on a minimalist and very steep path. Arteaga knew where to lead Bingham. The exhausting four-hour climb yielded a stunning array of ruined buildings constructed of finely cut ashlar. As Bingham (1948:155) wrote, "it fairly took my breath away. What could this place be? Why had no one given us any idea of it?" His fond notion that Machu Picchu had been abandoned did not mean that the site was unoccupied. Two peasant families were growing crops on some of the terraces and the young son of one of them showed Bingham around the ruins.

Fig. 7.4 The Canyon of Torontoy which H. Bingham entered in 1911 looking for ruins. Here the Urubamba River cut through a batholith that is part of the Cordillera de Vilcabamba

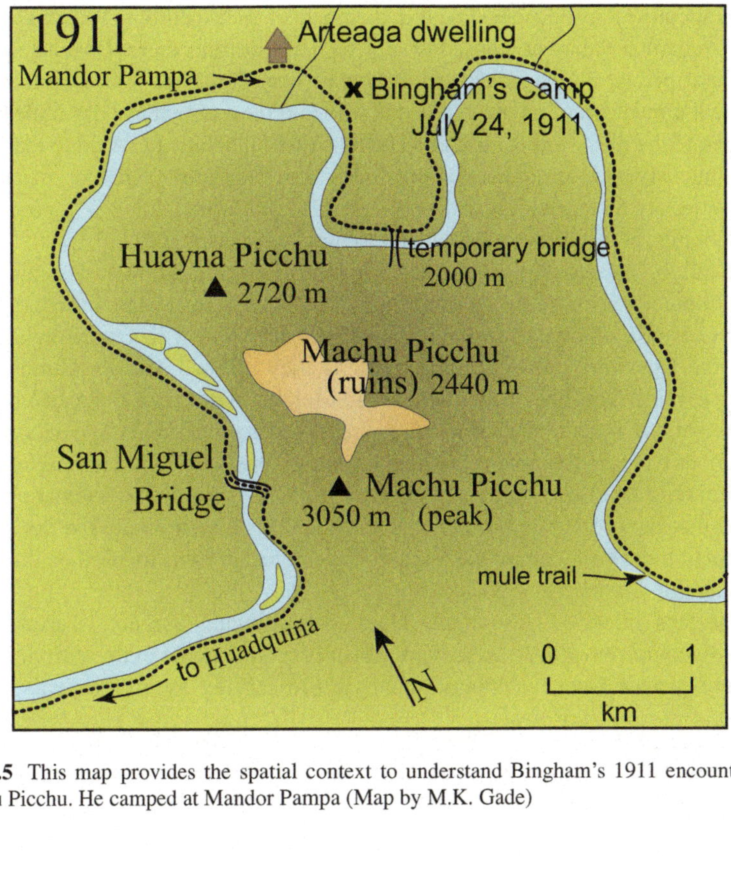

Fig. 7.5 This map provides the spatial context to understand Bingham's 1911 encounter with Machu Picchu. He camped at Mandor Pampa (Map by M.K. Gade)

7.4.2 The Counter-Narrative of July 24, 1911

It took half a century before it was revealed that Bingham's description of a chance encounter on that July day dissembled the real sequence of events. He neglected to say that at his meeting with Albert Giesecke in Cusco he had learned of the existence of ruins high above the western side of the Urubamba gorge. Though Giesecke had not seen the site himself, he knew where Bingham should spend the night and with whom he should speak in order to be led to them. It was Bingham who first asked Arteaga about ruins. Giesecke had given him Arteaga's name, knowing him to be the crucial contact person for information about ruins in the vicinity of Mandor Pampa. No publication mentioned that crucial conversation during Bingham's lifetime. Five years after Bingham's death, Giesecke spoke to a journalist from the Lima newspaper *El Comercio* describing for print for the first time what actually had happened (Montoya and Giesecke 1961). In that same year, Giesecke (1961) published similar remarks in the Cusco archaeology journal. That information seems not to have changed the interpretation of the "discovery" event; for example, the author John Hemming (1970:482) described the Bingham-Arteaga meeting as serendipitous, which certainly was not the case.

Giesecke remarked in his newspaper article that "Before leaving Cusco, Bingham went frequently to my house as did other members of the expedition. We spoke mainly of the Urubamba Valley and possible sites for ancient ruins there. Fortunately I had made this trip as far as Hda. Echarate downstream from Santa Ana (today Quillabamba) at the invitation of Braulio Polo y la Borda who told me during that 7 day mule trip the traditions of the place and the life and customs of its inhabitants.[2] Among these data, he continuously pointed out sites on hillsides where ruins are said to have existed and now covered up by forests. This was not first hand, but from talking with peons who, when looking for lost cattle, mentioned it to him" (Giesecke 1961).

The Polo-Giesecke trip on horseback occurred during the relentless and often torrential precipitation of the rainy season in this the wettest part of the Urubamba Valley. Giesecke recounted the trail conversation he had with Don Braulio about Paititi, the legendary pre-Columbian city. As they reached Mandor Pampa, Polo, keeping in mind their line of trail talk, asked Melchor Arteaga in Quechua about the existence of any ruins in the surrounding area. Arteaga replied that the only ones he knew about were up the nearby mountain on the other side of the river. Drenching rain, high river levels and furious turbulence made any crossing problematical. Polo decided that the propitious time to check out that lead would be in the dry season. Like that of most Peruvians at the time, Polo's interest in ruins focused on the dream of finding pre-Columbian gold.

Giesecke's interview also made clear that bilingual Sargento Carrasco, not Bingham, conferred with Arteaga who, like most rural people in this part of Peru at that time, was a Quechua monolingual. While serving as a Quechua-speaking

[2] Braulio was the son of Tomás Polo y la Borda who in 1897 acquired Haciendas Echarati and Potrero and in 1902 bought Hacienda Sahuayaco with Luis Gonzalez Willys.

intermediary, his authority as a policeman was also meant to exact cooperation from potentially refractory natives. The departmental prefect, J.J. Nuñez, insisted on Carrasco's service to Bingham for another reason. From the prefect's standpoint, Carrasco's most important role was to report back about where the North Americans had gone and what they had found. The policeman certainly would have told the prefect that, in his 4 h at the ruins, Bingham took photographs and wrote descriptive comments in his notebook, but did not make any effort to look for buried treasure.

7.4.3 Aftermath of the Big Find

Bingham's plan to trek into the Vilcabamba to find other ruins explained his short stay at Machu Picchu. With no authorization from the prefect to make any excavations, he decided to return back to the site the following year with official permission. To ensure that the site would not be ransacked in the meantime, Bingham sought to keep Machu Picchu a secret from *cusqueños*. Treasure hunters looting a site before a careful study could be made was a legitimate concern. But Bingham's motive for concealment was also personal. He did not want anyone else to produce a description of Machu Picchu before he did.

Months later, Bingham sent two Indians to clear enough vegetation for Lanius and Tucker to photograph and map the site. When Bingham got to Lima on his way back to the United States, he personally urged President Leguia to authorize him alone to undertake excavation. That proposal seemed arrogant, for although no one in Cusco at the time had enough archaeological training to make a competent dig, Bingham himself had never taken part in an excavation. As became apparent during Bingham's several expeditions to Machu Picchu, the patient work of archaeological recovery did not interest him. When local scholars learned via the grapevine that Bingham tried to keep them from knowing about the site, they saw themselves victims of a conspiracy.

Outside Cusco, Bingham was not so hesitant about letting the world know of his feat. On August 3, 1911, he wrote to J.S. Keltie at the Royal Geographical Society that he "found the ruins of a wonderful Inca city…" (Martin 1980:41). In the 1912 report of his expedition in the *Bulletin of the American Geographical Society*, Bingham nonchalantly wrote: "A few archaeological discoveries were made by the Director. Among them are the ruins of a number of Inca or pre-Inca cities, including Machu Picchu, a city probably built by the 'megalithic race' who preceded the Incas. The ruins are on an almost inaccessible ridge, 2000 ft. above the Urubamba River" (Bingham 1913b). Of interest in that paragraph is the remark that the discoveries were made, not by the expedition team, but by "the Director." Also in 1912, the *National Geographic Magazine* published a preliminary report on Bingham's 1911 expedition that disclosed "the discovery of eight Inca and pre-Inca cities and temples" (Anonymous 1912). Two photographs of the stonework at Machu Picchu are labeled "discovered by Hiram Bingham." It is not clear when or even if this April 1912 issue of the magazine reached the Sociedad Geográfica de Lima, the one public entity subscriber in that country.

Bingham's 1912 expedition focused on clearing the Machu Picchu site of its jungle. Whereas two expedition members spent several months there, Bingham himself spent only 2 weeks at the site. In Lima, Guillermo Billinghurst by then had replaced Augusto Leguía as president and Bingham lost his powerful advocate.[3] Discontent in Cusco with "el Doctor" came to a head when several cusqueños publicly leveled the accusation that the North American had surreptitiously taken gold artifacts from Machu Picchu for his own gain. No one then or later was able to verify that charge. Later it became known that Bingham purchased a collection of 366 ancient objects and bribed Peruvian customs officials to get them on board ship without documentation. The person who sold the collection to him was the son-in-law of Carmen Vargas de Romainville, Tomás Alvistur, who Bingham (1913b:530) described only as an "enthusiastic amateur archaeologist." Readers never learned about Alvistur's role in providing antiquities to Bingham.

Animosity held by many Peruvians at that time had less to do with misappropriating national patrimony than it did with the festering resentment that a clever foreigner profited by taking artifacts that should have gone to them. Looting of buried treasure was then still generally accepted in Peru and, even today, pre-Columbian sites are not safe from depredation. Bingham's standard response to the charge that he had removed archaeological materials from Machu Picchu was that a Peruvian government decree had given him the needed permission to take out of Peru what he had found for further study. But that was only half the story. Bingham had manifested a cynical attitude toward Peru's claim to antiquities when he made a spectacular charge to the president of Yale about Max Uhle, the director of the National Museum in Lima. Uhle, he claimed, only pretended to increase the government's archaeological collection; he actually smuggled nine-tenths of the objects he discovered out of the country (Letter HB to AH 10-14-1912). Such an assertion may have been made to justify Bingham's own behavior in acquiring his personal set of antiquities.

Bingham acknowledged his own removal of antiquities for the Peabody Museum. Two problems associated with that transfer led to much debate over the years. President Leguía's authorization had ignored an earlier Peruvian law prohibiting the export of antiquities. Secondly, the agreement between Leguía and Bingham specified that those materials, once they had been analyzed, would be returned to Peru. Neither Bingham in his lifetime nor a succession of directors of the Peabody Museum at Yale honored that agreement until 2012. Some of the criticism directed at Bingham reflected visceral opposition to Leguía and his autocracy. Jose Gabriel Cosio (1913) led the charge. Charges of misappropriation mounted during Bingham's 1915 expedition, by which time he had gained a reputation as a thief. The accusations from Cusqueños cut him to the quick, and he reacted by cutting off contact with Cusco scholars. Another point of contention was that Bingham did not

[3]Augusto Leguía y Salcedo (1863–1932) served as President of Peru in two different administrations: the first, from 1908 to 1912, was initiated by a coup; the second, from 1919 to 1930, ended in a coup. A member of the oligarchy, Leguía ruled essentially as a dictator. A saving grace for many was the responsibility he took to modernize the country.

supply cusqueños with copies of his work (Pardo 1961:23). Personality also played a part. Bingham's attitude of superiority rankled many educated Peruvians. Furthermore, he did not observe the traditional patron-client relationship that expected powerful individuals who receive rewards of wealth and status to look after the needs of underlings, the system by which haciendas and all levels of government functioned.

Although Giesecke never accused Bingham of impropriety, neither did he vigorously defend him. Bingham's sparse, cool and perfunctory epistolary exchanges with Giesecke reflect Bingham's sense that Giesecke, his own countryman, had betrayed him by not coming to his rescue. In 1926, Bingham wrote Giesecke asking if the rumors were true that the University of Cusco had rescinded his honorary Litt.D. degree so that "he could adjust his entry in *Who's Who* accordingly…" (HB to AG 3-8-1926). Giesecke would not have let withdrawal of that award happen. Much of the decline in postal contact with Giesecke came as a result of Bingham's heavy involvement in congressional responsibilities. Years went by in which he had little time to think about Peru. Only when he failed to win reelection to the Senate did Hiram again think and write about his Andean experiences. In 1948 he published *Lost City of the Incas* and gave translation rights to a Chilean, not a Peruvian, publisher. Over the decades, sales of the book reflected the wide popular appeal of the Bingham account, though most readers had no idea how flawed that interpretation was.[4]

Bingham's decision not to acknowledge Giesecke's indispensable assistance rested on a three-sided pique. First, Giesecke did not keep Bingham's find secret, yet Bingham tried to withhold the information even from Giesecke—the very man who had told him where to go. Late in 1911, when Giesecke went to Lima for medical treatment, he read in the Lima newspapers a report that Bingham, before he returned to the United States, had delivered a talk at the Sociedad Geográfica de Lima on his expedition. That public lecture in Lima was not published, but some facts about it appeared in a note in the *Geographical Journal* the very same year (Markham 1911). The journal editor, C. Markham, had close ties to Peru. It is reasonable to assume that Bingham believed that word of his presentation in Lima would not find its way to Cusco. At that time, the coast seemed a world away from the southern highlands. On the basis of that journal note, Giesecke wrote his friend at the University in Cusco, José Gabriel Cosio, in October 1911 that Bingham had found ruins at Machu Picchu. Subsequently each went to Machu Picchu before Bingham returned in June of that year. In January 1912 Cosio went, taking along as his guide, Enrique Palma who had been there already in 1902. A few weeks later Giesecke and several students explored the site more thoroughly than had Bingham in 1911. Both Cosio and Giesecke wrote and published what they saw at Machu Picchu before Bingham's big article on it appeared in 1913 (Cosio 1912). Bingham could not have been pleased.

---

[4] Between 1948 and 2002, this book, so full of erroneous interpretations, went through seven facsimile editions from North American and British publishers. The Spanish-language edition, translated as *La ciudad perdida de los Incas*, went through four editions.

7.4.4 *Hiram Bingham's Interpretations of Machu Picchu*

Bingham never divulged in print or lecture the exact story of his encounter with Machu Picchu. In spite of the aura of Calvinist rectitude inherited from his forbearers, Geltung overwhelmed transparency. Enhancing his reputation as an explorer dominated his thinking. Over the decades Hiram received an unusual amount of praise, even though as the site became better known, archaeologists began to doubt Bingham's interpretations of its age, identification, function and population. A flagrant assertion was Bingham's (1913b:22) opinion that the Incas built Machu Picchu "probably 2,000 years ago." This view grew out of a belief, apparently derived from Clements Markham, about Inca origins. Bingham connected the three-windowed building found at the site with the three-cave legend of Inca origin. In spite of all evidence to the contrary, Bingham persisted in the belief of an early construction even as it became known that the Incas did not predate 1100 AD. Perhaps most egregiously, Bingham considered Machu Picchu to be the site of the Vilcabamba la Vieja mentioned in the chronicles. He could not bring himself to revising his early views. He maintained that as the "discoverer" of the site, his opinions were the correct ones. The public assumed he knew what he was talking about it; the cognoscenti knew better. The archaeologist Philip Ainsworth Means (1931:254–255), who had correctly deduced that Pachacutec had built Machu Picchu in the early fifteenth century, had a much more solid knowledge of Inca culture history than Bingham.

Bingham's assessment of Machu Picchu's function also was far off base. From the start, he viewed it as having primarily a religious character, citing the ruins of a putative temple and a monument to the Sun ("Intihuatana") carved from living rock. To promote that view he gave fanciful names such as "high priest's house" and "sacred plaza" to different components of the site. Based on an early bone study (Eaton 1916), he claimed that most of the inhabitants of Machu Picchu were "virgins of the sun," evoking a kind of nunnery in the service of the state religion. Subsequent osteological analysis of the skeletons at the site found them not to be primarily female (Verano 2003). Religion implies pilgrimage; if Machu Pichu had that function, it would not have been abandoned. Hedging his bets, Bingham also gave Machu Picchu a military fortress role. His 1930 book used the word 'citadel' in its title implying a need to ward off jungle Indians coming up the Urubamba Valley. On another point, Bingham (1916) grossly miscalculated Machu Picchu's total population. He called it a "city" and estimated that at its height 10,000 people had lived there. Yet dwellings at the site could not have sheltered more than 1000 inhabitants and probably no more than a third of those were permanent residents.

Even when reviewers of Bingham's 1930 monograph challenged his hypotheses, he was undeterred. He repeated the same rejected ideas in his last book, *Lost City of the Incas* published in 1948. His ego would not let go of the story he had created. Besides, readers loved it. That the book remains in print testifies to Bingham's skill in weaving an effective tale of adventure as well as in defending his discoverer role. The Swiss scientist Arnold Heim (1957:136) visited Machu Picchu in 1946 and simply repeated Bingham's ideas as the truth.

The definitive shift in the interpretation of the site's function came in the late 1980s when John Rowe (1990) convincingly pieced together from the Cusco archives that Pachacutec had built Machu Picchu as his country estate in the mid-fifteenth century. Rowe's first clue about Pachacutec was in an archival document cited in a book by Glave et al. (1983). Previously Rowe (1944/1945:313) had interpreted Machu Picchu primarily as a shrine site. Unlike Bingham, who is not recorded to have changed his ideas about Machu Picchu, Rowe did not hesitate to do so when new evidence emerged. Rowe concluded that the site had been part of a vast expanse of property extending 30 km to the north and 10 km to the south. The high quality of construction reflected the skills of planners and builders who followed Pachacutec's design. Like people of high privilege today, Pachucutec sought a setting of beauty and inspiration in a good climate. The warm but not uncomfortably hot weather and lush vegetation of this zone would have been especially appealing in June, July and August when the imperial city of Cusco, then as now, has cold nights and brown landscapes.

7.4.5 *The Geltung of Events*

Bingham's faulty interpretations of Machu Picchu left him with one essential achievement: the claim of discovery. During his lifetime, his status as the discoverer went unchallenged in the United States where he made his fame. Not until after his death did it become apparent that his claim required calculated scheming. First he kept from his expedition cohorts what Giesecke had told him. He then arranged to be the only member of the expedition to explore the ruins on that emblematic day of the 24th of July. If any of his expedition companions had gone with him, he could not have been the sole literate witness. By sending Isaiah Bowman farther down the valley Bingham eliminated him as an expeditionary rival.

Isaiah Bowman was the first North American geographer to write a dissertation based on field research in Peru. His record of publication enhanced the intellectual authority to someone who already had an authoritarian manner. Bingham's son claimed his father was envious of Bowman (Bingham 2000). If Bowman had accompanied the director to Machu Picchu, the latter would have had to share the spotlight. Indeed, given his intellectual bona fides on Andean matters, Bowman may have received the lion's share of the credit. Scientific discovery often follows the "Matthew effect," in which major attribution accrues to the best-known person associated with a find, rather than to the one who does most of the work or has the most original idea about it. Thus, Bowman's own suggestion that he work on mapping in the Urubamba jungle and take William Erving and Paul Baxter Lanius with him met entirely with Bingham's approval. Bowman wanted to collect first-hand data about the tropical forest, an environment he had not studied before, so that he could include that information in the book on the Andes he was planning. Beyond that, both parties sensed that physical separation would help to keep their touchy relationship from bursting into conflict. Later, when Bingham (1913b:408–440) wanted a map made of Machu Picchu, he called on Tucker and Lanius, not Bowman, the skilled field mapmaker. As for Erving, Bingham did not have an easy relationship with him either and felt it best if he went with Bowman.

The geographer Neil Smith (2003:78) erroneously claimed that Bowman felt he had been maneuvered out of the Machu Picchu discovery. Smith (2003:76) quoted from Bowman's notebook that "I knew then why Bingham wanted me to go to Santa Ana as far as I could and wait for him there." From that remark Smith extrapolated that "Bowman would have dearly liked to have collaborated in the discovery of Machu Picchu." If that was true, Bowman kept his feelings well hidden. In a letter dated August 1, 1911 written from Santa Ana, Bowman wrote Bingham about Machu Picchu: "good for the ruins and good for you. They told us at Huadquiña that there were some near the bridge of San Miguel. Also of others known to occur in the village. Duque also told us that those at San Miguel were much better than Ollantaytambo" (Smith 2003:78). Bowman's remarks indicate that those two hacendados knew about Machu Picchu before Bingham.

Bowman's (1916:viii) lack of resentment is suggested by his gracious forward in the Andes book: "It is difficult to express the gratitude I feel toward Professor Hiram Bingham, Director of the Expedition, first for the executive care he displayed in the organization of the expedition's plans, which left the various members largely carefree, and second, for generously supplying the time of various assistants in the preparation of results. I have enjoyed so many facilities for the completion of the work that at least a year's time has been saved thereby. Professor Bingham's enthusiasm for pioneer field work was in the highest degree stimulating to every member of this party. Further, it led to a determination to complete at all hazards the original plans." In my view, these were genuine sentiments, for Bowman was pleased about his research results on that trip, did not begrudge Bingham's finds, even though they represented another magnitude of achievement than his own.

Of Bingham's two companions at Mandor Pampa, Harry Foote and Kai Hendricksen, neither was prone to confrontation nor did they probably think of the stakes involved. One or both of them would logically have accompanied Bingham just in case of trouble, if they had not been dissuaded. Bingham wrote that the two men had "chosen" not to accompany him to look for ruins—Foote wanting to collect insects near the river and Hendricksen needing to wash his clothes. Neither Foote nor Hendricksen ever proffered in print their own reasons for not accompanying the chief on that day. In Bingham's eyes, these two literate and educated men represented possible competitors in a discovery narrative. He calculated that his two Peruvians companions, Carrasco the policeman and Arteaga the guide, offered no competition for the glory he wanted for himself. In Bingham's view, they would not qualify as co-discoverers for they could not appreciate the intellectual importance of the find. Bingham's entire Calvinist upbringing about honesty and righteousness withered when faced with the opportunity for international fame.

Geltung appeared also in the subtle shift in the reported facts. Initially Bingham acknowledged that Agustín Lizárraga ("a local muleteer") had previously visited the ruins and had scrawled his name in charcoal on the stones of the "temple". (Bingham 2000:19). Hiram's son and biographer, Alfred Bingham, pointed out that his father had also written in his journal "Agustín Lizárraga is discoverer of Machu Picchu and lives at San Miguel bridge…" (Bingham 2000:19). Bingham reported that Lizarrága had visited Machu Picchu in 1904. In fact, however, Lizárraga, together with Enrique Palma Ruiz, mayordomo of Had. Ccollpani and Gabino Sanchez, hacienda peon, came to Machu Picchu on July 14, 1902, not in 1904 as Bingham reported. An hacienda owner, Justo Zendón Ochoa give the three men material support. In his book on that expedition, Americo Rivas Tapia (2011) wrote that Lizárraga was not the "indio ignorante" that Bingham implied, but a well-read man who worked as a road administrator between Pachar and Echarati. That graffiti was evidence to Bingham that Lizárraga, though a peon at Hacienda Ccolpani, could write his own name and thus was potentially in a position to relate his own account. However, that threat vanished in 1912 when, during the heavy current of the rainy season, Lizárraga drowned trying to cross the Urubamba River on that makeshift bridge. Subsequently Bingham had a workman spend 2 days "…erasing from the beautiful granite walls, the crude charcoal autographs of visiting Peruvians" (Bingham 1913b:452). With graffiti erased and the muleteer dead, the main challenge to Bingham's hegemonic desire to be "the" discoverer disappeared.

Through more than three decades, Bingham embellished his claim to precedence. His 1922 book mentioned Lizárraga as a previous visitor, but elsewhere he wrote that "…with the possible exception of one mining prospector, no one in Cusco had seen the ruins of Machu Picchu or appreciated their importance" (Bingham 1922). That differed from the letter he had previous written to J.S. Keltie at the Royal Geographical Society on August 3, 1911, in which he claimed that "so far as I can discover only three Peruvians have seen it (except a few Indians)" (Smith 2003). In his 1930 scientific treatise, Bingham dismissed Lizárraga as a "treasure hunter," a label, justified or not, clearly meant to disparage the man's legitimacy (Bingham 1930:108). He also mentioned Peruvians who discredited the significance of Lizarraga's arrival at the ruins. Bingham's 1948 work mentioned Lizárraga not at all. Scholars in Cusco never forgot Lizárraga. José Gabriel Cosio (1912:44), who had seen Lizárraga's name on the wall in January 1912 when he visited Machu Picchu himself, proclaimed in print that Dr. Bingham had been "scooped." But Lizárraga, unlike Bingham, left no images or written descriptions, and so no wide debate about precedence ensued. Several decades later, however, when the world recognized Machu Picchu as an icon of pre-Columbian America, Lizárraga's name was frequently invoked on websites about Machu Picchu, though he was almost certainly not the first visitor. Jorge Flores Ochoa (2011), whose maternal antecedents had an hacienda near Machu Picchu, cautioned that the Machu Picchu ruins received many visitors and that it made no sense to elevate Lizárraga or any one particular person as the "discoverer." Oral history of the region held that generations of people had visited that area, some treasure hunters, others not.

Bingham (1913b:427) also felt compelled to make related discovery assertions, claiming that his team found the Inca ruins on Huayna Picchu on his 1912 expedition. That assertion ignored the fact, published in Cusco in 1912, that A. Giesecke and his group a month earlier had climbed up the steep trail and found the remains of buildings constructed there (Giesecke 1912). When Hiram Bingham's Peruvian trips are analyzed, it becomes clear that for him being a "discoverer" overshadowed everything else. Having made one "discovery," he searched for others.

Bingham's (1916) reference to Machu Picchu as a "lost city." by which he meant "...with the exception of a few local Indians, no one in Peru was aware of its existence," was cleverly designed to portray him as discoverer. He wrote "...the professors at the University of Cuzco know nothing of any ruins down the valley," Yet Giesecke, president and professor, had told Bingham where to find it. Bingham learned neither the site's pre-Columbian nor its post-Columbian history. In the archival records, Francisco Pizarro, in 1539, granted "Picchu" to his brother Hernando Pizarro as an encomienda in which coca, chile pepper, and fruit were the tribute items. By the seventeenth century, Machu Picchu had fallen in the hands of the Augustinian Convent in Cusco, though the property was not registered in their archives (Glave et al. 1983:181, 247). They were in conflict with a Cañari Indian, Francisco Poma Gualpa, cacique of an ayllu of Cañari *chasqueros* who were in charge of the royal messenger service between Cusco and Arequipa. The Cañaris claimed the land until their ayllu was dissolved in 1849. Because the area had been Pachacutec's personal property, few Indians lived in the zone and labor shortages hindered crop production. In the mid-seventeenth century, the Inca path through the bottom of the canyon disappeared. The 1650 earthquake felt in Cusco and beyond may have triggered landslides that destroyed the path as well as many small agricultural plots.

A notarized property transfer dated 1776 identified "Picchu, Machuy Piccho and Huana Picchu" owned by Antonio Ochoa who, in 1812, sold it to Marco Antonio de la Cámara y Escudero (Mould de Pease 2003). The property that included Machu Picchu changed hands several more times after that. The name "Picchu" or variants thereof appeared on several maps, evidence that it was a well-known landmark to those who knew the area. The 1865 map of Peru by Antonio Raimondi correctly indicated the peak of Machu Picchu, but since Raimondi never traveled through the gorge, an informant must have told him about it. In Ollantaytambo in 1875, the French explorer Charles Wiener (2010) heard mention of Huayna Picchu and "Miccho Picchu" as places in the valley within the District of Ollantaytambo and placed them on a map. Bingham eventually learned that Machu Picchu was part of Hacienda Cutija owned by Mariano Cerro. After that, its ownership was transferred to the Núñez family. In 1943 the Peruvian government expropriated it without compensation and in subsequent years acquired additional land to protect the environment and especially the landscape views of and from the site.

Ignorant of the documentary record, Bingham confidentially invented Machu Picchu as a place previously unknown to the Spaniards. He assumed that since brush and trees had overgrown parts of it, that the jungle had swallowed it after the Incas had departed and for centuries it had been in an inviolate state. More than four centuries of visitors left little documentation. Among possible pre-Bingham explorers of Machu Picchu were three nineteenth-century Germans. The engineer Herman Göhring (1877), who wrote an official report for the Peruvian government on the eastern valleys, mentioned the fortresses of Chuquillusca, Torontoy, and Picchu. Yet nothing indicates that Göhring actually ascended to the ruins above the floor of the canyon or that he even traveled through this part of the Urubamba Valley. Decades later Georg (aka Jorge) M. Von Hassel (1907), also an engineer with an interest in archaeology, explored the eastern valleys. A stronger case surrounds a third German, Augusto R. Berns, who had both the motivation to explore the ruins and who was in the right location to easily accomplish that (Greer 2009). In 1867, Berns acquired 25 km^2 of land in the Urubamba canyon across the river from Machu Picchu. There he set up a saw mill to produce crossties for the Southern Railway, then in the early stage of construction. That business venture eventually failed and the machinery was left to rot, so that the site became known as "La Máquina." Berns then set up a company to attract investors in a goldmine and, in 1887, another company to plunder antiquities in complicity with the Peruvian government. Sheer proximity and dark motive suggest that Berns may at some point have reached Machu Picchu and saw it as such a treasure trove that it merited designation as a business. But it is not known what, if anything, Berns may have taken from the site. That Berns was a liar and swindler does not foreclose his finding treasure and staying quiet about it.

Evidence of other pre-Bingham visitors is mainly circumstantial. Daniel Buck (1993), an amateur historian of Peru, has sifted the evidence concerning Thomas Payne, a British Protestant missionary at Urco near Calca. Payne's relatives and associates asserted that he reached the ruins in 1905 either alone or in the company of Stuart McNairn, also a missionary, who had lived in Peru from 1904 to 1911. Relatives also recounted to Buck that in 1911, Payne had supposedly told Bingham precisely where to go to find Machu Picchu. Thomas Payne was a plausible candidate for such an adventure, yet if he had actually reached the ruins, he would have written about it in letters. If he did, those letters have apparently not survived.

More certain is that people who lived in the zone entered the site of Machu Picchu. Paolo Greer (2009), an Alaskan amateur historian with long experience in Peru, believed that decades before Bingham, *huaqueros* had pillaged the ruins and that Doña Carmen Vargas, owner of Hda Huadquiña, had a collection of antiquities that may have come from Machu Picchu (Gänger 2014). Her son-in-law, Tomás Alvistur, sold 385 pieces of Inca ceramics to Bingham for US$2500.00. Given Alvistur's knowledge of that area close to his mother-in-law's estate, it is feasible that some of those pieces came from Machu Picchu. Taking the long view, Greer claimed that plunder of the site may go back to the Spanish Conquest of 1532. The chronicler Juan de Betanzos (1996:138) asserted that a gold statue of Pachacutec was placed on top of his tomb, which would have been at Machu Picchu. Greer hypothesized that the statue was removed as part of the ransom the Incas paid to release Atahualpa at Cajamarca and that the Spaniards later melted it down. In its place native people erected a stone statue of Pachacutec there where it stood until around 1880 when treasure hunters removed it.

The Cusco elite had been aware of the mercenary value of Inca objects since the nineteenth century and this interest stimulated looting and forgeries (Gänger 2014). Ana Maria Centeno (1817–1876), together with husband Pierre aka Pedro Romainville, were passionate antiquarians. In 1868 they and others formed the Peruvian Archaeology Society in Cusco. Centeno had a well-known cabinet of curiosities in her Cusco house that included Inca artifacts. After her death, they were sold to the ethnographic museum in Berlin. Pierre's son, Adolfo, sold antiquities to foreign merchants in Cusco (Gänger 2014:86).

Bingham's self-proclamation as discoverer without qualification meant that the crucial piece of information from Albert Giesecke that had led him to Machu Picchu could not be specifically acknowledged. The closest Bingham came to admitting a connection appeared in the 1913 *National Geographic* article. He thanked "the president and faculty of the University of Cusco who aided us in numerous ways and whose many courtesies included not only hospitable entertainment at houses of the professors, but assistance in finding interesting points whose whereabouts was not generally known" (Bingham 1916:445). At the time, few readers could have guessed what he was referring to in those eleven last cryptic words. Their meaning did not become clear until after Giesecke's 1961 revelation. Acknowledgements in the 1930 scientific account mentioned only Augusto B. Leguía, the President of Peru; Don Mariano Cerro, owner of the estate which included Machu Picchu; and Don César Lomellini, the Italo-Peruvian outfitter in Cusco. Bingham (1930:39), however, was correct in his statement that "the ruins remained unknown to other savants until 1911." In his heart of hearts, he knew by then that that was his real niche in the Machu Picchu story.

Bingham's 1948 book did not mention Giesecke either. An opportunity for rectification or deeper reflection by a man past middle age yielded instead to the Geltung. In that book he repeated his original story of that day on July 24, 1911 when he glimpsed the temple of the three windows. It was the most thrilling moment on his three expeditions and a discovery moment to savor. Hiram's reputation as an explorer rested on it. Relevant here is Lowenthal's (1996:146) reminder that once a self-chronicle is consigned to print, it is hard for its author to remember it in any other way. Public duplicity contrasted with what Bingham knew in private. In a letter written to Giesecke the year he returned to Peru, Bingham stated that "I cannot remember a time when I owed so much in the way of honor, hospitality and efficient help to me" (Letter HB to AG 11-2-1948).

Bingham's writings emphasized the operational aspects of the expedition, but it is important to contextualize the larger circumstances that facilitated his Machu Picchu encounter. An extraordinary convergence of serendipitous timing made it possible. Without the recently completed railroad from the coast, getting to Cusco with a great deal of gear would have been infinitely more complicated. Without the fairly recent mule path through the Torontoy gorge, Bingham's party would not have been able to reach Mandor Pampa. Several nineteenth-century European explorers, Francis de Castelnau in 1846, Antonio Raimondi in 1865 and Charles Wiener in 1875, did not have the advantage of that trail. Their route instead followed trails through passes, either above Ollantaytambo or Calca, and from there into the tropical valley far below Machu Picchu. Even if those stalwarts somehow had known the precise location of Machu Picchu, without a trail through the Canyon of Torontoy canyon it is unlikely any of them could have taken their pack mules to where they could reach the ruins.

Another happenstance of time and space for Bingham was the presence of Albert Giesecke in Cusco, who had arrived only the year before to take over the university rectorship. Equally to the point, Bingham was fortunate that Giesecke, ever the diplomat and facilitator, chose not to contest the details of Bingham's account during the latter's lifetime. Giesecke's own rather oblique reference in 1961 came after Bingham had died and on the occasion of the fiftieth anniversary of the find. Being in the right place at the right time was the basis of Bingham's success. Fortunate timing included the Peruvian president Leguía who had taken power the year before the expedition's arrival. Leguía had his own personal reason to make a deal with the man from Yale. He needed an advocate for his son who had gone to the United States to study and, as a quid pro quo, was willing to break the laws of his own country to give Bingham what he wanted.

Photographic technology was also a big factor in Bingham's success in telling his story to the world. At a time when photography was used mostly for taking portraits, he had the prescience to bring a camera, tripod and lots of film. The Eastman Kodak company had developed the camera ("Kodak A3") he used only about a decade earlier; the older technology would have been too bulky and complicated for such an expedition (Balm 2008). It was not simply that each photograph conveyed information that could not be expressed in words. The images provide an indisputable record of what he found and, for the doubters, a kind of irrefutable proof of his presence there. The cusqueño José Gabriel Cosio (1912) provided an excellent description of the ruins in his article, but its impact was less without photographs.

7.5 Deconstructing Bingham's Vanities

From the start Bingham saw himself as chief of the expedition. He set its agenda, chose the members, and solicited the funds. He also saw himself as the explorer, which is precisely what he listed as his occupation in his early entries in *Who's Who in America*. Clothing worn in the field, consisting of riding breeches, vest and fedora, conveyed a dashing image that became part of his expeditionary persona. Hiram the Explorer also had a controlling nature. As chief of operations he screened information that came out of these trips so that he would not be scooped. Bingham had expedition members sign a legal contract that required them to clear with him anything they wrote relating to the Peru trip before publication. In that way Bingham could prevent any of his cohorts from trespassing on topics he considered his domain. Isaiah Bowman (1912) published a short article on an Inca wall without first clearing it through Bingham. That transgression initiated a sharp rebuke from Bingham who regarded the Incas a topic that only he could divulge in print (Letter HB to IB 1-7-1913).

Hiram Bingham was a striver and maneuverer of the first order. He wanted fame as an explorer, but realized the importance of money for that. Meeting his goal required material assets well beyond those of his youth. His grandfather Hiram Sr., heavily involved in implanting Christianity in the Hawaiian Islands, had dedicated himself to a religious calling. His father, Hiram Jr., fully expected Hiram III, his only offspring, to carry on missionary work in the Pacific. As a student in New England, first at Andover and then Yale, he was introduced to alternatives to piety, personal sacrifice, and genteel poverty. By the time he got to Yale, Hiram no longer accepted the canon of beliefs that his parents had tried so fervently to instill in him. In 1900 he married Alfreda Mitchell, heiress to the Tiffany jewelry fortune. She was raised with a secular attitude and she moved Hiram away from religion. The pair seemed to have few shared interests, but her family's financial resources assured him of both wealth and social status. They made possible several of his early trips and facilitated his fieldwork. He could pay Indians to guide him, find and deliver artifacts, and to submit to 38 measurements of their bodies, which was one of the anthropological activities of the first expedition. He knew that Peruvian peasants were so poor he could buy their cooperation. Bingham saw it as payment for services rendered, others saw it as bribery. He monetized his dealings, even at the Machu Picchu excavation in 1912, by offering a sol ("silver dollar") for every burial cave found. His workmen found eight caves and asked for eight soles. On a trip to Vilcabamba, he gave one sol to a local who led him to one ruin and in that way found a site he otherwise would have missed.

His financial cushion facilitated his transition from professor to conservative Connecticut politician. His political ascent began in 1922 when he was elected Lieutenant Governor of that state. Subsequently he became Governor and then U.S. Senator, a political position that ended in 1932 when a congressional censure for an ethical violation involving the use of a lobbyist to draft legislation caused him to lose his bid for reelection. Reprobation was prompted not only by his wrong doing, however, but also by his egotism and elitism which aggravated his colleagues (Miller 1982:153–154). He returned to government service between 1951 and 1953 when President Truman appointed him head of the new Loyalty Review Board during the tumultuous period of the McCarthy witch hunts. Bingham's hand in several controversial dismissals from the Civil Service reinforced the dark view that many people took of him. In 1937, Hiram divorced the mother of his seven children and later married the younger woman he had taken as a mistress. He left his own sons out of his inheritance (Bingham 2000).

Hiram Bingham's wealth, education, social status and political roles supported the view he had of himself as an *Übermensch* living a life of creative adventure. But his oversized ego led him to actions and stances that sullied his achievements. His unbridled determination to control the Machu Picchu discovery narrative to his advantage has greatly affected his place in the history of exploration. As the fame of Machu Picchu grew through the decades, Bingham attributed his find to his organizational skills and tenacity. By framing his Machu Picchu exploit as a grand adventure full of perils, it conveyed a sense of reward for dangers incurred. To enhance the image of the heroic explorer, he made exaggerated assertions that a dense jungle covered the ruins and that vipers lurked everywhere in the undergrowth. Given the presence of the trail to Cusco, the Urubamba was reasonably well connected, countering Bingham's remark (1913b:473) that Machu Picchu was in an "extremely inaccessible part of the Andes." If the ruins were to be meaningful, some reconstruction was necessary. Bingham was opposed to any such work at the site, for he wanted Machu Picchu to remain as he personally remembered it (Fig. 7.6).

Fig. 7.6 In 1921 Martin Chambi photographed Machu Picchu before restoration had started (Photograph by Martin Chambi)

Bingham's refusal to change his interpretations about Machu Picchu even in the face of professional opinion hindered the advancement of knowledge. Many lay people could not imagine that the discoverer could possibly be wrong.

Other personality clues came from reading his comments on the broader Andean landscape. They tell us little about what he actually saw, but much about the fact that his was a special view that had not been seen before (Poole 1998:128). He imposed his own toponymy on features that already had traditional Quechua names. Three ice fields in the Cordillera Vilcabamba were given names: "Grovesnor Glacier" [for the head of the National Geographic Society, who published his work]; "Morskill Glacier" [for the British manager of the Southern Railway who gave the expedition free passage]; and "Alfreda Mitchell Glacier" [for his dear wife, the money bags of the family]. The act of naming was more a calculated way to curry favor than to express gratitude.

For almost 40 years, Bingham made many and varied requests to Albert Giesecke. They started in 1911 when he asked Giesecke to alert the departmental prefect, J.J. Núñez, of his impending arrival in Cusco. Following that first visit, Bingham prevailed upon Giesecke in 1912 to send him a large fossil found near Cusco, a major logistical undertaking. In 1913 he asked Giesecke to provide him with abundant documentation of Cusco Valley in order to construct a modern map. Reciprocity did not occur in this relationship. Bingham turned down Giesecke's requests one after the other. In 1915, Giesecke wrote to ask if Yale could start an archaeological training program at the University. Bingham replied that it was an "impossible request," possibly because he did not want Peruvians carrying out excavation at Machu Picchu that might contradict his findings. In 1948, he asked Giesecke to tell him about the impact in Peru of his *Lost City of the Incas* book, wanting to know "...if the intellegentsea (sic) has shown any real interest in my theories re Machu Picchu" (Letter HB to AG 1948). In 1950, after a devastating earthquake destroyed two-thirds of Cusco, both ex-president Bustamente and Giesecke asked Bingham to lend his prestigious name to head a relief committee. Hiram declined without offering a reason. Then Giesecke in 1951 solicited help in a subscription campaign for restoration work at Machu Picchu. Bingham replied that he was adamantly opposed to the whole idea of any reconstruction at the site. The last negative response came in 1955, a year before Bingham died, when Giesecke wrote to ask permission for his name to be used for a conservation fund for the site. Innocent as that request was, Bingham turned Giesecke down on that as well.

In spite of multiple rebuffs, Giesecke remained over five decades Bingham's big promoter in Peru (Giesecke 1961). During his third expedition in 1915, Giesecke initiated Hiram's induction as a member into the Instituto Histórico del Cusco and, at the same time, honored him with a doctorate *honoris causa* from the University. Friends and relatives of Hiram Bingham visiting Peru received Albert's attentive hospitality. Giesecke got involved as a consultant for radio shows and a film about Bingham in Peru. Several invitations Giesecke sent to Bingham to visit Peru were turned down until, finally, in 1948, he accepted one. Having recently finished his *Lost City of the Incas* book, Hiram saw a return as an opportunity to promote its sale and influence and, for sentimental reasons, wanted his second wife Suzanne to see the place that had been so central to his life before politics. Arriving as a distinguished grey-haired gentleman, Bingham received deference and honor. The Peruvian government featured Bingham at the inauguration of the vehicle road from the valley floor to the ruins of Machu Picchu above. This "Carretera Hiram Bingham" replaced the mule trail that had opened in 1934 (Fig. 7.7). Known locally as "el zigzag," the new hairpin road henceforth carried busloads of visitors from the train station to the entrance to the ruins. Albert Giesecke, at the time the civil attaché at the U.S. Embassy in Lima, had arranged everything connected with Bingham's visit. Attaching the name of Bingham to the road was Giesecke's idea, and indicated that on cultural matters, he could be very influential in the Peruvian government.

7.6 Albert Giesecke (1883–1968) as the Key Figure in the Machu Picchu Story

The focus on Hiram Bingham and the reticence of Albert Giesecke to tell his own account of the events of 1911 long distorted the Machu Picchu discovery narrative. Giesecke looked for the best side in people, an attitude that went hand in hand with his concern for and involvement in the common good. Quite in contrast to Hiram Bingham's self-promotion, Giesecke's focus was on others rather than on himself.

Fig. 7.7 The "Hiram Bingham Highway" links the railroad station with the ruins 400 m above the floor of the canyon. Bingham came to Peru in 1948 for the inauguration of this unpaved road (Photograph by DW Gade)

He did not advertise his formidable achievements, which explains why many are poorly known. Some were done behind the scenes, and others got the credit. Though he lived in Lima much longer than he did in Cusco, his name is closely tied to the Imperial City. Giesecke not only put the University of Cusco back on track, he had a profound impact on the city. Three times elected mayor, Giesecke essentially pulled the city out of its medieval stupor and toward what a city of the twentieth century should be. He himself planned how and where water and sewer pipes could be laid. He planned the arrival of electricity in his municipality and moved the central market away from the main plaza to a hygienic building elsewhere.[5]

Giesecke also revolutionized local Peruvian attitudes about and understanding of the Cusco region. His 1922 coined term "*cuscología*" covered all subjects and topics that focused on the regional (Aparicio-Vega 2000). Giesecke was the first to appreciate the role of archaeology in promoting the economic development of the department. Giesecke's own contributions to the university, municipality, and society should be part of any discussion of early twentieth-century Cusco (Super 1994; Krüggeler 1999; de la Cadena 2000).

Albert Giesecke had several attributes that explain many of his accomplishments (Fig. 7.8). Perhaps no other North American has understood the Peruvian point of

[5]As in medieval towns, raw sewage ran uncovered down the middle of Cusco's narrow streets and the daily markets held on the two main plazas generated unclean spaces.

Fig. 7.8 Albert A. Giesecke (1883–1968), here seen in 1960, was a North American who spent most of his life in Peru. He first met Hiram Bingham in 1911

view as well as Giesecke did. Soon after he moved to Lima in 1923 to work for the Peruvian government, he became the most knowledgeable person in the country on cultural matters. Perhaps because he never became a Peruvian citizen, he kept his role in the background and mostly out of the media. He later became the cultural attaché at the U.S. Embassy in Lima. A diplomat in the deepest sense of the word, Giesecke had a gift of dealing with people known in Spanish as "don de gentes." In 1937 he founded the Peruvian-North American Cultural Institute (ICPNA) that served as a model for those that started in other Latin American cities. Generations of Peruvians learned English in the ICPNA classrooms. He never returned to the United States to live, and Peruvians with experience of Yankees have considered Albert Giesecke as "the most Peruvian of North Americans" (Rubio Correa 2007).

Giesecke was instrumental in starting the university archaeology museum in Cusco by encouraged an hacendado, José Louis Caparó Muñiz (1845–1921), to hand over his collection of 2000 artifacts to the University and by lobbying the government in Lima to buy it. Giesecke used his powers of persuasion with Victor Larco Herrera who, in 1924, gave his artifact collection to the state as part of a property exchange. Many objects salvaged from illegal exportation were later added to the new institution. In 1938, Giesecke, supervising the excavation of Pachacamac, a major oracle site near Lima, unearthed intact the wooden idol that Hernando Pizarro erroneously claimed to have broken into pieces in February 1533 when the Spaniards sacked the sanctuary. Giesecke was the key intermediary between the Wenner-Gren Foundation and the University of Cusco in funding a Department of Anthropology with John Rowe as its first director.

Albert Giesecke's effectiveness in Peru owed much to his personality, but also to his early experiences and advanced training (Giesecke 1963; Gade 2006). Unusual for the period, he had accumulated knowledge of South America while still a school boy in Philadelphia's Central High School. A teacher, Cheeseman Herrick, sparked Albert's interest in the history and economy of this part of the world. In addition to 3 years of German and 2 years of Latin, he also took 2 years of Spanish. Growing up speaking German at home with his immigrant parents made Albert comfortable communicating in different vernaculars. His international outlook was reinforced when, as a college graduation present from his parents, he went to Europe and learned French. At the University of Pennsylvania, he took archaeology from J. Alden Mason, geography from Paul Goode, and governments of Latin America from Leo Rowe. Rowe, who later became director of the Pan American Union, recommended Giesecke to the president of Peru as the best candidate for a job in the Ministry of Education. He also studied with Emory Johnson of the Wharton School, who had been head of the team that recommended construction of the Panama Canal.

Three years of research and study at Cornell University led him to a Ph.D. in economics. Thus Giesecke came to Peru in 1909 to work at the Ministry of Education and in 1910 to Cusco as rector of the university. Well prepared as a scholar, he had a decent knowledge of the country and a language facility that enabled him to become rapidly fluent in Spanish. In 1910, the central government in Lima appointed him president of the University of Cusco. Though in his 30s, Giesecke became a fine administrator, defusing the hornet's nest that had roiled the institution until his arrival. Albert served as an exemplar to his Cusco colleagues, a scholar who was as much interested in research as teaching. Throughout his life, he initiated all manner of civic engagement, and played important roles in the development of social capital in Cusco and Lima over six decades.

7.7 The Geltung of History

As the geographer J.K. Wright (1947:15) opined, "the most fascinating *terrae incognitae* of all are those that lie in the minds and hearts of men." To those who seek to move beyond the myth, Bingham's account of Machu Picchu will always remain problematical. His case, however, is not unique nor the most egregious. In 1519, Pedro de Balboa marching across the Panamanian Isthmus ordered his companions to turn around and go back so that he would have no other contenders as discoverers of the Pacific. Several other explorers were anxious to be discoverers without honoring the facts. Father Louis Hennepin in the late seventeenth century made the false claim that he had explored the Mississippi River to its mouth. James Bruce in 1772 asserted he had found the source of the Nile River in Lake Tana, even though the Portuguese Jesuits had preceded him by 150 years. David Livingstone also failed to grant credit to the Portuguese whose explorations in 1616 in Southern Africa had far predated his own. In a case of shameless fraud, Frederic Cook (1865–1940) claimed to have reached the North Pole in 1908 ahead of Robert E. Peary.

Hiram Bingham could be compared to the explorer Christopher Columbus as an explorer analogy. Both the Admiral of the Ocean Sea and the Professor/Senator claimed discovery for themselves, even though both America and Machu Picchu had previously received outside visitors. And, of course, people were already living in both of the places ostensibly "discovered". Columbus and Bingham were each dead wrong about what they said they had found: the New World was not India and Machu Picchu was not Vilcabamba. Both men failed to convince the respective authorities, governmental in the first case and scientific in the second, of the correctness of their interpretations. The royal court of Spain dismissed Columbus for his grand geographical illusion that he had reached Asia. Bingham's unsustainable and now quite bizarre hypotheses about the site above the Urubamba River did not pass muster in the court of scholarly opinion even well before 1930. Yet during their lifetimes Columbus and Bingham refused to revise their opinions. Both men sought to turn their exploits into increased status and privilege for themselves. Neither chose to share glory with others, for that would have diminished their accomplishments: Columbus failed to give Vicente Yáñez Pinzón any recognition for his important role; Hiram Bingham did not publicly acknowledge Albert Giesecke. In 1966 the travel writer Selden Rodman interviewed Giesecke in Lima about the latter's crucial tip to Bingham in 1911. Giesecke told him that Bingham replied: "Very interesting, but it's not what I am looking for" (Rodman 1967:159). Waisbard (1974) wrote that when she asked Giesecke after his lecture in 1961 why Giesecke did not correct the Bingham narrative, he refused to explain his reasons. We may surmise that Giesecke did not want to jeopardize Bingham's achievement, which, if exposed, would have revealed Hiram to have been a liar.

7.8 Machu Picchu in Perspective

The scientific unfolding of Machu Picchu has over a century had multiple impacts of cultural and economic significance (Burger and Salazar 2004). The publicity that Bingham generated at Machu Picchu awakened cusqueños to the significance of their pre-Columbian past. At about the same time, the abrupt end of the rubber boom dampened the El Dorado complex that Amazonia was a reservoir of bottomless wealth. Local intellectuals began to read chronicles, not for glorious accounts of the conquest, but for descriptions of Inca achievement. For the first time, the government attempted to protect archaeological sites from despoliation. Giesecke involved students in learning about where they lived rather than focusing on theoretical abstractions. Initiated by Giesecke and continued by Jorge Cornejo Bouroncle and others at the university, students in Cusco wrote term papers about various facets of peasant life and local history that were part of their own experience. In July 1912, he led a group of students to see Machu Picchu for themselves.

Giesecke, trained as an economist, detected a tourism vocation for Cusco before anyone else had. He ardently promoted construction of an airfield for Cusco, which was the starting point for local recognition of the potential of commercial aviation. Albert prepared the first English travel guide to the wonders of Cusco Department and included a section on Machu Picchu (Giesecke 1924). He was the major force, much of it behind the scenes, for developing the infrastructure needed to open

Machu Picchu to tourists. Construction in 1928 of a rail line through the Canyon of Torontoy, an on-site government hotel at the ruins, and the vehicle road from the train to the ruins were all projects in which Giesecke had a voice.

In 1929, the Peruvian government passed a law designating state authority over archaeological sites. With that law, ownership of Machu Picchu passed from Hacienda Cutija to the state. The Peruvian government at the time did not intervene, however, when the owner of Hda Mandor sold off 3482 ha to individuals and companies. The shambolic town of Aguas Calientes was made possible by that parcelization. A burgeoning collection of hotels, restaurants and shops crowded into this part of the canyon where the train stops at the station. No town should have been authorized. Settlement in the bottom of the canyon, with 70 % slopes on both sides, is vulnerable to annihilation from debris flows or landslides. Buildings close to the river are subject in certain years to flooding.

Belatedly it was realized that Machu Picchu was not just about the ruins, but also about the beauty of their natural surroundings. In 1968, the protected area was expanded to create a national Archaeological Park encompassing 10,724 ha; in 1981, it officially became an Historic Sanctuary, covering three times the area of the park. In 1983, the Sanctuary became a UNESCO World Heritage site, an important honor, but also a designation that introduced a certain amount of international control over the area. Pressure from world bodies has checked development plans favored by the Peruvian government. Biotic inventory and monitoring have become important activities (Galiano Sánchez 2000).

By the end of the twentieth century, Machu Picchu received an influx of more than 200,000 visitors a year. Since the 1908s a small portion have arrived on foot via the Inca Trail (Angles Vargas 1999). Its nationalistic symbolism has grown along with its importance to the tourism business. Politicians embrace Machu Picchu to further their own aspirations. In 2001 the day following his official inauguration, Alejandro Toledo took the presidential oath of office at Machu Picchu. The act was seen as a way to reclaim it for the indigenous of Peru. Though the ruins are in the public domain, the real economic beneficiaries of Machu Picchu are a few entrepreneurs who operate the rail line and hotels. Neither the government nor the Peruvian travel industry have wanted to restrict visitor numbers. Yet the pressure of evermore increasing number of visitors degrades the ruins. As an alternative to the zigzag highway, entrepreneurs proposed a cable car to carry people from the canyon floor to the ruins. First proposed in 1975, it became a serious proposal by 1998 and was approved by the state. But concerned constituencies protested on the grounds that a cable-car (*teleríƒico*) would be visually intrusive and would compromise the integrity of the site. Only in 2003, when international voices threatened to remove Machu Picchu from the list of World Heritage Sites, did Peruvian officials desist in this plan. Aside from the mode of access, concerns are expressed about increases in numbers of people that could be brought to the site. In 2013, visitors, two-thirds of them from outside Peru, numbered more than a million for the first time (Mincetur 2014). On any given day, between 3000 and 4000 pedestrians daily clamber through the ruins causing deterioration with simple wear and tear. Furthermore, Machu Picchu is vulnerable to seismic movement, solifluction and infiltration (Bouchard et al. 1992).

7.9 Conflict of Attribution

When heavy political commitments took him away from his still unpublished Machu Picchu materials, Bingham contracted with Philip Ainsworth Means to write up the data as the "big" book about the ruins. Means, an accomplished scholar and noted Andeanist in his own right, invested time and energy in the project. But when Bingham received the page proofs from the Yale University Press, he eliminated without explanation Means as co-author of the book. Means (1930:901), reviewing this book for the American Historical Review, mixed commendation with pointed criticism: "Why, Dr. Bingham has clung to an exploded theory, for which there is not an atom of serious proof, instead of going on to show the real history of the site, I cannot say." Advanced age did not soften his egotism. A disingenuous omission to his last book, *Lost City of the Incas* published in 1948, can only be described as self-serving in its duplicity. In it, Bingham quoted Diego Rodriguez de Figueroa's 1565 chronicle, but left out the critical sentence in which the chronicler referred to "Picho," i.e., Machu Picchu (Rodriguez de Figueroa and Pietschmann 1910:94). Bingham (1948) knew that, if included, the reference would have undermined his previous assertion that Spaniards did not know about Machu Picchu in the colonial period.[6]

For more than five decades, Hiram Bingham's adventure was part of the standard Machu Picchu narrative. His name is inextricably bound to it. Indeed, part of the charm of the site for the average visitor was the 1911 account of serendipity that either they had read or had recounted to them by a guide. The mystique of Machu Picchu has entailed more the name of Hiram Bingham who found it than that of the Inca Pachacutec who built it. Bingham's spelling of the site became the internationally accepted orthography. Bingham inspired other explorers to seek hidden pre-Columbian cities in the back country of South America. In the annals of South American exploration, the name Percy Fawcett (1860–1925) stands as one who became entranced with the idea of lost civilizations garnered from the Machu Picchu account. Fawcett disappeared while on one of his trips into the Brazilian wilderness and was never heard from again.[7]

As Machu Picchu gained iconic stature as a symbol of Peruvian identity, a strong nationalist reaction against the Bingham legacy appeared in Peru. In the 1990s, Bingham's controversial place in the discovery narrative exploded. Rather than a hero, he is often presented as a duplicitous schemer who meticulously planned out his fame and stole unknown objects of the national patrimony. Peruvian historian

---

[6]The exact words of Rodriguez de Figueroa are worthy including: "Esta noche durmí al pie de un cerro nevado en un pueblo despoblado Condorama, donde avia una Puente en tiempo antiguo, que pasavan el rio de Vidcos para ir a Tambo y a Sapamarca y a Pichu que es de la tierra de paz" (Figueroa and Pietschmann 1910: 94).

[7]Fawcett had read a manuscript in Rio de Janeiro that evoked a ruined city of stone in the interior of Brazil. In 1925, the American Geographical Society contributed $1000 to an expedition. But the search in the State of Mato Grosso for that supposed city had tragic consequences: Fawcett, his son and his son's friend died, presumably killed by Indians.

Mariana Mould de Pease (2000, 2001) labeled Bingham "the discoverer," intertwining the falseness of his claim with the controversy over the artifacts he took to Yale. Although the contract with the Peruvian government called for all the pieces to be sent back, a century passed without these objects even being archaeologically analyzed. For Peruvians, Bingham's prominence in the narrative became increasingly incongruous for a place infused with *peruanidad*. Peru had itself changed. Now even Peruvians of European descent such as Señora Mould de Pease could rally around the transcendental meaning of Machu Picchu for their cultural identity. That a foreigner should get credit for discovering a site that some Peruvians had always known about was unacceptable. Long ignored or overlooked, the statement of cusqueño José Gabriel Cosio (1912:20) that Hiram Bingham was not the discoverer of Machu Picchu eventually resonated with Peruvians. Others have since ridiculed Bingham's assertion (Aimi 2011).

Nevertheless, as flawed as he was, another issue also warrants examination. Hiram Bingham had a precious attribute, that of intellectual curiosity, then in short supply in Peru. While inquisitiveness for its own sake drove many Europeans and North Americans to learn about the past, Peru had mainly treasure hunters even in its educated class. If Bingham had not gotten there, it is very probable that within 6 months, Albert Giesecke, another foreigner, would have reached the ruins; it was on his docket of things to do. Peruvian society as a whole lacked a spirit of inquiry, a shortcoming that could not be reduced to the difference between a developed and an undeveloped society. Giesecke sensed that systemic failure to ask questions when he came to Cusco in 1910. An approach of scholasticism dominated the universities that focused on the truths inherited from the European past. Recognizing that, Giesecke sought to turn the university away from that stifling mode of thinking and its endless theorizing and ideological posturing and replace it with striving to understand what lay right under everyone's noses: Peru, a land of exceptional environmental diversity, glorious culture history of several civilizations, and a contemporary way of life rich in traditions.

Anti-Bingham rhetoric, however, also leads to a sober consideration that might, at some point, trigger serious debate. Yes, Peruvian scholars over the past century have created much new knowledge about many aspects of their fascinating national territory. Such outstanding *amautas* as Fortunato L. Herrera, Cesar Vargas Calderon, Javier Pulgar Vidal, Jorge Flores Ochoa, Franklin Pease, and Horacio Villanueva Urteaga are national treasures for the roles they have played in understanding Peru. Yet from a larger world perspective, intellectual curiosity has not been valued by Peruvian society. As Bingham mentioned, even the hacienda owners, who formed the elite of the Urubamba Valley, had little interest in wanting to understand the pre-Columbian ruins often on their own properties.

The Peruvian university was, in the past, a place where many students avoided immersion in their studies in favor of political activity. Antonio Raimondi (1874:I:8) admonished Peruvian students to "dispense with politicking and devote yourselves to getting to know your country and its immense resources." Sheer inquisitiveness, not economic gain, inspired Raimondi who became Peru's greatest explorer. He acknowledged the power of wanting to know: "…stimulated by curiosity,

I retreated to the densest of forests as if I was fleeing human footprint in order to put myself directly in this marvelous world" (Raimondi 1874:I:8). With that attitude, discoveries never cease.

References

Aimi A (2011) Scoperte Machu Picchu: la versione di Bingham. Archeo: Attualità del Passato 317:56–67

Angles Vargas V (1999) Machu Picchu and the Inca Road/Machu Picchu y el camiono Inca. Lima

Anonymous (1912) Explorations in Peru. Natl Geogr Mag 23(4):416–422

Aparicio-Vega MJ (2000) Cuscologia y sus origenes. In: Millones L, Tomoeda H, Fujii T (eds) Desde fuera y desde adentro: ensayos de etnografia e historia del Cuzco y Apurimac. National Museum of Ethnology, Osaka, pp 95–121

Balm R (2008) Discovery as autobiography: the Machu Picchu case. Terra Incognitae 41:102–113

Bingham H (1913a) The discovery of Machupicchu. Harper's Monthly Mag 26:709–719

Bingham H (1913b) In the wonderland of Peru. Natl Geogr Mag 24:387–574

Bingham H (1915) The story of Machu Picchu: the Peruvian expeditions of the National Geographic Society and Yale University. Natl Geogr Mag 27(172–186):203–217

Bingham H (1916) Further explorations in the land of the Incas: the Peruvian expedition of 1915 of the National Geographic Society and Yale University. Natl Geogr Mag 29:431–473

Bingham H (1922) Inca land: explorations in the highlands of Peru. Little Brown and Co, Boston

Bingham H (1930) Machu Picchu: a citadel of the Incas. Yale University Press, New Haven

Bingham H (1948) Lost city of the Incas: the story of Machu Picchu and its builders. Duell Sloan and Pearce, New York

Bingham AM (2000) Explorer of Machu Picchu: Portrait of Hiram Bingham. Triune Books, Greenwich

Bouchard JF, Carlotto V, Usselman P (1992) Machu Picchu: problemas and conservación de un sitio Inca de ceja de selva. Bulletin de l'Institut Français d'Etudes Andines 21(3):905–927

Bowman I (1912) A buried wall at Cusco and its relation to the question of a pre-Inca race. Am J Sci 34:497–509

Bowman I (1916) The Andes of southern Peru: geographical reconnaissance along the seventy-third meridian. Henry Holt and Company, New York

Buck D (1993) Fights of Machu Picchu. South Am Explorer 32:22–32

Burger RL, Salazar LC (eds) (2004) Machu Picchu: unveiling the mystery of the Incas. Yale University Press, New Haven

Cosio JG (1912) Una excursión a Machupiccho ciudad antigua. Revista Universitaria 1(2):2–22

Cosio JG (1913) Informe elevado al supremo gobierno sobre los trabajos de la expedición científica de la Universidad de Yale. Revista Universitaria 2(5):2–34

Cruz Ccorimaya N (2009) Christian Bües, biogeografía y recopilación de obras. Centro Cultural José Pio Aza, Lima

de Betanzos J (1996) Narrative of the Incas. In: Hamilton R, Buchanan D (eds and trans) University of Texas Press, Austin

de la Cadena M (2000) Indigenous Mestizos: the politics of race and culture in Cuzco, Peru, 1919–1991. Duke University Press, Durham

Eaton G (1916) The collection of osteological material from Machu Picchu. Connecticut Academy of Arts and Sciences, New Haven

Flores Ochoa JA (2011) El "descubrimiento de Machupiccho: tradicion oral y etnohistoria. Revista El Antoniano 117:3–14

Gade DW (2006) Albert A. Giesecke (1883–1968)—a Philadelphian in the land of the Incas. Expedition 48(3):27–33

Galiano Sánchez W (2000) Situación ecológico-ambiental de Santuario Histórico de Machu PicchuL Una aproximación. Programa Machu Picchu, Cusco

Gänger S (2014) Relics of the past: collecting and pre-Columbian antiquities in Peru and Chile. Oxford University Press, Oxford

Giesecke AA (1912) Una excursión a Machupicchu. Boletín de la Sociedad Geográfica de Lima 28:294–300

Giesecke AA (1924) Guide to Cuzco: the mecca of South America. Editorial Garcilaso, Lima

Giesecke AA (1961) Breves apuntes de la vida y obra de Hiram Bingham. Revista del Museo e Instituto Arqueológico 19:11–35

Giesecke AA (1963) The reminiscences of Albert A. Giesecke. Columbia University Oral History Collection, Part 2, No. 71. Microfilming Corporation of America, Glen Rock

Glave T, Miguel L, Remy MI (1983) Estructura agraria y vida rural en una región andina: Ollantaytambo entre los siglos XVI-XIX. Centro de Estudios Rurales Andinos "Bartolome de las Casas," Cusco

Göhring H (1877) Informe al supremo gobierno del Peru sobre la expedición a los valles de Paucartambo en 1873. Imprenta del Estado, Lima

Gram D (2005) The lost city of Z. Doubleday, New York

Greer P (2009) Machu Picchu antes de Bingham. El Antoniano (Cusco) 19(114):32–39

Heim A (1957) Wunderland Peru. Verlag Hans Huber, Bern

Hemming J (1970) Conquest of the Incas. Harcourt, Brace and Jovanovich, New York

Krüggeler T (1999) Indians, workers and the arrival of "modernity.": Cuzco, Peru (1895–1924). The Americas 56(2):161–189

Lowenthal D (1996) Possessed by the past: the heritage crusade and the spoils of history. The Free Press, New York

Markham CR (1911) Mr. Bingham in Vilcapampa. Geogr J 38(6):590–591

Martin JG (1980) The life and thought of Isaiah Bowman. Archon Books, Hamden

Means PA (1930) Review of: Machu Picchu, a citadel of the Incas by Hiram Bingham. Am Hist Rev 35(4):899–901

Means PA (1931) Ancient civilizations of the Andes. Charles Scribner's Sons, New York

Miller C (1982) Fathers and sons: the Bingham family and the American mission. Temple University Press, Philadelphia

Mincetur (Lima, Peru) http://www.mincetur.pe. Accessed 10 Sept 2014

Montoya G, Giesecke A (1961) El misterio de Machu Picchu. El Comercio (Lima), June 21 to July 24

Mould de Pease M (2000) Apuntes interculturales para la historia inmediata de Machu Picchu: Las funciones de Hiram Bingham y Albert A. Giesecke. Revista del Archivo Regional del Cusco 15:133–147

Mould de Pease M (2001) Un día en la vida peruana de Machu Picchu: avance de historia intercultural. Revista Complutense de Historia de America 27:257–279

Mould de Pease M (2003) Machu Picchu y el código de ética de la Sociedad Arqueológica Americana: una invitación al diálogo. Fondo Editorial Pontificia Universidad Católica del Perú, Lima

Pardo LA (1961) Machupicchu (Una joya arquitectónica de los Incas. Editorial H.G. Rozas, Cuzco

Poole D (1998) Landscape and the imperial subject: U.S. images of the Andes, 1859–1930. In: Joseph GM, LeGrand CC, Salvatore RD (eds) Close encounters of empire: writing the cultural history of U.S.-Latin American relations. Duke University Press, Durham, pp 106–138

Raimondi A (1874) El Perú, 4 vol. Lima

Rivas Tapia A (2011) El gran descubridor de Machu Picchu, Agustin Lizárraga. Agencia Andina, Cusco

Robledo L (1899) La via fluvial del Urubamba. Boletin de la Sociedad Geográfica de Lima 8(8):417–450

Rodman (1967) The Peru traveler: a concise history and guide. Meredith Press, New York

Rodriguez de Figueroa D, Pietschmann R (1910) Bericht des Diego Rodriguez de Figueroa über seine Verhandlungen mit dem Inka. Titu Cusi Yupanqui in den Anden von Vilcapampa. K. Gesellschaft der Wissenschaften zu Göttingen, Göttingen

Romero CA (1909) Informe sobre las ruinas de Choqquequirau. Revista Histórica 4:90–95

Rowe JH (1944/1945) Review of: archaeological excavations in the cordillera Vilcabamba, southeastern Peru by Paul Fejos. Am Antiquity 10:312–314

Rowe JH (1990) Machu Picchu a la luz de documentos del siglo XVI. Histórica 14(1):139–154

Rubio Correa M (2007) Albert Anthony Giesecke Parthymueller: el mas peruano de los norteamericanos. Privately printed, Lima

Smith N (2003) American empire: Roosevelt's geographer and the prelude to globalization. University of California Press, Berkeley

Super J (1994) History, Indians, and university reform in Cuzco. The Historian 56(2):325–339

Verano JW (2003) Human skeletal remains from Machu Picchu: a reexamination in the 1912 Yale Peruvian scientific expedition, collections from Machu Picchu: human and animal remains. In: Burger RL, Salazar LC (eds) Peabody Museum. Yale University, New Haven, pp 66–117

von Hassel JM (1907) Informe del jefe de la comisión exploradora al Alto Madre de Dios, Paucartambo y Urubamba por la via del Cusco. In: Ultimas exploraciónes ordenades por la Junta de Vias Fluviales, Junta Fluviales, Lima pp 259–394

Wagner PL (1996) Showing off: the Geltung hypothesis. University of Texas Press, Austin

Waisbard S (1974) The mysteries of Machu Picchu. Avon Books, New York

Wiener C (2010) Pérou et Bolivie: récit de voyage [1880]. Librairie Hachette, Paris

Wright JK (1947) Terrae incognitae: the place of imagination in geography. Ann Assoc Am Geogr 37:1–15

Manuscripts

I. (archives of the Association of American Geographers)
Letter from Hiram Bingham to AAG 2 November 1948 (AAG No. 62.1374)
Letter from H. Bingham to Albert A. Giesecke 8 March 1926 (AAG Archives 3.44)

II. Yale University Archives (New Haven, CT)
Yale Peruvian Expedition Papers. General Correspondence. Group No. 664
Manuscript and Archives, Sterling Memorial Library, Yale University
Letter from Isaiah Bowman to Hiram Bingham 1 August 1911
Letter W. Farabee to Hiram Bingham 21 April 1911
Letter Hiram Bingham to Arthur Hadley 4 October 1912
Letter from Hiram Bingham to Isaiah Bowman 7 January 1913
Letter from Isaiah bowman to Hiram Bingham 17 January 1913
Letter from Albert Duque to Hiram Bingham 14 August 1911
Letter from W.C. Farabee to Hiram Bingham 7 April 1911

III. Albert Giesecke Archives (Lima, Peru)
Letter from Hiram Bingham to Albert Giesecke 1948

Chapter 8
Vilcabamba: Fabled Redoubt of the Urubamba Region

Abstract West of the main Urubamba River, a mountain fastness sliced through by two main tributaries, the Vilcabamba and the Cosereni, has a snowcap-to-jungle ecology. The fragmented topography of this remote area called the Vilcabamba provided a last refuge for the Incas after the Spanish Conquest of 1532. Spaniards gained control of this zone only in the 1570s, motivated by the presence of Inca mines of precious metals. Its isolation and hidden ruins gave Vilcabamba its reputation as being a land of mystery. Two accounts seek to capture some of its character. One is that of the extraordinary experience of an American geographer, Stuart White, who went to Vilcabamba in 1980 and stayed for 2 years. The other is my own 1964 expedition seeking the Inca site of Vilcabamba la Vieja. The log made from that trip is placed in a retrospective context and the future is discussed.

8.1 Isolation and Marginality

Like few other places, the name Vilcabamba evokes the geographical mystery of a region long poised on the border between the known and the unknown. Despite a linear distance from Cusco of only about 100 km, this rugged mountain fastness that drains eastward to the Urubamba is remote and difficult to penetrate (Fig. 8.1). From Cusco its main entry was controlled near Chaullay by a suspension bridge called Chuquisaca. By clipping the strands of this plant fiber bridge, the Incas could diminish the threat of invasion. To this secluded region of Vilcabamba, Manco Inca and his three sons retreated, seeking refuge from Spanish domination. After 1532, the year of the Conquest, four decades passed before the Spaniards imposed their will here. Although they had entered the region in 1534, they made little impact until 1570 when Augustinian friars sought to found churches and destroy Inca religious sites (Fig. 8.2).

Colonial control came in 1572 when Viceroy Francisco Toledo ordered the capture and beheading of the last king of the Vilcabamba dynasty. Spanish authorities had learned from a woman of the old Inca elite that Vilcabamba had rich veins of gold, silver and mercury opened during the reign of Inca Pachacutec (Regalado de Hurtado 1992). Viceroy Toledo wanted those riches and also wanted to convert the native people. He ordered the construction of a new agglomeration (*reducción*) called San Francisco de la Victoria de Vilcabamba, now often referred to as

D.W. Gade, *Spell of the Urubamba*, DOI 10.1007/978-3-319-20849-7_8

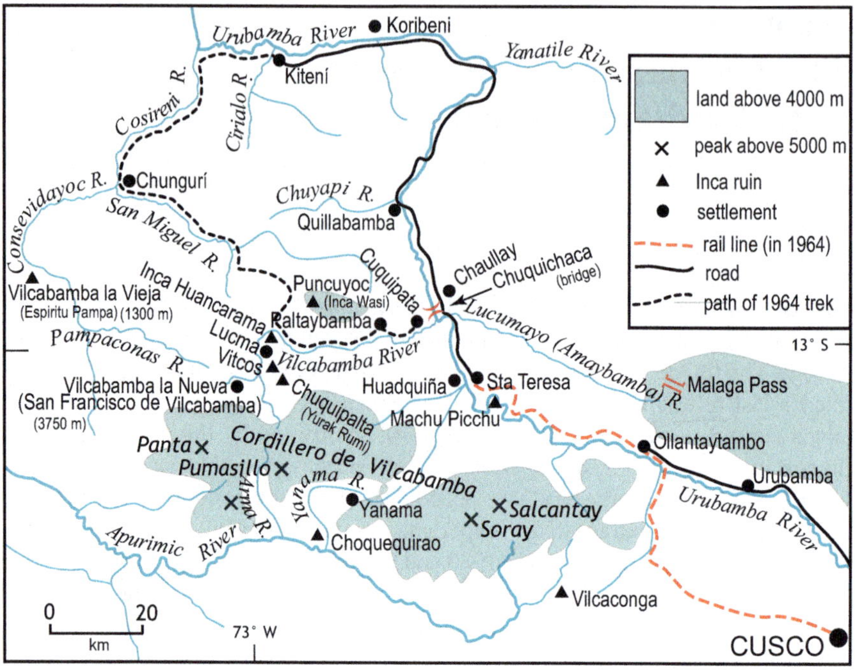

Fig. 8.1 The larger region of Vilcabamba (not just the present district) is drained by three rivers, the Vilcabamba which flows eastward into the Urubamba; the Apurimac to the south and west; and the Cosireni to the north, which gathers the Pampaconas, Consevidayoc and San Miguel Rivers, before it empties into the Urubamba

Fig. 8.2 The parish church at Lucma, was built in the 1570s on an artificially-leveled Inca platform (usnu). The stones used in construction of the church came from the nearby Inca site of Hauncarana

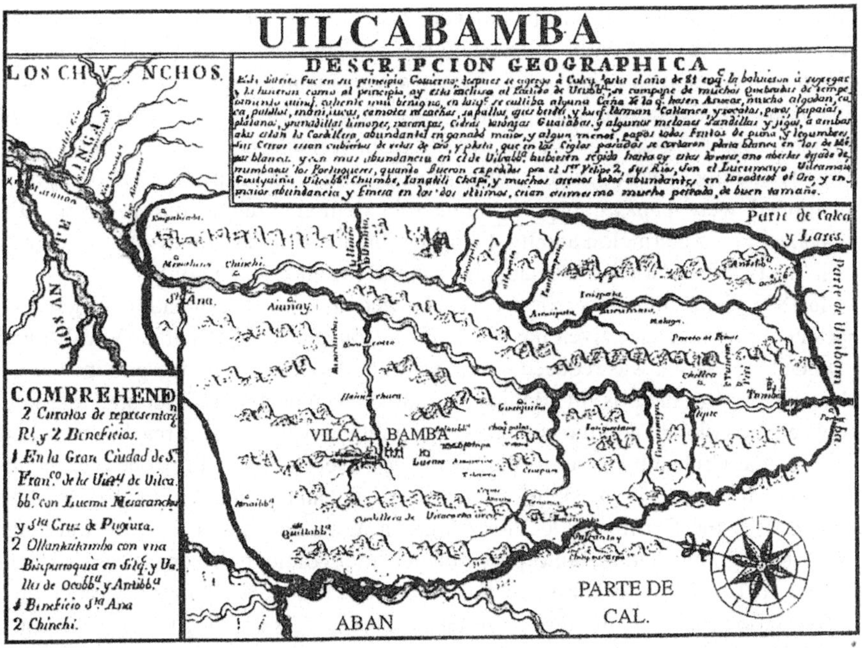

Fig. 8.3 The Partido of Vilcabamba in the late colonial period included most of the present-day Province of La Convención as far as Santa Ana (Quillabamba). Of the four parallel rivers on this rudimentary map, the Urubamba is the third from the top. (Source: Oricain 1906)

Vilcabamba la Nueva, to replace the pre-conquest Vilcabamba 10 km upvalley from Paltaybamba (Cobo 1956:II:20). The new Vilcabamba became the seat of a political entity known as a “corregimiento” that extended about 50 leagues north to south (one league = five kilometers), and controlled everything from Ollantaytambo to the frontier with the *indios infieles* of the forest (Fig. 8.3). The Spanish focused on the high country above 3600 m above sea level (asl) where the mines were located. The miners who came there, including many Portuguese, forced Indian laborers to exploit the veins over a radius of 30 km around Vilcabamba la Nueva. The prospect of becoming an important settlement prompted the establishment in 1586 of a Mercedarian convent. But output of precious metals did not meet Spanish expectations, and between 1590 and 1610 one mine after another closed. Population declined and in 1601 the Mercedarians left the town. By 1684 no European miners remained in Vilcabamba. Colonial authorities suppressed the corregimiento and transferred administration of the vast area to Calca. By the eighteenth century, little remained but a moribund Quechua-speaking village of potato farmers whose facial features betrayed interbreeding with European miners.

Below 1600 m asl the Vilcabamba corregimiento had had very few people in the early colonial period. The Toledan resettlement program of the 1570s laid out no towns in the hot country. Two Spaniards, with their highland peons and black slaves lived on haciendas Santa Ana and Guayanay and later formed incipient towns called "beneficios." An eighteenth-century vignette by Pablo Oricain (1906) described the Vilcabamba region as one of many canyons with mainly a warm climate in which Spaniards grew sugar cane, at that time for making sugar, not alcohol. Other warm climate crops were cotton, coca, palillo (a plant yielding a yellow dye in its root), peanuts, manioc, sweet potatoes, arracacha, squashes and capsicum pepper. Fruits mentioned were papaya, banana, granadilla, lemon, orange, citron, grapefruit, guava, melon, watermelon and fig. Fish were plentiful in the streams. Placer deposits in the two tributaries, the Chapi and Yanatile, yielded some gold.

In 1857 Vilcabamba was reorganized, changed in size and designation. Its territory became a province and given the name of La Convención. Thereafter, Vilcabamba, politically speaking, was reduced to a district of 5046 km^2 within the province of La Convención. But there is, in addition, a traditional "Vilcabamba": an area that falls between the colonial and the contemporary political definitions. It reaches west and south of the Urubamba River to the Apurimac River. It covers life zones extending from 600 m asl to 6271 m asl, from selva alta to glacial ice. South of the Vilcabamba Valley, rises the Cordillera de Vilcabamba, a mountain massif 250 km long and made up of granite and metamorphosed Paleozoic sediments. Twenty peaks surpass 5000 m asl; the two highest are Salcantay (6271 m) and Pumasillo (6070 m). Glaciers top a dozen mountains in this chain. Three rivers draining the area flow into the Urubamba: the Santa Teresa, Vilcabamba, and Cosireni.

Population of the whole district in 2012 was about 18,000 people, most of whom live above 2500 m asl and speak Quechua as their first language. Few haciendas were established in this part of the larger Urubamba region. The largest was Paltaybamba, formed in 1830 as a land grant ("amparo") from the Government of Peru as repayment for support during the Independence struggle. Paltaybamba subsequently absorbed Hda Mesacancha. In the 1920s, Paltaybamba, 45,527 ha in size, became the property of José Pancorbo, who had a reputation as a feudal baron. During the rubber boom Pancorbo received government concessions in the San Miguel Valley.

Two unpaved vehicular roads now penetrate the region from the Urubamba; they were not there in the 1960s. The main one, up the Vilcabamba Valley to Huancacalle, had been a mule path (*camino de herradura*) and before that an Inca trail. The other road extends from Kiteni on the Urubamba up the Cosireni River to Chunguri near the mouth of the Rio San Miguel. Many trails predating the Spanish Conquest are still used by mules. Many of these paths, sometimes carved out of bedrock, follow along the side of a mountain hundreds of meters above the floor of a valley (Fig. 8.4).

8.2 A Geographer in the Vilcabamba High Country

When geographers recount where they have gone and what they have seen, little bits of new knowledge about places are likely to emerge. When one has a lengthy living experience, deeper and more expansive insights emerge. I wish to highlight here one individual, Stuart White, who first came to the Vilcabamba in 1977 as a result of a

Fig. 8.4 Cliffhanging trails, common in the Vilcabamba region, are used by walkers and beasts of burden. This one, cut into rock, lies ca. 1000 m above the floor of the Yanama Valley. (Photo by John Servayge with permission)

sequence of happenstance decisions. The role of contingency in advancing knowledge has many examples of which this is one.

After graduation from Harvard, White became a Peace Corps Volunteer in an isolated rural community in Cauca Department, Colombia, where his assignment included taking care of livestock. He was not a veterinarian, but learned a good deal about farm animals while in that post. When, in 1973, his two and a half years of Peace Corps service ended, Stuart went to visit his compadre and his godson to help them make a land claim in the jungles of Putumayo. Having done what he could there, he caught a ride on a Colombian navy boat which took on civilian passengers and proceeded downstream to Leticia; from there he went to Iquitos by plane and to Pucallpa by boat. In Pucallpa he met Antonio Dávila, a woodcutter who invited Stuart to join his family and workers who were going upstream to Mishagua in the Urubamba drainage area, a trip of 3 days in two motorized boats. Stuart became interested in Antonio's cedar and mahogany logging operation on the Mishagua and stayed for 2 weeks. Then, with three Matsigenka, White canoed through the Pongo de Mainique and from there, followed a forest trail upriver. When he came to the compound of Fidel Pereira, he had arrived at the doorstep of the legendary mestizo landowner who had several Matsigenka wives and who ruled much Matsigenka territory outside the missions. They drank manioc beer (masato) together, and then Pereira showed Stuart the path to Kiteni farther upstream and told him about a hut in the forest where he would be safe from the jaguars that were known to roam the area. Stuart found the structure just as darkness approached and stayed inside overnight. The next day he hiked to Kiteni, at that time the terminus of the road, and the following day, took a truck to Quillabamba. Enroute, White noticed many Quechua colonists who had come from the Highlands. Why were they there and where had they come from? From Quillabamba, Stu White traveled to Cusco and from there flew to Lima and the United States.

As has happened with many sojourners, Latin America opened for Stuart White a plethora of questions and prompted him to seek more formal education in order to answer them. In the fall of 1973, he faced decisions concerning what discipline and which program he should pursue. He considered epidemiology at Johns Hopkins, ethology at Cornell, and geography at UC Berkeley. He then visited the Wisconsin Geography Department where Professor Bill Denevan was very supportive of his nascent scholarly interests. White enrolled in the geography program and for his Master's thesis returned to Peru for 2 months to make a formal study of Dávila's logging operation. In that time, he learned how Matsigenka Indians cut wood without modern machinery. Two high-value species, cedro (*Cedrela odorata*) and mahogany (*Swietenia macrophylla*), monopolized most logger's attention. By then, large specimens of these species could be found only far up the Urubamba tributaries. His information on logging became the subject of a published article (White 1978).

The next step in his program was to select a research topic for a dissertation. White proposed returning to the Department of Cusco to study how Peruvians who had migrated from the Highlands to the jungle adapted to their new environment. To carry out such a study, White decided he first needed to learn more about serrano life by settling into a rural community. On a whim, he got in a truck headed for the Vilcabamba Valley. When the rudimentary road ended, he trekked several hours to the old mining town of Vilcabamba (la Nueva). No longer the capital of Vilcabamba District, the town was essentially an Indian hamlet with no services. Most people spent days in their potato fields, which was why the town seemed deserted at midday. Late in the afternoon, Stu spied an elderly peasant couple sorting potatoes on their front patio. Stuart—this *q'ana runa may karusmantachus jamun* ("white man from distant parts")—asked them in Quechua for a place to stay. At 3660 m asl, the town was very cold at night. Thanks to his year-long course in Quechua at the University of Wisconsin, Stuart and the monolingual Quechua-speaking couple could easily communicate.

One overnight led to another; Stuart stayed with Benito and Nicolasa Uchupe a year (Fig. 8.5). Their grown children had left the hearth and were no longer in the area. The old couple and White developed strong personal ties. Accepted by the whole community, Stuart lived the life of an Andean villager. He participated in the daily routines of herding sheep, looking for lost cows in the heights above the village, bringing salt to livestock, rounding up horses, planting and harvesting crops, making chuño, building fences, and serving as godfather to several children. He castrated bulls,

Fig. 8.5 Benito and Nicolasa Uchupe, monolingual Quechua couple, with whom Stuart White lived for a year in 1980–1981 in San Francisco de Vilcabamba (aka Vilcabamba la Nueva). (Photo by Stuart White with permission)

rams, and stallions, a skill learned in the Peace Corps in Colombia. The peasant folk of Vilcabamba benefitted from his knowledge of how to remove parasites from livestock. He also helped in the annual corvée of thatching the roof of Vilcabamba's colonial church, long without a resident priest. White traveled beyond the village to bring salt from salt mines in the area and to collect firewood, and he made several treks to visit families or places he had heard about.

Settled into a routine, White decided to spend a second year in Vilcabamba. Renting a stone hut on the village fringe, he planted his own chacra. During that second year, a life-threatening case of infectious hepatitis forced him to return to the United States for tests, medication and recovery. Not totally cured, he nevertheless returned to Vilcabamba. Few scholars have so completely entered peasant life in such a remote locale. Living in Vilcabamba provided Stuart deep insights into paleotechnic farm practices, native cosmology, Quechua as a language, peasant psychology, and the dynamics of the Andean landscape. For his dissertation, he prepared what was essentially a novel based on his Vilcabamba living experience and the people he met. Although he included a long chapter on the meaning of landscape, it was the first and still only doctoral dissertation in American geography of that sort (White 1981).

Two topics became the basis for subsequent articles. Noticing the annual use of controlled burning of highland vegetation, White (1985) extrapolated from that how so much of the Andes had been converted from a forest to grass cover.

His second article was archaeological, based on his 1978 explorations of the Puncuyoc Range north of the Vilcabamba River. A stone-inlaid trail led Stuart to a spectacular complex of Inca ruins that no one had previously described (White 1994/1995). He returned there seven times, sometimes with others, to survey the site, make maps, and record observations. Surrounded by craggy heights of up to 4412 m asl, the complex, which White called Puncuyoc after the mountain range that loomed above it, was often shrouded in cloud and mist. Chief among the dozen or more stone structures was a high-gabled main building, often called "Inca Wasi" (i.e. house), constructed of finely cut ashlars. With its trapezoidal niches and protuberant stone pegs, White surmised it to be a late Inca construction, perhaps instigated by Manco Inca after he had fled Cusco and Spanish colonialism. Going to the historical record, White determined that the oracle site which the Spanish chronicler Diego Fernandez had described in his 1557 account was, in fact, this structure.

Campesinos in the area knew about these ruins, but few outsiders had visited them. The indefatigable German agronomist Christian Bües trekked there in 1934 and included it on his map, but wrote only that "la ruina incaica de Idma Huasi de dos pisos, bien conservado sin techo que cierra al abra de Idma Huaicco." (The Inca ruin of Idma Huasi at the pass of Idma Huaicco has two floors in a good state of preservation but without a roof) (Bües 1939). In 1952, Victor von Hagen led an expedition to Peru and mentioned that his team had "discovered" an Inca site in the heights above the Vilcabamba Valley. Even though he had in his possession a copy of the Bües map on which the site was marked, Von Hagen (1958:118–119) claimed credit for its "discovery," without himself having gone near the region. Hugh Thomson (2003:240), an explorer, wrote that "Stuart White was the first explorer to have seen Inca Wasi before he wrote about it." MacQuarrie (2007:433) incorrectly attributed the discovery of the ruins of Puncuyoc to Von Hagen's expedition. Also, Vince Lee got his interest in Puncuyoc from the work of White, not that of Von Hagen. MacQuarrie made no mention of Stuart White. Vincent Lee (2000:214), acknowledging White's pioneering role in describing Inca Wasi, later interpreted Puncuyoc as a ceremonial site related to the Inca calendar. Most recently, Bernard Bell (2011:108) interpreted Puncuyoc as a solar observatory by noting that on the June solstice, the first sun rays stream through the eastern door of the Inca Wasi.

8.3 The Lost City of The Incas and Other Ruminations

Exploration has always been part in my own interest in places. It is a way to learn. To be an observer and a witness underpins my idea of what it is to be a geographer. Trekking has its own strong appeal. I have never understood those who call themselves geographers but are uninterested in exploring new places. In 1964, a Cusco newspaper announced in a sketchy article that an American adventurer, Gene Savoy (1927–2007), had reached a place called Espiritu Pampa in the District of Vilcabamba and that this location was the fabled "Lost City of the Incas." It was a reminder that another gringo 53 years earlier had claimed to have discovered two Inca sites,

Fig. 8.6 1964 Vilcabamba expedition members from left to right: Stephen Klinge, Rainer Mosebach, Ronald Arias, Michael Beede and Daniel Gade. (Photographer unknown)

Machu Picchu and Vilcabamba la Vieja. That person, Hiram Bingham, then confused one with the other by misreading the historical record. Savoy, indeed, had reached Vilcabamba la Vieja and shortly later, *Time* magazine recounted the Savoy exploration (Peru 1964a).

The challenge of contributing to this discussion galvanized me to organize my own expedition to Espiritu Pampa in 1964. Three companions, Ronald Arias, Michael Beede, and Stephen A. Klinge, were Peace Corps Volunteers from California. A fourth participant was Rainer Mosebach, a German we met in Cusco who impressed us with his account of a previous trek through the Paraguayan Chaco (Fig. 8.6). [Pouring over the undated map prepared by Christian Bües, we agreed that the shortest and thus the best way to reach Espiritu Pampa—a place named on that map—was to trek much farther north along the Urubamba River and then hike up its tributary, the Cosireni, a stream called Consevidayoc in its upper reaches. Savoy and his group had taken a different route, one that went up the Vilcabamba Valley, over the pass, and down into the misty, wet, and forested Pampaconas Valley. After several weeks of assembling gear, our expedition took off. The reportage in the Cusco daily newspaper, *El Comercio*, hopelessly scrambled the purpose of the expedition, the route chosen, and it described me as an "estudiante peruano de botánica" (Peru 1964b)].

8.3.1 Log of the Trip

The elliptical account that follows of this week-long trip comes from my record of incomplete thoughts jotted down at convenient moments.

8.3.1.1 September 3, 1964

Day of departure from the city of Cusco after several days of intense planning. Loaded down with 23 kg packs, we set off for the Urubamba Valley on the narrow-gauge railroad to Huadquiña. From there we rode a truck to Quillabamba, where we spent part of a day buying supplies and sharpening our machetes for the jungle trek ahead. Another truck took us across the river to the Dominican mission at Koribeni, then the end of the road. The river was low, and a boatman was only willing to go 25 km downstream to a place called Sirialo. Rapids and shallows below this point forced us to start our overland hike 40 km sooner than we had counted on. At Sirialo, we were delayed several hours because the trail was crossed by brush fires set by slash-and-burn agriculturalists, highland Indians who had moved down into the tropical valleys in search of land and a better life. These Quechua speakers had been gradually pushing out the Matsigenkas, the forest Indians who had traditionally inhabited the tropical part of the Urubamba Valley. Several hours of walking and it was beginning to get dark, so we pitched camp near a small stream and ate a meal of oats-and-milk, oats-and-soup, and coffee. Swarms of insects made us glad we had taken along mosquito netting.

8.3.1.2 September 4

On the trail early, we followed the winding Urubamba River, at times dropping to its edge, then climbing to avoid a steep precipice. As the morning wore on, the vegetation started to close in, the heat grew oppressive, and walking became progressively harder, requiring more frequent rest stops. The temptation was to eat as much as possible to lighten our packs—especially the "Quaker," the word Peruvians use generically for oats instead of the Spanish word "avena." By late afternoon, we reached a blackwater stream where we took a swim, then continued on to the mouth of the Rio Cosireni where we spent the night.

8.3.1.3 September 5

We rose at daybreak, eager to move on the trail that more or less paralleled the Cosireni. According to calculations, we could reach our goal, Espiritu Pampa and its ruins, in 3 days of steady hiking. This schedule fell apart when the trail suddenly vanished in dense undergrowth. As we hacked at the vegetation to open a path, stinging red ants fell on our exposed necks. Confused, we backtracked and came upon some wood-cutters who set us on the trail which, however, went back to the same river where we had gone swimming the day before. From there, the trail climbed steeply to the top of a knife-edged ridge, and then dropped gradually to follow the course of the Cosireni. By afternoon, we came to a farm called Selva Alegre where we were given the best refreshment available, *masato*, or beer made from manioc that had been chewed to start the fermentation.

Virtually no fruit was obtainable on our whole trip; bananas and papayas would not be ripe for another month. On our way again and a few miles farther on, we came to another farm where the peasant give us uncucha root (*Xanthosoma* to the initiated) and manioc to which we added oats, quinoa, Maggi soup mix, and water. The best meal of the trip. Here, too, I found growing jiquima (*Pachyrrhizus tuberosus*) a native root crop considered quite rare. That night we bedded down in the pig yard on some peccary and monkey skins on loan from the farm family, careful not to expose our toes to the bites of vampire bats. Before I had left for Peru, I had read an article on how common it was for vampires to be responsible for most rabies when dogs were not involved.

8.3.1.4 September 6

We continued up-valley on that trail through thick forest which gradually changed in floristic composition as our elevation increased. Soon the trail led us to the top of some very steep cliffs, affording spectacular views of the Rio Cosireni. By early afternoon we reached a chacra (farm) of highland settlers (colonos) called Yubeni, where we were told there was no trail to Espiritu Pampa. Not wishing to believe this even if it were true, we pushed on to a larger farm called Santa Rosa where we were again informed that no trail lay beyond. After a meal of oats and guavas and boiled corn supplied by the farmer, we slept in the granary, entertained by acrobatic mice in the rafters. From inside the house came the noise of a battery-operated shortwave radio playing raucous Indian music from Cusco, which at that time seemed a thousand miles away. The colonos all had guns, which they used to kill wildlife for food (Fig. 8.7).

8.3.1.5 September 7

Up early, uncertain what the day would bring. After refortifying ourselves with Quaker, we decided to push on. It turned out to be a long hard day of hiking up hills and down hills through brush. Finally we admitted that, indeed, past that point, there was no cleared way to get to Espiritu Pampa, not more than 35 km away. The realization sunk in that we had neither the energy nor the provisions to hack our way through the dense jungle to the ruins.

Fig. 8.7 One reason why wildlife in the selva of the Urubamba and tributaries has diminished is the widespread use of guns by most colonos. The animal skins on display are, on the left, a collared peccary and, on the right, a howler monkey. (Photographer unknown)

8.3.1.6 September 8

By morning, we were psychologically prepared to turn back. Choosing not to return by the same route by which we had come, we started slowly up the valley of the Rio San Miguel. In spite of its considerable isolation, some of the land was in haciendas, and I kept wondering how commercial agriculture could be successful in places like this where the road lay 3 or 4 days away by mule. By noon, we reached 1500 m elevation (5000 ft), passing coffee and sugar cane plantings. At Hacienda Miraflores, we were invited to stay overnight, but chose to continue up the valley in a light rain that soon got heavier. At Hacienda Florida, elevation 1780 m, we requested permission to spend the night under some kind of shelter. The mysterious owner never appeared, yet issued orders to his Indians to provide mattresses and to kill a chicken. We ate this tough bird with gusto at a table set with a white linen tablecloth that looked like it hadn't been used in years. Soon after dinner, we fell into a deep sleep, brought on by fatigue and knowledge that a tough climb still lay ahead. Lucma, the nearest town, was two long days distant on the other side of a 3700 m-high peak.

8.3.1.7 September 9

Crowing roosters helped us hit the trail at the crack of dawn. It soon got steep and about noon it got vertical. A steady drizzle began to fall. Although by this time we had consumed most of our food, the remaining 20 kg in my backpack felt like tons on my aching shoulders. Benumbed, I couldn't appreciate the spectacular view of the surrounding peaks and the San Miguel Valley below. Now the misty cloud forest with giant tree ferns and hanging epiphytes dominated and, with it, utter silence except for the squish-squish of our thoroughly soaked boots. Finally, at about 5 p.m., the top had been reached, the trail flattened out, and the treeless puna emerged. The rain, however, increased and the damp cold was much more miserable than the oppressive heat of the jungle. An hour of tramping through marshy puna up to our ankles brought us close to the point of utter exhaustion. Looming ahead in the darkness was a precarious open-sided shelter that mule-drivers had constructed. The altimeter read 3300 m above sea level. With a roof to keep off the rain, but no protection from the wind, we shared the place with the Indians. Luckily for us, they had brought dry wood or it would have been impossible to start a fire. The high elevation, rain, darkness, and the wind made the night bitterly cold. After some potato soup which the Indians gave us, we settled down to sleep as close to the hot embers as possible, uncertain about our hosts. It occurred to all of us that these mule-drivers might try to run off with our gear in the middle of the night. I remained awake most of the night.

8.3.1.8 September 10

Dawn approached, and the native mule-drivers started moving around, lighting the fire that had gone out. After some coffee we took off, still climbing for at least another 18 km up to the divide which, according to the altimeter, was 3750 m above sea level. Then the descent began, extending for about 21 km almost straight down to the Vilcabamba River. Compressed in this short distance were puna grassland, then a bush formation, and an elfin woodland. Finally about noon, we reached the bottom and some miserable Indian huts. One was a small store which had canned tuna fish (the only widely distributed canned food in Peru at that time), stale bread, and cola. Fortified with what seemed to constitute a real banquet, we pushed on to Lucma, capital of Vilcabamba District and found it to be quite a different kind of Peruvian pueblo. Set on a river terrace, the large open grassy spaces between the adobe huts reflected its origin as a town of mule-drivers (Fig. 8.8). The fields around the town were in natural pasture with very little land in crops. A priest came once a year for the village fiesta to celebrate Mass in the tumbledown little church. We stayed in the *almacén* (warehouse) of the mayor and during the night found that the principal thing stored there was fleas.

Fig. 8.8 The town of Lucma in 1964. Capital of the District of Vilcabamba, Lucma has a diffuseness explained by its history as a town of mule drivers. (Photo by D.W. Gade)

8.3.1.9 September 11

Up at dawn to pack the mule that would carry our gear the 50 km or so to the road head near the Urubamba River. Relieved of backpack weight, we fairly flew down the trail. The climate and crops became quasi-tropical below 1800 m with chacras growing manioc, coffee and bananas. In mid-afternoon, we reached Hacienda Paltaybamba, a 130-year old estate designed in the colonial manner to be virtually self-sufficient. Paltaybamba was run-down and only partially functional in 1964 because of the scarcity of labor. Most of the land under cultivation was in sugar cane, converted into rum because of the high cost of transport.

Farther down the valley a large dog charged out at me. Picking up a rock to deter him, my aim was too good. The rock struck the animal straight on the forehead, killing him instantly. Better to compensate its owner than fear rabies infection in the middle of nowhere. In a light drizzle, we passed the town of Chaupimayo, spectacularly hung on the side of the mountain. It was known as the hideout of guerrilla leader, Hugo Blanco, who, in 1962, mobilized peasants against the tyranny of latifundistas. Captured by the military, he became a popular hero among the masses. Finally, we reached Cuquipata where the vehicle road began and where a truck eventually arrived and took us to Chaullay. So ended our journey beyond the cordilleras, which took us through some 180 km of jungle, cloud forest, and puna. Magnificent scenery and the chance to test ourselves against some of the wildest country in Peru partly compensated for our failure to reach the ruins. I treasure memories of a superb adventure at the border of the unknown.

8.3.2 *Reflections on This Trip Half a Century Later*

Because the 1964 exploration did not achieve its objective, my thoughts focus on the experience of the trek itself and the places along the way. The laconic observations recorded convey a basic truth of the epistemology of a reconnaissance. On a trip requiring great physical effort under difficult conditions, intense weariness makes it difficult to concentrate on an extended narrative. Antonio Raimondi's published accounts of his multiple trips hither and yon in the nineteenth century were often disconcertingly spare, probably for the same reason. Great fatigue best explains why many explorers write nothing down at all.

The trails and the areas through which they ran have changed greatly. In 1964, Kiteni, merely a named place on the Urubamba consisted of a couple of thatched huts; in 2013, it was a town of ca. 3000 people of highland origin with an airstrip. In 1964, Kiteni was accessible only by boat or on foot; now a road running north and south of Kiteni daily brings scores of dusty trucks, buses and cars through the town. A vehicular road from Kiteni up the Cosireni replaced the 30-km trail to the settlement known as Chunguri (also spelled on maps as Chohanguiri or Chunquiri) located near the mouth of the Rio San Miguel. In 1964, no trail went up the Upper Cosireni (Consevidoyoc) to the ruins at Espiritu Pampa. It was in that area that peasants had told us the only way to reach the ruins was by bushwhacking with a machete through two kilometers of forest, and after hacking for some time, we reluctantly turned back and trekked the 87 km trail from Chunguiri through the San Miguel Valley and up into the Puncuyoc Cordillera, over the Huarina Pass, and down to Lucma. From that town, the trail led down the Vilcabamba Valley to Chaullay on the Urubamba. Of the areas we went through, the San Miguel Valley was particularly attractive. It apparently had had no settlement during the colonial period, and the first records of it came when rubber gatherers, working their way up the Cosireni, created a trail.

8.3.2.1 The Story of Elvin Berg

A notable personage at that time was a Norwegian, Elvin Berg, who arrived in the area between 1902 and 1904 to exploit rubber. Berg founded the settlement of Yubeni (also Yuveni) as a rubber collection point. When the rubber boom ended, he stayed and acquired ownership of part of the San Miguel Valley, married a Quechua girl and raised a family. Before Berg died in 1919, Edmund Heller, an American biologist, collected birds and mammals there (Chapman 1921). In the 1930s Christian Bües found the valley mostly forested. Only 75 people then occupied the mid zone between 1400 and 1600 m asl, 20 of whom were Matsigenka (Bües 1939). Those Indians were refuges from the malaria epidemic that raged in the Urubamba between 1934 and 1938. As one moves up the San Miguel Valley, rain increases and the ceja de la montaña, full of tree ferns, takes over. In 1964 not one person lived in that ceja zone.

8.3.2.2 Christian Bües (1874–1948) and the Map

Working with and depending on Christian Bües' map increased my admiration for him and his cartographic gem of the Vilcabamba. Rather amazingly, in 1964 it was the only map available for the region; even the mapmakers from the Instituto Geográfico Militar in Lima had yet to map the area. The Bües map gave our exploratory team the idea that we could reach the ruins rather easily by going down the Urubamba in a boat and then hike the 40 or so km from the mouth of the Coserini (or Kushereni on the Bües map). As I found out later, Christian Bües became a cartographer when hacendados near Quillabamba hired him to make property maps of their estates which had never been surveyed. To make the Vilcabamba map, Ingeniero Bües spent several years off and on trampling through the region and asking locals the names of features. In some isolated locales, no one would have been around to ask, so perhaps he, himself, named some minor streams. Blank areas on the map, in the western third of the district, indicate where he did not go. But it is not entirely clear where Bües did go. He located Espiritu Pampa on the map, but not Vilcambamba la Vieja. He never mentioned any of these ruins in his writings, suggesting to me that he never saw them. Gene Savoy used the name Espiritu Pampa to refer to the area of Inca ruins he saw covered with jungle.

The original Bües map had no date, but the conjecture that Bües drafted the map in 1937 seems likely (Thomson 2003:264). Thomson mentioned that a person named A. Palma made a hand copy of the Bües map and that a German doctor, Enrique Berguan, living in the town of Urubamba, had then traced the Palma's copy and in 1952 passed it on to Victor von Hagen (Thomson 2003). The map I had was entitled "Sud Convención" and showed La Convención only as far north as the Rio Cosireni. Dated November 1958, it was apparently drafted in Cusco at the Tierras de Montaña office from the information on the 1937 map. When printing firms in Cusco got blue print machines, the map could be easily reproduced and copies of it distributed. My copy came from the Tierras de Montaña government land-claims office in Cusco.

8.3.2.3 Gene Savoy (1927–2007)

Gene Savoy was not the first to reach Vilcabamba la Vieja, but he was the first to identify it correctly as the "lost city of the Incas." His background for exploration did not involve much scholarly preparation. It was Hiram Bingham's (1911) article on Vilcabamba that brought Savoy to Cusco. Once there, Savoy got his hands on the Bües map, which enhanced his confidence to explore the zone. His team included Douglas Sharon, his cameraman, who later went on to become a noted anthropologist. Savoy assured himself of wide publicity by clearing just enough ruined structures to be able to photograph the stone construction. To some, Savoy seemed like a publicity seeker manipulating the media. At the time of his death, a New York Times necrology writer summed him up: "Gene Savoy," an amateur archaeologist whose success in finding some 40 Incan and pre-Incan ruins in Peru was matched by a flair for self-promotion that drew on his tales of peril in the jungle, his bandito mustache and Stetson hat, and a retinue of would-be explorers who paid to accompany him, died on September 11 at his home in Reno, NV at the age of 80 (Martin 2007).

Savoy's (1970) book, *Antisuyo*, presents his account of Vilcabamba la Vieja, which he called Espiritu Pampa from the place name found on the Bües map. Hiram Bingham had reached it more than half a century before Savoy, but did not successfully explore it. Savoy noted that when workmen removed the vegetation, remnants of round buildings were mixed in with those of the Inca rectangular plan (Savoy 1970:95). Knowing that Matsigenka built ovoid-shaped huts, he suggested that these structures reflected a jungle influence on Inca architecture. The writer Kim MacQuarrie (2007:445) also erroneously assumed that "local indigenous groups" who lived there had built the cylindrical stone structures. In fact, however, roundish constructions at that site tangibly indicate the presence of the Wari civilization that preceded the Inca in time. The Wari cultural development, centered across the Apurimac near Ayacucho until ca. 800 AD, also reached into what is now Cusco Department. The Incas at Vilcabamba la Vieja built their town on top of the Wari settlement. The 2010 excavation of a pre-Inca grave covered with two flagstones dramatized the Wari-to-Inca cultural sequence. A person of noble status had two scepters of palm wood, a silver mask and pectoral, 234 silver plates, and gold armbands (Salcedo 2011). This "Señor del Wari" stunningly verified Wari influence in the Vilcabamba centuries before the Inca expanded into their territory. Clearly the infrastructure already in place at Vilcabamba la Vieja explained why the Inca chose to put their refuge there. This discovery was the culmination of an archaeological excavation begun in 2002 under the auspices of the Cusco office of the Instituto Nacional de Cultura (Fonseca Santa Cruz 2010).

8.4 Vilcabamba as a Magical Territory

All explorers of Peru fantasize about finding a "lost" ruined pre-Columbian city somewhere in the jungle. The quest to discover a fabled city of yore has lured hundreds of adventurers into what can be called the "Paititi syndrome." The name Paititi refers to a real place lying far distant from Cusco in the savanna lowlands of Eastern Bolivia (Combès and Tyuleneva 2012). Scholars have known since the early 1960s that in the swamps of the Llanos de Mojos there had existed a civilization of notable cultural development. William M. Denevan (1966) was the first to provide a serious study of the cultural-historical meaning of the Llanos de Mojos. Clark Erickson (1995), who had been Denevan's graduate student, outdid his mentor in shedding even more light on the area. Levi-Bacci (2010) recently provided an overview of the history of this area.

The Incas knew Paititi as a location in the Province of Musu and they may have sought to conquer it. In the mid-nineteenth century, the Franciscan priest Julian Bovo de Revello (1889), who had unusual knowledge and insight into the lands east of Cusco, suggested that this mythical place called Paititi was in the Mojos.

Garcilaso de la Vega (1960:434–435) wrote that Inca Yupanqui undertook a military expedition to Musu by way of the Madre de Dios (Amarumayo) River and along the way, they subdued forest Indians before arriving at this zone of sophisticated agricultural achievement.

The British explorer Percy Fawcett (1953:186) sought to emulate Hiram Bingham's "romantic discovery," by finding another lost city thought by him to lie in the Amazon of Western Brazil. Over three decades, he pursued, fruitlessly and ultimately tragically, his chimera. Alberto Flores Galindo (2010:32) called the idea of Paititi an "imaginary geography, a sort of Conquistador oneiric horizon." The Paititi idea has continued to motivate dreamers to undertake expeditions into the forested valleys beyond the cordilleras. One Peruvian explorer described Paititi in quite precise terms as lying in the headwaters of the Rio Pantiacolla which flows into the Rio Alto Madre de Dios (Iwaki Ordóñez 1975). Gregory Deyermenjian (1999), an indefatigible American explorer, looked for Paititi over more than a decade in that same area. Earlier, a three-man expedition led by the American Bob Nichols, whom I knew in Cusco in 1963, was killed by the Matsigenka along the Pantiacolla while searching for a lost city.

8.4.1 *Choquequirao*

North of Cusco, the zone between the Urubamba and Apurímac has been prime territory for exploration. In the nineteenth century, two Frenchmen moving west via the Urubamba entered Choquequirao, an Inca site overhanging the eastern side of the stupendous Apurimac Canyon, 1700 m above the river. Though politically lying outside the district, Choquequirao is often considered to be part of the Vilcabamba realm. In 1834, a French diplomat, Eugène de Sartigues (1809–1892), reached Choquequirao by going up the valley of the Rio Santa Teresa on land then owned by Hda. Huadquiña. The visitor found the ruins overgrown, with trees sprouting inside roofless buildings. Reading between the lines of the Sartigues account, most of his 8 days there were spent searching for buried treasure. He erroneously thought this site a refuge of Manco Inca from the Spaniards. This Frenchman also entered the nearby Yanama Valley where a silver mine was worked on the slope and a sugar cane operation existed on the valley floor. So dense were the biting flies ("horribles petites bêtes") that he could not eat or drink without swallowing quantities of them (de Sartigues 1851:1036). The rum local people drank day and night inured them to the insects.

In 1847, another Frenchman, Léonce Angrand, traveled through the same fractured Vilcabamba landscape to Choquequirao. This *arqueólogo improvisado* seems also to have been motivated by the prospect of treasure, and to have wrongly concluded that Choquequirao had been the last Inca refuge (Rivera Martinez 2010). The assumptions made by both of these Europeans may have persuaded the Italian explorer Antonio Raimondi (1874) to himself proclaim Choquequirao to be the Inca refuge. Then, in 1909 Hiram Bingham (1910) reached that site, not from Huadquiña as the Frenchmen had, but by crossing the Apurimac River on a raft at a place called Pasaje. He enlisted the help of the prefect in Abancay who accompanied Bingham to Choquequirao, probably in the belief that the American had some special knowledge about where buried treasure could be found at the ruins. None of Bingham's writings about the site give any hint that he recovered valuable artifacts. Later Bingham read the document written by Carlos Romero (1909) and published in *Revisita Historica* in Lima which asserted that Choquequirao was not the last Inca refuge. Catherine Julien (2001) revealed its subsequent history by retrieving archival information indicating that Choquequirao was part of the encomienda that in 1539 Francisco Pizarro had granted to his brother Hernando. In the 1580s, colonial authorities created two new reducciones, Cachora and Huanipaca, on the west side of the river in order to nucleate the still disseminated native people of that zone. Old documents suggested that Choquequirao remained inhabited into the seventeenth century (Duffait 2005). Beginning in 1993, a French archaeological team made the first site survey and excavations. Mindful of the pioneering visit of the Eugène de Sartigues, the French government has provided an endowment to improve the tourist infrastructure.

For many years, the arduous climb from the canyon floor and the lack of a bridge made the site virtually inaccessible to visitors. Now tourists may take an organized 4-day round trip with mules, guides, and equipment, starting at Huanipaca or Cachora, both of which have foot bridges across the Apurimac. Hiram Bingham's curiosity drove him to undertake a rigorous trip. Since then, Choquequirao has become a trekker's destination. Its structures are not of the quality of Machu Picchu's, though geometric patterns with a llama motif are so unusual that von Kaupp and Fernandez Carrasco (2010:104) suggested the builders were mitmaqkuna whom the Incas had brought from Chachapoyas. Stone bank terraces built on 70–80 % slopes are remarkable feats of engineering. Two intrusive proposals, one is for a cable car stretching from Huanipaca or Cachora up to the ruins and the other is for a heliport at the ruins, have met much resistance.

8.4.2 *Vilcabamba la Vieja*

Vilcabamba la Vieja, also known as Espiritu Pampa, lacks Choquequirao's spectacular siting on a spur 1700 m above the river. Hiram Bingham is a principal figure in this story, too. In 1911, soon after he left Machu Picchu, he visited Vilcabamba la Vieja on his own by following the trail up the Vilcabamba Valley, going over the pass and down into the Pampaconas. No expedition members accompanied him so that he would not have to share discovery bragging rights with anyone else. Less than 4 months later, Clements Markham (1911) of the Royal Geographical Society in London published Bingham's hastily prepared piece on that site. Bingham, very soon after his return to the United States from that 1911 expedition, had sent pictures and commentary to Markham. Bingham wanted his achievement to be rapidly disseminated in the Society's *Geographical Journal*. Bingham's earliest comments and accompanying photographs of Machu Picchu, Vitcos, Chuquipalta and Vilcabamba la Vieja were published in the *Geographical Journal*. Although the comments were spare and contained factual errors, this appearance in print scooped the *National Geographic* whose first article on those finds did not appear until 1912. It is not in the published record how the National Geographic Society, which had first rights to publishing the results of the expedition they sponsored, reacted to having those notices appear first in a competing journal.

In his notes to Markham, Bingham equivocated about which site, Machu Picchu or Vilcabamba la Vieja, was the "lost city of the Incas." The explorer Gene Savoy argued that the label fit only Vilcabamba la Vieja, yet that site was not historically "lost" any more than Machu Picchu was. An anonymous 1783 manuscript mentioned the "pueblos antiguos de Vilcabamba" and mentioned Vilcabamba la Vieja being "en la tierra de adentro habitación principal de Inga…" ("…in the jungle, the main dwelling place of the Inca…") (Anonymous 1783). Around the same time, Oricain (1906:393) wrote that "ay tradicion de haver una magnifica poblacion nombrada Hatun Vilcabamba y otros no menos considerables." ("There is a tradition of a magnificent town called Hatun Vilcabamba and others just as impressive.") In 1976, the historian Edmundo Guillén (1994) from Lima trekked through the valley twice to locate all the places found in the chronicles and noted that Savoy had missed many details. Guillén, who left Cusco on June 1 and reached Vilcabamba la Vieja on June 6, spoke with the 90 year-old owner of a finca called Vista Alegre, who remembered Hiram Bingham stopping at his father's property. Guillén returned to Cusco from Vilcabamba la Vieja by trekking down the Cosireni to Kiteni on the Urubamba. In the 12 years since my expedition, a trail had been cut to make that possible. A month later, Guillén returned to Vilcabamba with two Polish filmmakers.

Vincent Lee (2000), a practicing architect from Wyoming, first came to Vilcabamba in the 1980s and made subsequent return trips. Lee not only provided a superb account of his explorations, but also made exquisite line drawings of sites and of each of the buildings on them. Lee also provided detailed maps of the places he explored. Two British explorers, Hugh Thomson (2003) and before him, John Hemming (1970) recounted their Vilcabamba trips. Additional explorers who have published Vilcabamba adventures include the Englishman John Beauclark (1980), the Spaniards Santiago de Valle Chousa (2005) and Maria Carmen del Martin Rubio (1988), and the Americans Gregory Deyermenjian (1985) and Robert von Kaupp (von Kaupp and Fernandez Carrasco 2010). Kaupp (1925–2012), who had a Ph.D. in anthropology from the University of Massachusetts, told a reporter that looking for lost ruins was one of the primary purposes of his life (Barnes 2012). Other exploration parties have entered Vilcabamba over the past half century without leaving retrievable trails of their journeys. Beginning in the late 1980s, Sendero Luminoso terrorists quelled the visitor flow for almost a decade. Increased presence of the military and police did not encourage travelers when word got out that the forces of law and order had themselves harassed foreign travelers. In 1987, another terrorist group, the MRTA, sought to establish itself in the valley of San Miguel, as if the bad guys had agreed to carve out different territories.

Publicity surrounding the various rambles to the Vilcabamba, Machu Picchu and the Lower Urubamba stimulated cinematographers to treat these adventures as a modern quest for the El Dorado myth (Scorer 2011). Several well-received films brought this part of Peru to the attention of an international audience. In 1972, "Aguirre the Wrath of God" by Werner Herzog appeared to much acclaim. The film started with the heights of Machu Picchu as the location of one lost city and descended into the jungles and the entangled myth of El Dorado as another lost city. Always hidden beyond the next bend of the river, El Dorado was never reached in that film. The arch-romantic idea of the stones of an ancient ruined settlement enveloped by luxuriant vegetation was also integrated into both the Walter Salles 2004 film, "The Motorcycle Diaries," and Steven Spielberg's 1984 film, "Raiders of the Lost Ark." If the theme of discovering a lost city in the South American rainforest began as a colonial legend, its modern inspiration is most directly attributable to the Hiram Bingham adventure. The lesson that can be drawn from the mystique of lost cities in the jungles of eastern Peru is that only by exploring can one discover. Behind exploration looms the driving force of intellectual curiosity.

8.5 Vilcabamba and the Face of Changes to Come

Since the early colonial period, the Vilcabamba region has held a geographical aura of mystery. Its isolation now, however, is less than it once was. The main entrance to Vilcabamba near Chaullay is within 5 h of Cusco, a city which can be reached within 24 h of travel from most points in the world today. As a locus of exploration, Vilcabamba's appeal combines visible ruins with a partially known historic record that dates from the early colonial period. Together they present the past in an intelligible way but new details can be added. For example, Corihuayrachina (also spelled Qoriwayrachina) was a remote Inca settlement outpost established to produce food and llamas for the silver mines near Vilcabamba (Frost 2004). On the basis of early Inca ("Killke") pottery sherds found there, an archaeology buff postulated a much earlier Inca presence in this area than that held by conventional wisdom (Ziegler and Malville 2013:211–212). The notion that Inca occupance of the Vilcabamba region came after Pachacutec conquered the Chanka people in 1438 begs for reexamination. Corihuayrachia also introduces new ideas about land use in the semi tropical Yanama Valley. In modern times, no coca has been cultivated there, but in the Inca period the floor of the Yanana Valley may have been in coca plantations with the leaf sent to the mines.

Vilcabamba is poised for tourist development. Its diverse landscapes from snowcaps to jungle wilderness enchant those who enter. This sub-region still exudes mystery about its pre-Columbian past. Improving the road in the Vilcabamba Valley and constructing an extension over the pass to the Pampaconas Valley will bring many visitors. A road to Puncuyoc by way of Lucma will make that site accessible. Authorities will also be under pressure to make the site of Choquequirao more accessible than it has been. Promoters argue that transportation improvements will also benefit agriculture and perhaps lead to the resuscitation of mining. For many trekkers, these developments are not welcomed. But modernization is inevitable, given the economic realities of Peru, the will of local inhabitants, and the world of mass tourism.

References

Anonymous (1783) Memorial acerca de cuatro ciudades inkas situadas entre los rios Urubamba y Apurimac. Biblioteca Nacional (Lima) sección manuscritos. Cuaderno No. 83, No. C1082. 13 ff

Barnes B (2012) A local life: Robert E. Von Kaupp, 86, independent scholar who searched for lost Inca capital. Washington Post, 25 February 2012

Beauclark J (1980) La cordillera Vilcabamba: el último refugio de los Incas. Boletín de Lima 1(4):78–81; 1 (5): 18–30

Bell BW Jr (2011) Decoding an Inca observatory: solar alignments at Puncuyoc, Nawpa Pacha. J Andean Archaeol 31(1):101–116

Bingham H (1910) The ruins of Choqquequirau. Am Anthropol 12:505–525

Bingham H (1911) Across South America. Houghton Mifflin, Boston

Bovo de Revello J (1889) Brillante porvenir del Cuzco o exposición de las esperanzas de engrandecimiento (1848). Imp. Calle de Quiilca, Lima

Bües C (1939) El valle del Río San Miguel, Cuenca del Urubamba. Revista Universitaria (Cusco) 77:42–56

Chapman FM (1921) The distribution of bird life in the Urubamba valley of Peru. Smithsonian Institution, U.S. National Museum Bulletin 117, General Printing Office, Washington

Cobo B (1956) Historia del Nuevo Mundo, 2 vol. Ediciones Atlas, Madrid

Combès I, Tyuleneva V (eds) (2012) Paititi, Ensayos y documentos. Itinerarios Editorial, Cochabamba, Bolivia

de Sartigues E (1851) Voyage dans les républiques de l'Amérique du Sud. Ecrit sous le pseudonyme de Lavandais E de. Revue des Deux Mondes 21(10):1019–1059

de Valle Chousa S (2005) El misterio de Vilcabamba: el hallazgo de la capital Inca perdida. Primerapersona, La Coruña: www.monografias.com/trabajos32.expedicion-Vilcabamba/expedicion-Vilcabamba.html. Accessed 23 Jan 2013

del Martin Rubio MC (1988) En el encuentro de dos mundos: los Incas de Vilcabamba. Ediciones Atlas, Madrid

Denevan WM (1966) The aboriginal cultural geography of the Llanos de Mojos of Bolivia. Iberoamericana 48. University of California Press, Berkeley

Deyermenjian G (1985) Vilcabamba revisited. South Am Explorer 12:4–11

Deyermenjian G (1999) Glimmers of Paititi: searching for a lost Incan refuge. Mercator's World 4(1):42–49

Duffait E (2005) Choquequirao en el siglo XVI: etnohistoria e implicaciones arqueológicas. Bulletin de l'Institut Français d'Études Andines 34(2):185–196

Erickson C (1995) Archaeological perspectives on ancient landscapes of the Llanos de Mojos in the Bolivian Amazon. In: Stahl P (ed) Archaeology in the American tropics: current analytical methods and applications. Cambridge University Press, New York, pp 66–95

Fawcett PH (1953) Lost trails, lost cities. Funk & Wagnalls, New York

Flores Galindo A (2010) In: Aguirre C, Walker CF, Hiatt W (eds and trans). Search of an Inca: identity and Utopia in the Andes. Cambridge University Press, New York

Fonseca Santa Cruz J (2010) Las nuevas evidencias de Espiritu Pampa, Vilcabamba, Cusco. Dirección regional Cusco Instituto nacional de Cultura Cusco

Frost P (2004) Mystery mountain of the Inca. Natl Geogr 209(2):66–81

Garcilaso de la Vega EI (1960) Comentarios reales de los Incas. Universidad Nacional del Cuzco, Cusco
Guillén E (1994) La guerra de reconquista Inka. historia épica de como los incas lucharon en defensa de la soberanía del Perú o Tawantinsuyo entre 1536 y 1572. R.A. Ediciones, Lima
Hemming J (1970) The conquest of the Incas. Harcourt Brace Jovanovich, Inc, New York
Iwaki Ordóñez R (1975) Operacion Paititi. Editorial de Cultura Andina, Cusco
Julien C (2001) Inca estates and the encomienda: Hernando Pizarro's holdings in Cusco. Andean Past 6:229–275
Lee VR (2000) Forgotten Vilcabamba. Empire Publishing, Jackson Hole
Levi-Bacci ML (2010) El Dorado in the marshes (trans: Ipsen C). Polity Press, Malden
MacQuarrie K (2007) The last days of the Incas. Simon and Schuster, New York
Markham CR (1911) Mr. Bingham in Vilcapampa. Geogr J 38(6):590–591
Martin D (2007) Gene Savoy, flamboyant explorer of Ruins, dies at 80. New York Times, 19 September 2007
Oricain P (1906) Compendio breve de discursos varios sobre diferentes materias y noticias geograficas (1790). In: Martua V (ed) Juicio de límites entre el Perú y Bolivia. Imprenta de Heinrich y Comp, Barcelona, pp 319–377
Peru (1964a) The lost city. Time Magazine, August 28, p 25
Peru (1964b) Nueva expedición a Paititi inician hoy voluntaros del Cuerpo de Paz. El Comercio (Cuzco) September 2
Raimondi A (1874) El Peru, 4 vol. Imp. del Estado, Lima
Regalado de Hurtado L (1992) Religión y evangelización en Vilcabamba (1572–1602). Pontifícia Universidad Católica del Perú. Fondo Editorial, Lima
Rivera Martinez E (2010) La obra peruanista de Leonce Angrand (1834–1838, 1847). Fundación M.J. Bustamente de la Fuente, Lima
Romero CA (1909) Informe sobre las ruinas de Choquequirao. Revista Histórica 4:87–103
Salcedo JV (2011) Increíble pero cierto: hallan Señor Wari en el Cusco. La Republica (Lima), February 24
Savoy G (1970) Antisuyo: the search for the lost cities of the Amazon. Simon and Schuster, New York
Scorer J (2011) "X" never, ever marks the spot: Latin American lost cities on screen. Iberoamericana 11(44):101–116
Thomson H (2003) The white rock: an exploration of the Inca heartland. Overlook Press, Woodstock
Von Hagen V (1958) Los caminos del sol (trans: Cosla R). Editorial Hermes, Mexico City
von Kaupp R, Fernandez Carrasco O (2010) Vilcabamba desconocida: exploraciones 1997–2003. Editorial Grafica Rivera, Cusco
White S (1978) Cedar and mahogany logging in eastern Peru. Geogr Rev 68(4):394–416
White S (1981) Moments in the narrative landscape of highland Peru. PhD Dissertation, University of Wisconsin, Madison
White S (1985) Relations of subsistence to the vegetation mosaic of Vilcabamba, southern Peruvian Andes. Yearbook Conf Lat Am Geogr 11:5–12
White S (1994/1995) Preliminary site survey of the Punkuoq range, southern Peru. Nawpa Pacha 22–23:127–160
Ziegler GR, Malville JM (2013) Machu Picchu's sacred sisters: Choquequirao and Llactapata. Johnson Books, Boulder

Chapter 9
Highland and Lowland Peoples in Contact in the Tropical Urubamba

Abstract Before and after the Spanish Conquest, a stretch on the Urubamba River between 730 and 650 m above sea level (asl) became a contact zone where forest tribes encountered people of highland origin (Incas, Quechua Indians, mestizos, and Spaniards). Matsigenka Indians settled in that zone, but the other forest tribe, the Piro, annually came upriver from a long distance to trade. In the Inca period, contact rested on trading coca, woods, medicinal plants and ornaments from the jungle for salt and other goods from the highlands. Except for a small coca-growing fringe, the Incas did not permanently settle in the hot country. They had negative perceptions of the jungle environment. After the Conquest, Spaniards first took over Inca coca plantations on the fringe and then established coca and sugar cane estates through the colonial period. As highlanders pushed down the Urubamba below 800 m asl to start haciendas, they came into contact with Matsigenka. Catholic priests started a mission to save souls, and haciendas used Matsigenka as a source of labor. An annual trade fair with forest tribes survived into the first decade of the twentieth century. Further highlander penetration down the Urubamba occurred during the cinchona bark and rubber booms in the late nineteenth century. Later, a new land policy of the Peruvian government distributed parcels in the jungle to highlanders. In the 1960s, road construction brought a migratory flow of highland peasants into the lower Urubamba. Engulfing the selva, highlanders cut forests, eroded the soils and depleted fishing and wildlife. A natural gas boom has led to further change in the lower Urubamba. The Matsigenka and Piro survive with their own blocks of land, but their aboriginal culture has largely disappeared.

9.1 Traditional Cultural Distributions and River Navigability

A major cultural-historical cleavage long divided the Urubamba Valley into sedentary highland people to the south and the semi-nomadic tribes of the tropical forest to the north.[1] Before and after the Spanish Conquest, a delineable zone of contact brought people of these two cultural configurations together for trade, labor and, later,

[1] I use the spelling Matsigenka as the now preferred orthography for the tribe. Other forms—Machiguenga, Machigenga, Matsigenga and Matsiguenga—have not disappeared.

D.W. Gade, *Spell of the Urubamba*, DOI 10.1007/978-3-319-20849-7_9

colonization. The kind and degree of contact that the Incas had with forest tribes is scarcely known, because a reliable historical record began with only the European arrival. In the twentieth century, acculturation, migration, and technology blurred distinctions that once were so strongly imbedded in the cultural geography of the Urubamba. The contact zone is passé, yet the historical geography of cultural encounters reveals much about the dynamic interplay of people. A whole succession of valleys on the Andean eastern front experienced these encounters but no one pattern fits all of them. The assumption that the edge of the tropical forest marked the cultural divide between groups cannot be supported; instead the contact zone represented a fluid domain where two cultural entities came into regular contact and settlement frequently shifted up or down (Lyon 1981).

Where along the Urubamba River the settlements of the forest dwellers—called Antis or Manaries and later Matsigenka—historically ended and those of the sedentary highland peoples began can only be approximated (see footnote 1). A supposition that the Matsigenka occupy the same territory today as they did at the time of the Conquest may be true only in the general sense that their homeland remains along the Urubamba. Exact boundaries fluctuated as one would expect of a socially decentralized, semi-nomadic group. No forts built by highlanders separated the two groups, suggesting both the lack of any set boundary and the absence of long-term hostilities between the Inca and the Matsigenka. That forest people sporadically pushed up the valley in their search for hunting grounds is likely, though an assertion that the Matsigenka once occupied the Urubamba almost as far up valley as Machu Picchu is not supported by either archaeology or ethnohistory (Renard-Casewitz 1972:209). Given the nature of their social organization, it is unlikely that historically the Matsigenka expanded into zones of traditional highland settlement. Forest Indians, however, visited the highlands not as invaders, but as guests of the Inca. For their part, the Incas settled the fringe of the hot country in the Urubamba drainage. During the reign of Tupac Yupanqui (1471–1493) coca cultivation began in the Amaybamba Valley. The Inca failure to settle more of the forested part of the valley constitutes one of the conundrums of Andean culture history. Lines of evidence for that void can be deduced from the rarity of settlement below 1500 m asl and comments made in the sixteenth-century chronicles.

After the Conquest, a small population of highland origin lived in the tropical Urubamba from 1500 to 700 m asl. A late colonial document about people living around present-day Quillabamba referred to them as "Ante-yngas, those who … maintain the dress and language of Cusco and cultivate fields in an organized manner…." (Oricain 1907:372). Although Matsigenka wore long tunics, they did not speak Quechua nor cultivate their fields in the ordered fashion of highland chacras. After the Tupac Amaru revolt of the 1780s, the colonial government tried to establish a town called Guananay four leagues (=20 km) upstream from a settlement at the headwaters of canoe navigation called Cocabambilla. Its purpose, never explained in colonial records, may have been intended as bulwark against possible incursions of forest tribes. Guananay, however, never actually developed into a town, perhaps because the threat never materialized. How high the Matsigenka occupied the valley before the Conquest is not clearly understood. Place names provide clues to the

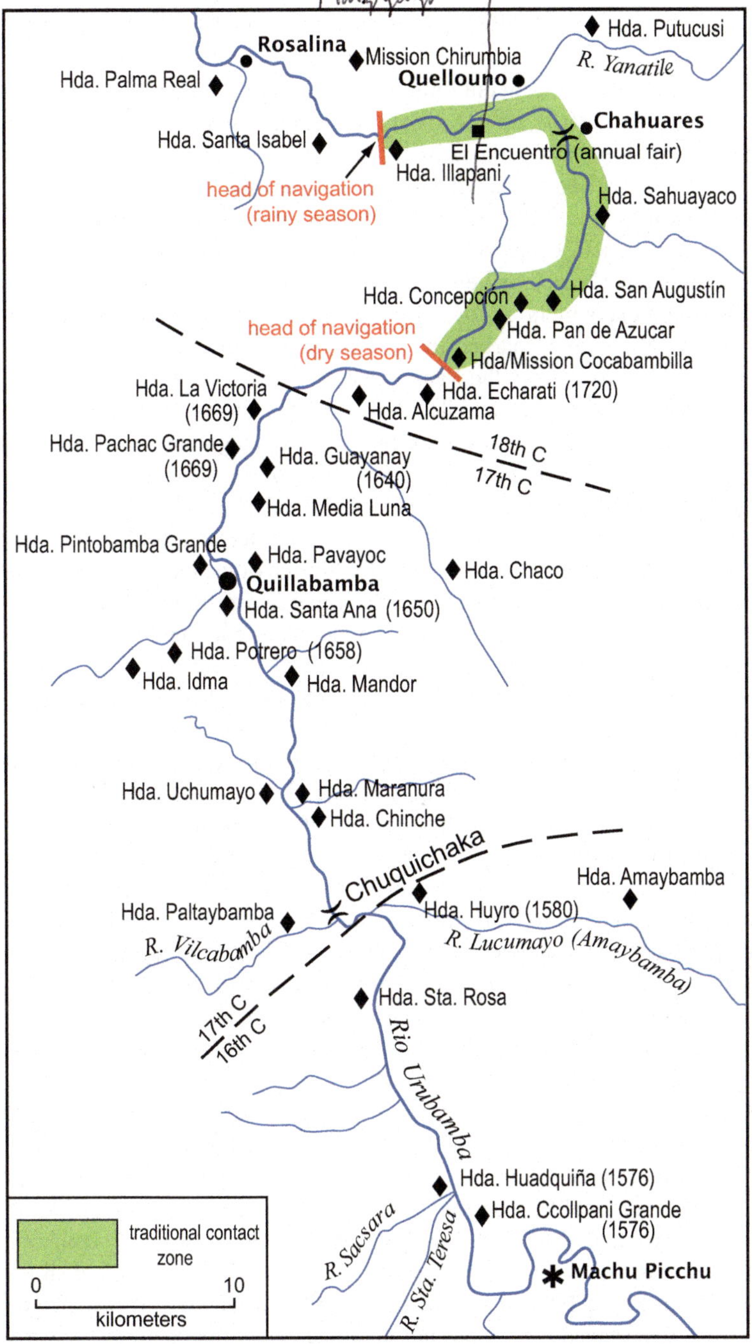

Fig. 9.1 Beginning in the sixteenth century, Spaniards and others from the highlands established haciendas in the main Urubamba Valley between 1800 and 650 asl and its tributaries. Tropical products focused especially on solid sugar and, beginning in the nineteenth century, on rum. North of Echarati hacendados encountered Matsigenka. A contact zone (*shaded area*) existed into the twentieth century between highlanders and tropical forest tribes. Now both the contact zone and the hacienda system have disappeared

ancient location of spheres of cultural influence. On the Urubamba upstream from its juncture with the Yanatile, all river names are in Quechua.

The contact zone, also an area of ecological transition between the dry tropical forest and the selva alta, included two kinds of movable boundaries (Fig. 9.1). One was the "advance limit" at the headwaters of canoe navigation to which Matsigenka moved their camps. Cocabambilla was a mission and a hacienda, but considering its location, may have been long before the Conquest a longstanding place of lowland-highland trade encounters. A second boundary was the juncture of the Urubamba with the Yanatile, which marked the normal upper limit of persistent Matsigenka settlement. Downriver from the Yanatile, all the streams have a Matsigenka name, suggesting the Urubamba-Yanatile confluence as the upper edge of traditional Matsigenka territory. The opposite extension, that is highlanders pushing down the river to settle, does not predate the nineteenth century. The Incas may have been most hindered in penetrating the forest by their lack of knowledge about boats.

Defined in its broadest sense, the contact zone between highland and lowland peoples in the Urubamba comprised a 50 km stretch of the main river. Cocabambilla was the crucial upvalley limit of the zone. Upstream from that point, rapids made river travel difficult and hazardous either going up or coming down. Two sixteenth-century expeditions that started at Quillabamba failed. One, led by Martin Garcia de Loyola that descended the river on rafts and in canoes, had its watercrafts capsize (de Ocampo 2013). In 1582, another Spanish expedition set out from near the same place. Three rafts carrying food and clothing supplies capsized and the following day, two more rafts broke up when they crashed into an overhang.

The problem was not only with the river but also the people. Highlanders were inept boatmen compared to forest tribes who deftly negotiated swift currents that capsized the canoes of outsiders. Seasonal differences in the river also affected navigability. The December to April rainy months bring down so much water from the Highlands that the Urubamba becomes a raging torrent. But even in the dry season, navigation is hazardous. Where lateral streams flow into the main river, the combination of shallows, rapids and whirlpools creates "malos pasos" which have to be carefully negotiated. Much farther downstream the Pongo de Mainique is a 3 km long narrows of fast-moving water and whirlpools (*remolinos*), where the Urubamba flows through a rock-faced canyon lined with waterfalls gushing out from the perpendicular limestone walls (Fig. 9.2). Drownings, ruined boats and lost cargo have been the price of many attempts to move through the Pongo. Successful negotiation of the passage is more likely during the low-water period. In August 1897, the North American J. Arton Kerbey (1905) went through the Pongo with two companions on a raft. During the rainy season this stretch is particularly hazardous, even for experienced boatmen. In February 1932, an expedition led by the British geologist, J.W. Gregory, portaged canoes around the most dangerous sections (Coverly-Price and Wood 1933; Leake 2012). Returned to the churning stream, their canoes, hit by waves, capsized. Eight of the ten people swam to shore, but Professor Gregory and a Matsigenka Indian were swept into a whirlpool and disappeared. Once the formidable Pongo is successfully negotiated from south to north, navigation on the placid Urubamba downstream has few hazards. This major geographical barrier to naviga-

Fig. 9.2 Pongo de Mainique is a treacherous narrows with whirlpools that hinders navigation on the Urubamba River. Samanez y Ocampo (1885:55) who traversed this stretch of water in 1884 wrote the following: "There are no words to describe this extraordinary pongo with its sublime and enchanting beauty mixed with horrors and dangers that freeze the blood with fear"

tion explains in part the difficulty that highlanders had in penetrating the tropical forest section of the Urubamba. For their part, forest Indians have routinely negotiated the Pongo.

9.2 Defining the Inhabitants

For many centuries the Urubamba Valley in its totality consisted of two ethnic territories: highlanders above and forest tribes ("Antis") below (Renard-Casevitz et al. 1986). Besides linguistic, religious and social differences of the first order, farming methods set the two groups very much apart. In the highlands, sedentary farming enabled Andean people to achieve food surpluses. The Antis lived as subsistence-oriented swidden agriculturalists, hunters and fishermen. The Highland Incas extended their hegemony into the tropical Urubamba to perhaps near Quillabamba. Yet no pre-Columbian settlements identified as permanent can be found in the main depression between Huadquiña at 1500 m asl and the Yanatile. Through that area, 105 km long, the Incas surely derived resources even if they did not settle there. Near the Yanatile, the Incas came in contact with forest tribes. Without permanent settlement, would the Incas have built trails down the Urubamba as far as the Yanatile? Trails were reconstructed as far as Chaullay, but no evidence for such a pre-Conquest road can be found.

Notwithstanding the contact zone, a clear division separated the people of highland culture, including the Spaniards after the Conquest, from the forest tribes. Some of that division was based on fear and misperception. Lack of a lingua franca and unfamiliarity deterred each population from entering the domain of the other. Highlanders called the forest Indians "chunchos," that is, barbarians. People of the jungle differed from highlanders in their use of ambush as a method and the bow and arrow as a technology. They also had in the Highlands an undeserved reputation as cannibals. The Neo-Inca Sayri Tupac, perhaps aiming to frighten the Spaniards, claimed that his chuncho allies were "eaters of human flesh" (de Matienzo 1967). In this far western part of the Amazon drainage, only the Conibo tribe on the Ucayali has had an anthropophagic history. Cieza de Léon (1959) and Garcilaso de la Vega (1960) were other chroniclers who repeated this assertion that resulted from an overgeneralization.

The Spanish chronicles mention nothing about the Inca as settlers in the Urubamba selva. The closest comment in that regard recorded that Tupac Yupanqui had made an incursion there (Santacruz Pachacuti Yanqui 1968). The American anthropologist William Farabee (1922) related an oral tradition that the Incas, in conjunction with the Piro forest tribe as their vassals, built Tonquini at the southern entrance to the Pongo de Mainique. Finnish archaeologists Siiräinen and Pärssinen (2001:46) argued that Tonquini was proof of an Inca "control zone" over the forested Urubamba. The absence of archaeological record of Inca presence anywhere in the lower Urubamba makes that interpretation unpersuasive. More remarkably, even in the higher tropical zone between 1500 and 700 m asl, the lack of Inca presence is puzzling. Killke-style pottery, collected mostly by C. Bües has been retrieved from the Urubamba as far downriver as Quillambamba. However, its presence is not necessarily evidence of an early Inca occupation of that part of the Urubamba.

9.3 Pre-Conquest Trade

The culture history of coca in this valley remains an enigma. Its reconstruction may be advanced if one brings together what is known in order to develop plausible assumptions. In the Urubamba below 1000 m asl many species of *Erythroxylum* are present, but only one of them, *E. coca*, is now thought to have been selected by human intervention. Other species are described as "wild," though it is unknown if some of those species were once cultivated by forest tribes ("Antis" in general). In the Urubamba, the Matsigenka is the forest tribe with a tradition of growing coca. This group may have cultivated the shrub for their own needs, and at some point, perhaps as early as the Wari period (550–900 AD) began to trade the leaf to highlanders. When the Incas, who arrived after the Wari, consolidated their hold on the Cusco highlands, they first depended on trade to get the leaf. As the Inca polity expanded and demand increased, this trade supply no longer sufficed. In the late Inca period, highlanders started to grow coca themselves perhaps, following the Matsigenka example. One kind of coca—*E. coca*—attracted their attention and they laid out plantations of this species using the labor of forest Indians who had become their vassals and mitimaes brought from different places in the Empire. An early trade pattern was implied by the assertion in Viceroy Francisco Toledo's *vista* of the 1570s that Anti people were suppliers of coca to the highland (Cook 1975). *E. coca* was far superior to other kinds of coca. The plant not only tolerated the cool nights at elevations above 1000 m asl, but the leaf quality better than wild coca, how understood in term of flavonoids and high alkaloid content. *E. coca* also sets seed at those higher elevations. When grown below 600 m asl, *E. coca* does not normally produce seed and thus can be grown only by using cuttings. Moreover, at low altitudes, annual leaf production is low and the plant short lived.

One may assume that the Matsigenka, whose ancestors lived on the Urubamba for several millennia, had a key role in coca cultivation before highlanders got involved in it. Matsigenka have their own calcium substance to extract alkaloids from the coca leaf. Whereas highlanders have used mainly the ash of quinoa (*Chenopodium quinoa*) for that purpose, the Matsigenka prepare an ash called *tocara* derived from the branches of a jungle tree. The Matsigenka also use *chamiuro*, a woody plant yielding a bark that sweetens the taste of the coca leaf. Some highlanders knew that plant too, but knowledge of it seems to have come from the Matsigenka habitat on the forested eastern slope. Matsigenka, like highlanders, have used coca as an indispensable accompaniment of work. They take periodic coca breaks just as highland workers have done. However, the geography of coca use in the Urubamba suggests highland influence in the custom. A clue is that north of the Pongo, the Matsigenka have not used coca at all, whereas south of the Pongo—the area closest to the highlands—coca has a long tradition. In the early colonial period, the main center of coca production was not the Urubamba, but the valleys of the Callanga, Toiama and Tono Rivers in the Madre de Dios drainage.[2]

Interests in the jungle could normally have been satisfied by trade rather than settlement. Products they sought included fine woods, medicinals, dyestuffs, ornaments, and specialty foodstuffs. Chonta (*Bactris gasipaes*), a spiny palm with very hard but also light and elastic wood, was used for spears, war clubs, maces, fishhooks, and staffs of authority. Balsa (*Ochroma* spp.) wood was valued for making rafts. The Inca elite also sought brightly colored bird plumes for the sumptuous feather work they wore. Resins for embalming and medicine were collected from jungle trees. Honey from wild bees was valued. Plant dyes from the lowlands included genipa (*Genipa americana*), annatto (*Bixa orellana*) and palillo (*Escobedia scabrifolia*). The Incas produced cotton, but to meet their needs they may also have acquired cotton textiles from the forest tribes.

[2] Two German engineers, Herman Göhring and Jorg Von Hassel were the first to describe the terraces and ruins of Inca dwellings of the zone. By then its importance as a major coca producing zone into the seventeenth century had been forgotten. Its center is in a 70-km radius around the settlement of Lacco (formerly an hacienda), which holds stone terraces, a ceremonial center and a necropolis. One of the sites, Hualla Mocco, may be the source of the colonial name of this region known as Hualla.

Tropical plants of medicinal value in the Inca pharmacopoeia included purgatives and aphrodisiacs. Wanarpo (*Jatropha ciliata*) was a well-known aphrodisiac that came from this zone (Santacruz Pachacuti Yanqui 1968:289). Inca shamans may have journeyed to the jungle to obtain materia medica in a long established trade channel (Urton 1981:177). The Incas may have used chamairo (*Mussatia hyacinthina*) as a sweetening additive to chewing the coca leaf. What foods passed into highland trade channels from the lowland people is unknown. The Incas themselves cultivated manioc, uncucha, peanuts, guavas and cherimoyas in warm intermontane valleys, thus trade in these items would not have been necessary. Perishability of these crops made it impractical to bring them long distances to Cusco.

9.4 Inca Perceptions of the Hot Country

9.4.1 *Sinister Elements*

Inca perception of the selva as a sinister place was based on how different it was from the highlands. The hot country's impenetrable forests contained fearsome creatures. The jaguar (*Leo onca*) was twice as large, powerful and stealthy a feline as the puma that the highlanders knew (Fig. 9.3). The vampire bat (*Desmodus rotundus*),

Fig. 9.3 Seventeenth century drawing of an Anti (forest Indian) facing a jaguar with bow and arrow. Forest dwellers were collectively called chunchos meaning barbarians, reflecting the disparaging view that highlanders held about people of the forest. (From Guaman Poma de Ayala 1980)

which extracted blood like a ghoul from its victims, was a lugubrious creature to people of the heights. The Andean highlands have no snakes, venomous or not, and highlanders have always had irrational fear of these reptiles. Below about 2300 m asl an array of deadly serpents appear in the valley. The common snake of the ceja de la montaña is the Andean lancehead (*Bothropsandianus*). Below 1400 m the feared lance or jergón (*Botrops atrox*), known for its fast-acting and lethal hematoxins, appears. Some vipers live in trees; one of them is the small but deadly parrotsnake or *loro machucuy* (*Bothriopsis bilineata*). Rattlesnakes (*Crotalis* spp.) inhabit the fields. A 100 km farther down the Urubamba from the Yanatile juncture, the fearsome bushmaster (*Lachesis mutus*) delivers lethal power with one bite. Lying in wait in the streams are the black caiman (*Melanosuchus niger*) and anaconda (*Eunectes murinus*), the latter of which is called amaru in Quechua and is part of highland folklore about the jungle.

As in the present, encounters with large animals were episodic, but biting, stinging, and burrowing insects became inevitable companions of all who enter the tropical forest. Some are venomous, such as centipedes in the genus *Scolopendra* and scorpions, especially those of the family Buthidae. Several species of ants afflict unwary humans with their bites. Chief among them are the *pungara*, a large black cobra ant (*Pachycondyla villosa*) and several kinds of giant hunting ants, especially *Dinoponera gigantea* and *Parapondera clavata*. The last species, known as *isula*, has a bite 30 times more painful than the sting of a wasp. Excruciating in a different way is the human botfly (*Dermatobia hominis*), whose maggot develops under the skin and becomes increasingly unbearable as the larva grows. Chiggers (*Pulex penetrans*), gnats in clouds, and mosquitoes are inevitable. Another plague of the hot country that is unknown in the highlands, is the leaf-cutting ant (*Atta* spp.), unknown in the highlands, that overnight makes a shamble of a crop field. Tropical forest dwellers contended with these pests, but highland dwellers had few except fleas and lice. To those accustomed to the deliciously temperate and bug-free highlands devoid of dangerous predators, the hot country of the eastern valleys was an uncomfortable and sometimes perilous realm. It was also unhealthy: the warm valleys were strongly identified as the source areas of contraction of the particularly frightening disease, leishmaniasis.

9.4.2 Land Use Hindrances

Agricultural limitations and hindrances presented other reasons to avoid the jungle. The poor lowland soils could not be continuously cultivated without regular enrichment. Farming required a swidden regime of cutting and burning the primary forest before planting a crop. Two years later, poor yields forced abandonment of the plot. That sequence of cut, cultivate and abandon required a semi-nomadic way of life and did not support concentrated population clusters. In contrast, highland Inca practice permitted some soils to be cultivated indefinitely if nutrients were applied. Moreover, none of the familiar highland dietary staples of potato, flour maize, or quinoa could be successfully grown in the hot part of the valley. Llamas, the only Inca beast of burden, were out of their element in the tropical forest. They were poorly adapted to the heat and humidity and lacking plants that met their nutritional requirements. Precisely how the Incas reacted to the dangers, aggravations, and limitations remains unknown, but the virtual absence of stonework suggests that they did not integrate the hot country into their permanent settlement fabric. Yet, in the end, the void presents a conundrum. Modern Quechua migration to the jungles of the east shows that highlanders can adapt to that environment. The Incas, an intensely organized polity, may have found the jungle to be inimical to their way of life. Permanent settlement was not possible and food surplus was hard to come by.

9.5 Ethnographic Patterns in the Tropical Urubamba

Spaniards organized their exploration of the tropical valley some 40 years after the Conquest. The Viceroy Francisco Toledo, who arrived in Peru in 1569, had as his priority the defeat of the Neo-Inca Tupac Amaru in Vilcabamba. Incursions in this forested zone led to his capture in 1572, an event that also put Spaniards in contact with the "chunchos" for the first time. Led by Martin Garcia de Loyola, captain of Viceroy Toledo's personal guard and nephew of St. Ignatius, the expedition encountered forest Indians called Manaries i.e. the Matsigenka. Others called these aboriginals "Antis," an Inca term for people of the East. The Spaniards frequently made the distinction between "indios infieles" by which they covered all forest-dwelling people and "indios cristianos" i.e. native people of the highlands. Matsigenka used the Quechua terms *punaruna or puñaruna* for highlanders. Another expedition, in 1582, led by Martin Hurtado de Arbieto using canoes and rafts on the main Urubamba River reached the land of the forest Indians (de Matienzo 1967:vii). A Spaniard called these forest people "'yndios amigos'" who were friendly to our Spanish nation….and well and honestly dressed" (de Ocampo 2013). He was referring to the long tunics that the Matsigenka used then and well into the twentieth century. The Spaniards also encountered another group, the Pilcozones, who were described as much harder to deal with than the Mañaries. Everything points to the Pilcozones as the tribe that later became known as the Chontaquiro and, after that, the Piro.

9.5.1 The Matsigenka

Arawak speakers, the Matsigenka have been linguistically related to the Campa. Well into the nineteenth century, many travelers on the Urubamba referred to the Matsigenka as Campa (Samanez y Ocampo 1885). The America ethnologist Allen Johnson (2004:28) surmised that the Matsigenka had been in place in the Urubamba Valley for as long as 5000 years and had, through isolation, separated from the main mass of Campa people who lived much farther north. He assumed the Matsigenka occupy the same territory today as they did at the time of the Conquest. Their subsistence activities of swidden agriculture, hunting, and fishing differed little from other Amazon groups (Cenitagoya 1931). Each family had three kinds of gardens: one in current use in which manioc, maize, yams, and peanuts were grown; a second that was abandoned except for the fruit trees in them; and a third cleared of forest but not yet planted. Hunting with bow and arrow persisted almost to the twentieth century when the rubber boom brought shotguns to the Matsigenka. Fishing, a dry season activity when water volume is low, involved catching fish in weirs or stupefying them with barbasco. The catch was smoked and dried. In the traditional Matsigenka social organization no chief imposed his control; the basic unit consisted of an extended family in which a married daughter lived near her parents. Both men and women wore a slit open shirt (*cushma*) woven of cotton, a garment suggesting the desirability of protection from coolish nights (Fig. 9.4).

9.5.2 The Piro

The other forest tribe in the Urubamba was the Piro, now also called the Yine, a tribe that in the sixteenth century was probably the same as those called Pilcozone by the Spaniards. In the eighteenth and nineteenth centuries, the Piro were known

Fig. 9.4 The cushma, woven of cotton, was worn by Matsigenka men and women. Pre-Conquest use of this garment is unknown, but may reflect vestimentary influence from the highlands. Most Matsigenka have abandoned daily use of the cushma

as Chontaquiro. Their home territory north of the Pongo de Mainique centered on the con fluence of the Urubamba with the Río Miaría. The Piro made excellent canoes, were skilled boatmen, and their mobility gave them a reputation as "the people of the river." Unlike the Matsigenka, the Piro had a strong tribal organization with a hereditary chief. Clever traders, they expanded their influence by learning Quechua, Spanish, Matsigenka and Conibo. Mobility and commercial acumen enabled the Piro to spread elements of highland material culture far into the Western Amazon Basin.

Beginning in 1711, the Piro became known to Urubamba highlanders as the source of certain trade items (Scipión Llona 1903:97). Their role as traders may have been a carryover from the Inca period. Piro long-distance movements occurred in the dry season when their flotillas traversed the Pongo de Mainique. From there they rowed 100 more kilometers to the confluence with the Yanatile where they traded with highland people. Rowing against the swift current, the upstream trip required 3 months; the return downstream trip took only 15 days. On the way upstream, the Piro encountered the Matsigenka with whom they traded, but also from whom they stole. Most egregiously, the Piro kidnapped Matsigenka children, packed them in their canoes, and sold them as slaves to highlanders. During the rubber boom when agents needed laborers above all, this predatory practice became lucrative.

Opinions about the Piro varied. Whereas to the Harvard anthropologist William Farabee (1922), the Piro were the "noblest tribe on the Urubamba," others deemed them to be cunning and ruthless. Samanez y Ocampo who traveled up the Urubamba with the Piro viewed them as "gypsies" for their frequent displacement, deceitfulness, laziness and fickleness.[3] Matsigenka feared the Piro because of their frequent raids. They considered the Piro to be ignorant of the forest and less knowledgeable about plants and animals than they were (Gow 2012). Piro lack of timidity gave them more leverage in commercial exchanges with highlanders than the Matsigenka who were generally shyer and more reticent.

9.6 Highlanders and Tropical Haciendas

The climate of the Urubamba below Machu Picchu dramatically changes to a macrothermal realm of different vegetation and crops. North of the Machu Picchu canyon, the *ceja de la montaña* gives way to a dry tropical forest that reflects the sharp rainfall seasonality rather than the orographic effect of the Cordillera de Vilcabamba. The Incas grew coca in the lateral Amaybamba (now referred to also as the Lucumayo or Huayopata) Valley. Some coca workers there had been brought as colonists (*mitimaes*) from zones of similar climates elsewhere. Access to Amaybamba was by way of Ollantaytmabo above which lies the pass known as

[3] In the eighteenth century, a Piro Indian, Juan Santos Atahualpa, sought to organize an uprising of all Amazon tribes in eastern Peru against the Spaniards.

Pantiacolla or, after the Conquest, Málaga. The traveler coming through the pass from Ollantaytambo first moved through a puna, then sharply descended into a maize-growing temperate zone and below that into the quasi tropical elevations below 1800 m asl where coca was grown.

Spanish interest in most of this warm environment was tentative, although they took over the Inca coca fields within a decade after the Conquest. In 1540 it became an encomienda of Hernando Pizarro, which subsequently passed on to Arias Maldonado and then his son Diego (Julien 2001). Demise of the encomienda returned the zone to indigenous control for a time. As late as 1568, this coca zone was still owned by descendants of, the Inca nobility. Indians from around Ollantaytambo arrived as workers for 3-week stints to pick coca and then returned to their home ayllus. Spaniards soon saw sugar cane, at that time grown for the crystalline product, not alcohol, as more lucrative crop than coca. In 1569–1570, Hacienda Huyro, started by the Augustinian order, set up a sugar cane mill (*ingenio*) in which the power of horses moved the wheel (Glave and Remy 1983:125). Coca and sugar cane were grown on most properties and mules replaced llamas as the beast of burden. Whereas Indians did the work of coca cultivation and harvest, estate owners brought in African slaves as a dependable labor supply for sugar cane cultivation.

The main depression of the Urubamba began to receive attention during the 1570s (Menéndez Rua 1948). The last native suspension bridge crossed the river at 1200 m asl; below that, rafts were the only way to cross the stream but, during the torrential rainy season, rafting was dangerous. This isolation of one bank from the other explained why the expression "Valle de Amaybamba" included not only the right bank of the Amaybamba itself, but also the right bank of the larger Urubamba all the way down to the end of the colonized zone at Cocabambilla (Villanueva Urteaga and Macera 1980:77). Similarly, the "Valle de Quillabamba" referred only to the left bank of the Urubamba and included the haciendas Huadquiña, Santa Ana and Potrero. Early colonial interest of the upper reaches of the tropical Urubamba included its wood resource. The highlands had little wood and building material for Cusco came from this zone. For religious and manorial structures, builders preferred the fragrant and strong timbers from the cedro tree (*Cedrela odorata*). Highland Indians carried those large pieces of wood the 18–20 leagues (=90–100 km) to Cusco from the area around Chaullay (Egaña 1954:III:428). Farther down the Urubamba, supplies of wood from this species were vast, but the distances to Cusco were too great. The hacienda zone lost part of its wooded cover when sugar cane became the major crop and wood was needed as the fuel in boiling cane juice (Fig. 9.5).

Spaniards in Cusco became aware of the tropical Urubamba as a place where they could make money growing and selling sugar cane and coca. The Jesuits were particularly keen to start haciendas in order to financially support their schools. Though not arriving in Peru until the 1560s, the Jesuits soon perfected the economics of estate production and in time their profitability became the envy of many landowners. Hacienda Huadquiña, founded in 1576, was their first property in the main Urubamba depression (Fig. 9.6). Its success encouraged the Jesuits to expand their agricultural involvement farther down the valley either by purchase or bestowal.

Fig. 9.5 Sugar cane field on flat land near Quillabamba with boiling house on the right. Wood for fuel in these operations came from clearing the slopes of trees. Sugar cane cultivation requires an economy of scale which explains its disappearance with the breakup of the hacienda system

Fig. 9.6 Early twentieth-century view of Hacienda Huadquiña when Mariano Vargas owned the property. For more than 300 years, Huadquiña was the largest, most self-contained hacienda in La Convención. Beyond the chapel are the cane fields. Bells were used to call workers to their tasks (Source: Hardy 1919)

About 65 km farther down the valley, Hda Santa Ana (1050 m asl), founded in 1650, occupied a river terrace well suited to agriculture. The previous owner passed it to the Society in return for payment of Masses to be said in his name. In 1720, the Jesuits acquired Hda Echarati (750 m). All three estates produced sugar as their economic mainstay. At Huadquiña, the harvest occurred once every 2 years; at Santa Ana, warmer temperatures enabled the cane to grow faster and its harvest occurred every 18 months. At Echarati cane was cut and processed every 12 months. Echarati also initiated cacao cultivation in the Urubamba. As a substitute for the unreliable supply of Highland Indian labor, the Jesuits brought in African slaves. A fourth Jesuit property was Cocabambilla, which, being at the head of navigation on the Urubamba, gained designation as a "Puerto." When Matsigenka came there in canoes, the Jesuits realized that Cocabambilla could be not only a money-earning hacienda, but also a mission where forest Indians could be Christianized and as a launching point for trips to contact more *infieles* downstream.

Already in 1714, the Jesuits sent a Matsigenka boy to Cusco where he learned Spanish and religion; baptized, he was then returned to Cocabambilla the following year (Ferrero 1967:44–46). In 1715, the Jesuit Cristóbal de Quevedo launched the first missionary expedition, taking with him that Matsigenka boy as interpreter (Quevedo 1900). They traveled only ca. 50 km down river on rafts as far as Illipani where they found the first Matsigenka settlement. That expedition provided the first real information to highland people about the nature of tribal communities and locations for possible future farms. When the Jesuits were expelled from Peru in 1767, the property was thrown into a legal limbo until 1799, when Franciscan padres from Cusco took over the property principally as a logistical center from which to establish missions to the forest people. In 1804, the mission the Franciscans had founded near Chirumbia was soon abandoned. To evangelize, the priests persuaded natives to come to their Cocabambilla mission to settle. An original contingent of 44 Matsigenka was later complemented by 75 more. The Franciscans taught them religion and, not incidentally, used them as a labor source. As Viceroy Toledo had done in the sixteenth century in the highlands, the Franciscans adopted the reducción concept to relocate native people in order to most efficiently evangelize them.

9.7 Forest Indians and Highlanders in Contact

In the late colonial and early republican periods, the "chunchos" began to enter the consciousness of hacendados on two fronts. One was the growing importance of the annual fair that brought Piro and Matsigenka Indians up the river to trade every July for 6–8 days. The fair was held on a sandy floodplain stretch just below the junction ("El Encuentro") with the Yanatile. Although they lived north of the Pongo de Mainique, the Piro dominated the fair by their greater numbers, forceful personalities, and good instinct about trade items that appealed to highlanders. Arrival of the "salvaje" flotilla was a much anticipated yearly event for many people in that part of the Urubamba. Many merchants came from Cusco and stayed at the Cocabambilla mission until they received word that the Indians had arrived at El Encuentro. Taking 1843 as an example, 30 canoe loads of Piro folk arrived at El Encuentro (de Sartigues 1851). Seven years before that, 200–300 Piro, roughly the same number, arrived for the event (Miller 1836). Among Piro trade goods were hammocks, bags of cacao, and gold dust and in return they wanted metal objects and salt (Valdez y Palacios 1971:90). Since the coveted objects were less bulky than those brought with them, the Piro traded some of their canoes as well. The lack of a common vernacular between highlanders and lowlanders may have accounted for the observation that trading occurred in silence (Valdez y Palacios 1971:215).

Some forest Indians continued upstream to barter at Haciendas Sahuayaco and Echarati, the latter being reached on July 14 during the feast day of the hacienda's *fiesta patrona*, Nuestra Señora de Carmen (de Sartigues 1851). Festivities included large consumption of rum and chicha, which increased the possibility of violent confrontation. At Cocabambilla, the Franciscans did not welcome the obstreperous Piro en masse, but accepted delegations from that tribe if they deposited their bows and arrows before entering the mission. Tobacco was cultivated at the mission as a trade item mainly for forest Indians who preferred it to coca. Even greater highland-lowland exchange would have occurred at Santa Ana (Quillabamba) except the current that far upstream was too swift and full of hazards to negotiate in loaded canoes.

9.7.1 *Objects of Trade in the Nineteenth Century*

Both during the Encuentro fair and at other times, materia medica from the tropical forest was an important trade item in the late colonial and early republican periods (González 1833:50; Miller 1829:72–73). Some of these jungle remedies are still sold today in the highlands at stalls called *hampi katu*. Large quantities of cinchona bark (*cascarilla*) exchanged hands at El Encuentro. Sarsaparilla (*Smilax officinalis*, *S. aspera* and *S. aequatoriales*) had a double, seemingly contradictory use, as a libido stimulant as well as a treatment for syphilis (Schultes and Ruffauf 1990). Some of it got into the export trade. Two congeneric euphorbiaceous plants, *huanarpo macho* (*Jatropha macrantha*) and *huanarpo* (*Jatropha ciliata*), were favored aphrodisiacs, just as they had been in the Inca period. Alkaloids and saponins in the root accounted for their effectiveness, but their phallic appearance, related to the doctrine of signatures, was part of huanarpo's mystique. The foot of the tapir (*Tapirus tapirus*), was used to ward off epilepsy, a disease once believed to be contagious (Gade 1999, 2003). Beliefs brought from Spain about epilepsy persisted among peasant folk in the Andes until the early twentieth century.

An oily seed called *quina quina* was used as a flavoring, emollient and medicinal said to cure headaches. These seeds came from a tall jungle tree (*Myroxylon balsamum*). Another species, *Copaiferapauper*, was a healing agent and detoxifier. Church incense came from exudations from the base of the copal tree (*Hymenaea courbaril*) and the *Eupatorium bullatum* tree. Calaba oil, known in Peru as *aceite de*

Maria (*Calophyllum brasiliense*), was used as a febrifuge, pectoral and to heal wounds. Topicals for use on wounds or skin infections came from the bark or latex of *Bellaco caspi*, *Plumeria rubra* and *Hura crepitans*. Two flavorings traded to the highlands were wild vanilla (*Vanilla planifola*) and *cacao chuncho*, a semi-cultivated variety of cacao that evolved in the Amazon Basin. Since church altar candles could not be made from tallow, beeswax from native bees, though not considered good quality, was in demand. Both manatee fat as a cooking oil and turtle oil for lighting found their way to the highlands (Fernández Moro 1952:251). Previously tamed monkeys, parrots, and macaws (*guacamayos*) were traded as amusing pets for highlanders. Jaguar skins and bird feathers had ceremonial and decorative value.

Highlanders appreciated the virtues of certain woods brought from the tropical forest for the same reason that the Incas had valued them. They also acquired woven cotton as table cloths and bedspreads. Bows, arrows, seed beads and tree-fiber hammocks had value as exotic objects. Piro raiders on their way up the river kidnapped Matsigenka children and women and sold them to coca haciendas (Valdez y Palacios 1971:205). A child was traded for an ax. Wide trade contacts maintained by the Piro explained the presence in Cusco of decorated clay vessels made by tribes on the Ucayali who lived far to the north.

Highlanders knew what the forest people wanted most: metal axes, hatchets, knives, billhooks and cutlasses. The Collins machete, made in the United States of forged steel, became a highly coveted trade item after the middle of the nineteenth century. The tool revolutionized forest clearing in the jungle. Scissors, needles, and fishhooks were other desired trade objects. Decorative objects—mirrors, buttons, rattles, glass beads (*chaquiras*) and woven wool textiles—also appealed to forest tribal people. Matsigenka, though not the Piro, appreciated rum, cheese, hardtack, cakes, and onions. An important trade item to forest people on the Urubamba was salt brought from the highland town of Maras. Even for the Piro north of the Pongo, Maras salt was more accessible than that from the so-called Cerro del Sal in the central Peruvian montaña.

Around 1890, the annual trade fair between serranos and chunchos moved to Rosalina. A trail had extended there from Echarati and, for those who arrived in boats from downstream, that site presented fewer problems for canoe navigation than did El Encuentro. A decade later, the fair moved farther downriver to Sirialo. Dominican missionaries cleverly combined a religious feast day honoring Santa Rosa on August 23 with the trade fair (Camino 1978: 132). At an event under missionary surveillance, Piros could be prevented from selling Matsigenka children and women. That fair lasted until 1914, the last year in which the Piros came up the river to trade. By then, metal objects brought by rubber agents had become widely available. The Matsigenka increasingly received trade goods from the haciendas between Rosalina and Echarati. An unknown number of them, seeking protection from slave traders during the last decade of the rubber boom, became workers on haciendas in the labor-scarce area.

The Piro, whose home territory lied north of the Pongo de Mainique, had a quite different psychological profile from the Matsigenka. They initiated contacts with other groups and, at the same time, insisted upon setting the agenda in those relationships. In 1874, a Piro delegation journeyed far upriver from their home territory to Santa Ana (Quillabamba) to ask for Peruvian assistance in building towns and mule trails in their region (Buendia 1874:265). That unusual request persuaded three zealous Franciscan priests from Cusco to organize an expedition with the aim of setting up a mission for the Piro north of the Pongo where the Rio Miaría flowed into the Urubamba. At the time Piro territory extended north of there to Camisea (Samanez y Ocampo 1885). The Franciscans learned much about these people when they reached an ill-fated agreement with Piro boatmen to guide them down the Urubamba. In Fr. Sabaté's (1877) written account, the trip was launched at Illapani and started with much anticipation and good will. Missionizing fervor described "our beloved Piro full of a sweet inner peace and indescribable contentment." Less than 30 pages later, Sabaté replaced those charitable feelings with descriptions of the Piro as "barbarous criminals." Not only did the Piro boatmen harass the priests by incessantly demanding gifts, they stopped along the way to rob the Matsigenka of chickens, bows, arrows, and even canoes. After several days of high tension, the Piro and priests reached the Rio Miaría. Totally disillusioned by Piro behavior, the padres nevertheless still wanted to establish a mission. However, the Piro did nothing to facilitate their efforts, nor did they even give them food. Starving, the missionaries escaped their tormentors and in a series of harrowing experiences returned to Cusco.

The arrival of hundreds of forest Indians every year, first to the annual fair and then at the haciendas upriver from there, raised security issues among highlanders. Chunchos, as they were generally called at that time, were in Andean terms unpredictable savages. As highlanders began to eye the lands below Cocabambilla as sites for new haciendas, anxiety increased. Hacendados naturally supported the Franciscan effort to evangelize the Matsigenika as the chief means to their pacification. To be on the safe side, estate owners established a militia called the "comandante de fronteras" to guard against the advance of forest Indians and what that might mean for the well-being of their properties (Maurtua 1907:26). Hacendados, who viewed "salvajes" as an obstruction to bringing the selva into production, also discussed their possible extermination (Rénique 1980). Although in some forested zones of eastern Peru, the natives were aggressive, in the Urubamba that was not the case with the Matsigenka. The Piro had potential for belligerence, but their home territory, far from the contact zone, was not threatened by any land grab. No records exist of the Matsigenka destroying haciendas or killing highlanders.[4] More than almost any other Amazonian tribe, they had a pacific character.

[4]The American explorers Herndon and Gibbon traveling in the valleys east of Paucartambo in 1851 found that highlanders were fearful of accompanying them into the jungle. Since the early colonial period, the Huachapari and other tribes outside my study area have had a history of destroying haciendas and killing highlanders. The Urubamba had a different kind of contact history.

9.7.2 *Hacienda Expansion Down River*

Highlanders looked at Matsigenka territory as a potential appropriation. In 1804, estate owners opened a pack trail to Chirumbia to bypass the shallows in the Urubamba that made travel on the river so tenuous. This trail encouraged expeditions to launch their trips farther downstream from Cocabambilla. In 1847, the Count de Castlenau started his trip on the river Chahuares; in 1874, Padre Sabaté set out in canoes from Illapani; and in 1886, Carlos Fry shoved off from Rosalina. With the "port" function of Cocabambilla no longer viable, the Franciscans closed their mission there and set up a new one at Sepahua among the Piro.

The pack trail to Rosalina facilitated the establishment of new haciendas after Independence in 1821. The Peruvian government offered land in the zone to reward individuals for their role in the struggle against Spain or as payments for debts incurred. In that way, Pedro Polo, a criollo who had fought in the Battle of Ayacucho, convinced the new government in Lima to grant him a land concession in the Urubamba. His earlier acquisition of Echarati and other properties made him one of the richest men in the Department of Cusco. José Rueda, also received a huge land concession that he called Hda San Agustin, sections of which he first rented out and then sold (Raimondi 1898). These became the separate estates of Sahuayaco, Illapani, and Rosalina, the last two of which intruded onto lands where Matsigenka were living at the time. Only small areas ion these properties were actually farmed; Sahuayaco had almost 12,000 ha but only about 34 ha were under cultivation. Taking over Matsigenka land raised hacendados' anxiety, and in an attempt to mollify natives, gifts were given. Traveling in 1834, the Frenchman de Sartigues (1851) noted that estates gave cows, pigs and fowl to ensure good relations with the Matsigenka. In 1865, the explorer Raimondi (1898:190) noted the practice of gift giving had become a regional custom.

Through most of the nineteenth century, hacendados, expecting retaliation for their usurpation, armed themselves against possible attacks on their property and lives (Sabaté 1877). Although Matsigenka relations with estate owners were not always good, they conducted no raids on these Spanish-speaking interlopers. Matsigenka social structure largely accounted for their peaceful nature. Their traditional settlement pattern of isolated extended family groups physically distant from other groups did not encourage a leader to emerge who could mount a military campaign. Without a contingent of warriors to mobilize, the Matsigenka could not act on their hostilities. In addition, this tribe lacked a sense of territoriality that would have justified incursions on highlander haciendas. This non-violence and accommodation may have been a long tradition dating from before the Spanish Conquest. Rather than fight the Incas, the Matsigenka (then "Antis"), chose a *modus vivendi* that included becoming their vassals. They may have grown coca for the Incas or provided products collected from the jungle. They may also have formed a brigade of archers in the Inca imperial army.

A problem of all tropical haciendas in the Urubamba was finding sufficient labor to work the land. Serranos dominated the labor market in the old settlement part of the tropical Urubamba. By the eighteenth century, few African slaves remained there and scarcity of labor was a perennial problem for haciendas. Unlike in the highlands, there were no nearby peasant communities from which to conscript workers. Indians from the highlands became the principle workers on estates. But fear of mortal disease made highlanders reluctant to migrate to the hot country. The warm valleys in the Department of Apurimac where many Indians were accustomed to hot weather and had also acquired some immunity to malaria, became an important source of workers (Bowman 1916:78). Highland communities in Cusco Department could not be so easily enticed. When workers refused to come to the valley of their own volition, haciendas used goons called *guatacos* to forcibly bring them to the estate. To keep these unwilling workers on the property, hacendados paid a guard to prevent them from leaving (Macera 1968:85).

The nineteenth century brought a different tactic to entice workers to the estates of the tropical valley. Prospective workers were offered a wage at twice the rate of that in the highlands and the possibility of renting a parcel of land for their own use. If they chose the latter, the rent was paid in labor to the estate. Hacienda owners each owned thousands of hectares so renting parcels to highland peasants was no sacrifice. Besides subsistence crops, workers grew crops for sale. Hilly plots were often put into permanent crops which need good drainage. Estate owners kept flat land for sugar cane fields.

Haciendas below Echarati had major problems as commercial enterprises. One was the high cost of mule transport, making their products more expensive than those produced closer to Cusco. By way of example, at Hda Santa Ana producers sold an arroba (11.5 kg) of coca to middleman for four pesos; in Cusco, that same amount sold for 14 pesos. Coca from Hda Sahuayaco farther down the valley had to sell at an uncompetitive 20 pesos in Cusco in order to make a profit. Estates below 800 m also had a perpetual labor shortage. Not only were they distant from the home communities of peons in the highlands, they were also more unhealthy than haciendas in the Urubamba above 1000 m asl. Hacendados tried to fulfill their labor needs by enlisting Matsigenka as workers. Given the limited numbers of these people, as well as their low dependability as workers, that source was marginally satisfactory. Those who worked on estates were, in at least cases, not free agents. A secret slave trade of Matsigenka was said to have existed on the Urubamba until 1944 (Heim 1957:144).

9.7.3 Missionary Change

In contrast to the banks of the Ucayali where by 1791 the Franciscans had established themselves for good, nowhere along the Urubamba did they have long-term missionary success. All nineteenth-century attempts to establish missions at four different sites downstream failed. Disillusioned, the order sold Hacienda Cocabambilla around 1870 and withdrew from the region. Several decades of a missionary vacuum ended when, in 1900, the Dominicans (O.P.) received authorization

Fig. 9.7 View of Koribeni in 1968 showing the church and mission school and two kinds of dwelling. The oval-shaped palm-thatched structure is the traditional Matsigenka dwelling; the rectangular adobe buildings with tin roofs reflect highland influence. Dominicans from Spain founded the mission of San José de Koribeni in 1918. Since the 1990s, Matsigenka are reduced to an enclave surrounded by land taken over by colonos

from the Holy See to evangelize native peoples of the Urubamba and Madre de Dios Valleys. Their arrival from Spain coincided with the last decade of the rubber boom. Setting up their regional base of operations in the still emerging town of Quillabamba, in 1902 the Dominicans established their first Matsigenka mission less than 50 km downriver at Chirumbia. In 1918, the Dominicans founded another mission at Koribeni, 40 km downriver, which had been a cinchona bark collection point attracting Matsigenka settlement (Fig. 9.7). The missionaries were in periodic contact with Matsigenka farther down the Urubamba to provide religious and health services. After World War II, the Dominicans established three other missions: Sepahua in 1947; Timpia in 1953; and Kirigueti in 1961. Missionary strategy followed the reducción idea of encouraging Matsigenka to settle near a mission.

9.7.4 Matsigenka Acculturation

The Dominicans saw themselves as protectors of these people, for the brutalities associated with the rubber boom had not disappeared. Missionary contact had a substantial effect on the Matsigenka loss of their culture. One can argue that the impact of Western ways on the Urubamba Matsigenka began during the nineteenth century cinchona and rubber booms. More profoundly, Dominican priests and later nuns consciously and unconsciously imparted Western values to the Matsigenka while saving their souls. In writing about their missions, he Dominicans commented on the Matsigenka attitudes toward work, religion and family that reflected a view of superiority of European perspectives. One missionary characterized the Masigenka as "the personification of indolence itself" (Pio Aza 1928:53). The Dominican role in acculturation can be seen in dress. At Chirumbia, in 1902, the first year of that mission, all Matsigenka wore the traditional cushma; in 1954 none of them did so. By the 1960s, approximately 85 % of the Matsigenka in the Urubamba had been catechized. Many by then also spoke some Spanish, and wore Western clothing. Varying degrees of Dominican influence affected the remaining 15 % who continued to live in a dispersed pattern, wore the cushma and practiced slash-and-burn subsistence agriculture (Camino 1973).

I recall at Koribeni in 1968 the nuns teaching Matsigenka girls to embroider bluebirds on white pillowcases. Before the missions started their own schools, some Matsigenka boys and girls were sent to the Dominican boarding schools in Quillabamba. In 1940, the Dominicans had established an experimental farm and school called "La Granja" on land that had been part of Hacienda Santa Ana where Matsigenka learned about a kind of farming different from that practiced by their parents. A radio communications center in Quillabamba kept in frequent touch with missions down river. The Dominicans still control the parish church in Quillabamba, a town that has always been populated almost entirely of people of highland origin.

Another group of missionaries came to the Urubamba in 1945. The Summer Institute of Linguistics ("ILV" in Spanish), a Protestant missionary group from the United States heavily involved in Bible translation, started a mission center among the Piro. Later the Pereira clan, who favored the Protestants over the Dominicans, welcomed the establishment of a mission among the Matsigenka at Monte Carmelo near Rio Malanquiato (Camino 1979).

9.8 Economic and Demographic Changes in the Lower Urubamba

9.8.1 *Economic Booms on the Urubamba*

9.8.1.1 Cinchona Bark Collection

Even before the Dominicans, outside forces had major effects on the native forest people of the Urubamba. Two nineteenth-century booms based on collecting wild products somewhat overlapped in the Urubamba Valley. Between 1850 and 1890, the world avidly sought cinchona bark (*cascarilla*) which can be collected from several related species of cinchona that grew wild on the eastern front of the Central Andes. Quinine extracted from the bark had become the world's prime anti-malarial treatment. Relative accessibility made the Urubamba below 800 m asl one of the major Andean valleys from which cinchona bark was collected. In 1856, Cusco entrepreneurs formed a company to exploit it in this valley; later foreigners also started firms to do the same. Cascarilleros as they were called roamed the forest during the dry season looking for cinchona trees recognized by their smooth light green leaves. The trees, cut down with an ax, were stripped of their bark. Once dried and made into small bundles the bark was carried on the backs of Indians to collecting points on the river and loaded on to boats to be taken to Rosalina where a trail enabled mules to carry the bark to Hda Santa Ana. This new commercial activity provided the impetus for founding nearby a new town, Quillabamba. From Quillabamba, mules carried the bark to Cusco where it was processed and packed for export. Cultivation of the tree in Southeast Asia and India brought a sharp drop in prices and turned cinchona gathering in the hills back from the main river into a much less remunerative activity.

9.8.1.2 Rubber Gathering

Rubber collection started to gain some importance on the Urubamba after 1860, though the real boom years were between 1890 and 1912. Modest rubber collection extended into the 1920s. Two kinds of wild trees yielded rubber: *caucho* (*Castilloa elastica*) south of the Pongo and *jebe fino* (*Heveabrasiliensis*) north of the Pongo. Much of the labor force to tap, collect, and process the latex was supplied by peons of highland origin who, lured by money, left the valley haciendas to which they had been attached. Rubber agents also commandeered the Matsigenka to perform this work through violence, enslavement, and displacement. As elsewhere in the Amazon Basin, rubber gathering created social mayhem among the tribal people along the Urubamba. To escape rubber agents and the diseases they introduced, many Matsigenka took refuge in the high hills far beyond the main river. Where the lateral streams became unnavigable, they were generally safe from the clutches of rubber agents. In one case, Matsigenka sought protection from outside intruders by becoming part of the Pereira clan led by a powerful mestizo patriarch from Cusco who had Matsigenka wives. Controlling a section of the river as his fiefdom, Pereira ruled a complex social system that assured a safe haven in return for economic exploitation of Matsigenka labor. Decades after the end of the rubber boom, his son continued that paternalistic arrangement.

As rubber prices rose on the world market, a number of cusqueños sought to benefit from rubber exploitation and trade on the Urubamba. To overcome the hazards of Urubamba river navigation, in 1899 a group of Cusco businessmen led by Jesús Lámbarri undertook construction of a 142 km-long mule trail from Quellouno on the Yanatile River to the Timpia River. Completed in 1904, the Camino de Lámbarri facilitated rubber shipping to market by avoiding the dangers of upstream movement on the river. Cusco became a warehousing center for rubber waiting to be sent to the coast. The economic bubble motivated completion in 1908 of the Ferrocarril del Sur (Southern Railway) to Cusco from the coast. Then, however, world rubber prices fell sharply in 1912. Over a period of a decade the tempo of tapping declined and then stopped altogether. The forest repossessed the Camino de Lámbarri and outsiders with some exceptions departed the region. In the highlands, the railroad had enough other economic activity to be marginally profitable, even though the importation of cheap foreign goods on that line devastated the local economy. Interest in the Amazon in southern highland Peru withered until a revival occurred in the 1940s.

9.8.2 Twentieth-Century Colono Invasion of the Urubamba Jungle

Serrano movement to the Urubamba *selva alta* was spurred by a 1898 "ley de tierras de montaña" (law of jungle lands) opening new colonization opportunities. It assumed that the state owned the land east of the Andes as a *terra nullius*. Included in that concept is the notion that "salvajes", although they had lived there for millennia, had no legal rights to the lands they occupied. The law allowed a person or a juridical entity to lease a parcel of land ("*denuncio*") from the state in exchange for a modest payment. Failure to pay would supposedly cause the lease to lapse. At Sepa the Peruvian government had acquired a block of 37,000 km^2 taken over from a failed scheme to settle Polish immigrants. There the Peruvian government set up a penal colony for 500 with the double objective of rehabilitating them while also developing agriculture in the zone. In 1987, with its cost no longer justified, the Sepa penal colony closed. The land, though still owned by a government agency, later was used for cattle raising by private interests.

Impoverished highland peasants, some landless, others who had rights to parcels in their ayllus, took advantage of the law. In the first two decades of the twentieth century, a trickle of highland people entered the Lower Urubamba. The 1932–1933 malaria epidemic that struck the whole valley below Huadquiña, abruptly halted that movement. Resumption, even acceleration, of that migratory flow occurred with the convergence of several push and pull factors from the late 1940s into the 1960s. The high international price paid for coffee beans attracted highland peasants seeking cash income. Another impetus came from land hunger. Primary health care in the highlands reduced mortality and created population pressures that could not be satisfied in home communities. At the same time, pitiful agricultural yields, exacerbated by worsening soil erosion, negatively affected already impoverished Andean peasants. Then, between 1957 and 1960, a drought struck the highlands at the same time that malaria, which had been so devastating in compromising health and often lives, became a non-issue. Anti-malarial campaigns beginning in the early 1950s, during which dwellings were sprayed with DDT, eliminated mosquitoes. People also had access to prophylactic medicine to prevent contraction. All of these factors encouraged the movement of settlers, called colonos, to acquire plots of land in the jungle.

Lack of a sense of territoriality among the Matsigenka facilitated establishment of colonos in this part of the valley. Colonos acquired land used by Matsigenka for swiddens or for hunting without organized opposition. Extension of the vehicle road also benefited colono movement "adentro" ("inside," that is, farther down the valley). In 1963, the vehicle road had reached Sirialo and serrano arrivals began to take over land without even filing an official claim (*denuncio*). Whereas in 1806 a Franciscan expedition had found only "Antis" at Sirialo, in 1964 all of its inhabitants were Quechua speakers of highland origin. By a factor of four to one, Andean peasants by 1970 had overwhelmed the native population from Rosalina to the Cosireni River and Matsigenka then lived outside mission centers only far down the valley.

Settlement by highlanders, accompanied by dispossession of the Matsigenka, persisted during the Velasco regime (1968–1975) despite the fact that the Peruvian government recognized, for the first time, the territorial rights of aboriginal peoples of the forest. In 1974, the Law of Native Communities granted legal territorial rights to individual settlements of forest Indians, but not to all of the land claimed by them. The outcome of distributing those titles was to create islands of Matsigenka surrounded by a sea of farmers of highland origin who had acquired their land not by a legal process, but through simple invasion. The byword of the period was that land belonged to those who occupied it, an expression used to justify the breakup of haciendas in the highlands. Indeed, land invasion all over Peru became viewed even by the Velasco government in Lima as a form of social justice. Colonos, assuming correctly that the government would not remove them, continued to appropriate land in Matsigenka territory. Matsigenka soon organized themselves to protect their communal lands. To circumvent new restrictions, colonos often sought amenable Matsigenka individuals to file land claims on their behalf. More colonos and also more land speculation characterized the Alan Garcia presidencies (1988–1990; 2006–2011). Individuals connected to Garcia's APRA party got loans from the Agrarian Bank to acquire rights to lots in the lower Urubamba of from 5 to 400 ha without ever occupying those properties (Tamayo Herrera 1992:II:615).

Confrontations over land with the Matsigenka were freighted with the traditional attitude of Quechua-speaking migrants from the highlands claiming de facto authority over "chunchos" i.e. "barbarians." The discourse of "civilized" vs. "uncivilized," so deeply engrained in Andean culture history, dominated relationships just as they did in Inca times. Dominican missionaries were caught between their historic responsibility to the Matsigenka and the religious and other needs of newly arrived colonos who often usurped ancient hunting grounds for their fields. Many of these highlanders were pendular migrants who maintained seasonal residencies. From October to April, they cultivated their highland farm and in May went to the "valle" to harvest the crops they had earlier planted. In their half-year presence, they also managed to clear a bit more of the forest on their parcel. However much they disdained their tribal neighbors, people of highland origin in the valley adapted a bundle of traits from them. Not perennial crops, but subsistence crops that required swidden methods of cutting and burning were grown. Manioc replaced the traditional tubers of potato, oca, ullucu, and añu; upland rice replaced quinoa, wheat and barley. Maize was grown in both the highlands and lowlands, but the varieties and uses were entirely different. Masato made from manioc replaced the chicha of the highlands. The flinty lowland maize did not make good chicha. Bananas, mangos, and papayas substituted for the apples, peaches, and capuli cherries fondly remembered for their seasonal availability in the highlands. In clothing, colonos shed their thick woolen homespun and donned manufactured cotton textiles. Colonos sometimes constructed the characteristic adobe dwelling of the highlands; more commonly they built structures of wooden boards as a material better suited to heavy rainfall and high temperatures.

Commercial agriculture penetrated the valley below the mouth of the Yanatile (750 m asl) with the construction of roads. Coffee is the major perennial crop of that colonization zone as it had been for decades in the old settlement zone between 800 and 1500 m asl. For the Urubamba Valley as a whole, from its source to end, coffee now occupies twice as much land as does flour maize and three times as much as potatoes. In 1963, La Convención produced 6000 metric tons of coffee; in 2013, it exceeded 28,000 metric tons. Lower in quality, coffee from the lower elevation has to contend with lower prices than that grown above Quillabamba. The best quality coffee from the Urubamba reputedly comes from the Huadquiña Cooperative, which produces about 2200 m. tons between 1500 and 2000 m asl. Another commercial crop, achiote, is harvested over a wide swath of the Urubamba below ca 1300 m asl (Fig. 9.8).

Fig. 9.8 Achiote or annatto (*Bixa orellana*), a native plant of Amazonia, has long been valued as a source of dye contained in its seeds. Exports as a yellow coloring for dairy products began in the twentieth century. Growing as a shrub on the margins of fields, achiote provides income for colonos and tribal people all through the Lower Urubamba. Hand labor is used to remove the red seed from the prickly pods

9.8.3 Transportation, Demography and Development

The 1905 grandiose plan conceived during the height of the rubber boom to build a railway from Cusco, funded by a tax on rum and coca, to a point below the Pongo de Mainique never materialized for that distance. Still, part of the railway was built. Just as the rubber boom ended in 1913, rail construction began. By 1925, 61 km of the narrow-gauge (0.75 m track width) line from Cusco had reached Pachar and by 1928, the track had been laid through the canyon to km 101 below Machu Picchu ruins. Five years later, in 1933, a vehicle road from Quillabamba was completed up valley to the then-terminus of the railroad at Machu Picchu. Trucks and trains replaced mule transport for getting products to and from Cusco. In 1951, the rail line reached Santa Teresa at km 131 and in 1977, Quillabamba at km 171. Locomotives, pulling a maximum of 50 metric tons, carried from Cusco mostly potatoes and other agricultural products of the Sierra as well as bottled drinks. But frequent landslides plagued this route and in 1999 a huge mudflow below Machu Picchu destroyed a large section of track. Never rebuilt, the train to La Convención was replaced by a highway built over the Málaga Pass.

Downstream from Quillabamba, new roads provided access to and from a zone that previously could be reached only by canoe. Highlanders, never adept at navigating the river, campaigned for a vehicle road into the jungle (Fig. 9.9). In 1941 the road from Quillabamba reached Echarati and in 1945 Quellouno. In 1961, the Urubamba was bridged at Chahuares and in 1964 the road reached the left side of the river across from Koribeni mission and, soon after, Sirialo. In 1970, the road was extended to Kiteni and in the 1990s to Ivochote at the Rio Mantalo. From there, goods and people must be loaded onto motorized boats for locations farther down the valley. The trip between Ivochote and Quillabamba normally takes 7–8 h. The Camisea gas field is likely to stimulate the completion of a vehicle road through the remainder of the lower Urubamba Valley. By 2030 the Torontoy canyon between Cedrobamba and the Hydroelectric Station 10 km north of Machu Picchu is likely to be the only stretch of the Urubamba without road access. Transportation led to the creation of some towns and to the substantial growth of others. As roads reached Kiteni and Ivochote, those settlements became important nodes south of the Pongo. Settlements that had been missions attracted people, both Matsigenka and colonos. Other old Dominican missions have become towns. Sepahua is now a town of 7000 people of mixed ethnic character; it has a health clinic, secondary school, technological institute and radio station. Atalaya, now a provincial capital, has also become a regional center for a substantial stretch of the river. Camisea, the center of hydrocarbon exploitation, has grown into a sizeable settlement. In 1980 fossil fuel exploration teams from Shell Oil using seismic prospecting technology discovered a huge deposit of natural gas near Camisea. For two decades no development took place. A flu epidemic killed many Matsigenka and in 1988 the government pressured Shell Oil to leave the area. Finally, in 2000, the Peruvian government contracted with a consortium of companies led by Hunt Oil to develop the hydrocarbon reserves. Henceforth the Urubamba has become known as the natural gas center of Peru.

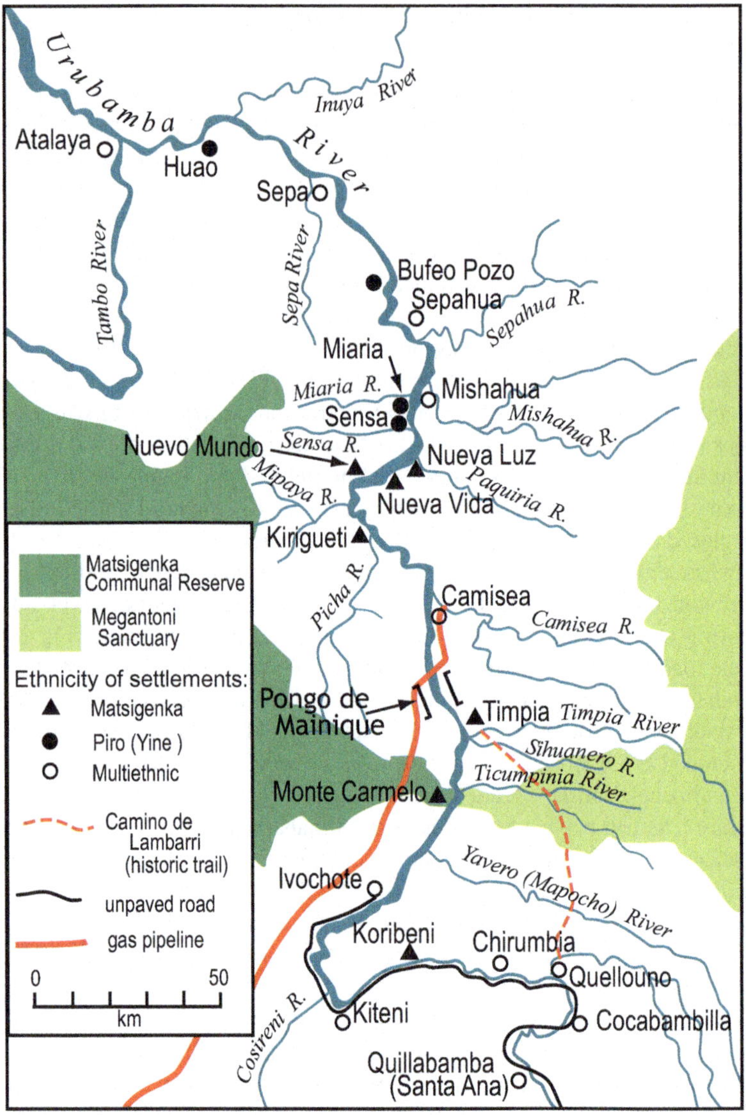

Fig. 9.9 The tropical Urubamba downstream from the old hacienda zone. Between Rosalina and Koribeni the dry tropical forest turns into the selva alta, which, in turn, changes to selva baja north of the Pongo de Mainique. Influx of colonos and exploitation of a natural gas deposit has resulted in new towns and new roads, as well as permanent agriculture and ranching. Tropical forest people have increased in numbers, but they have lost much of their culture. (Map by M. K. Gade)

A pipeline completed in 2010 carries natural gas to the Pacific Coast south of Lima. In liquefied form this gas moves to export markets.

Major construction projects at Camisea have, in the process, transformed livelihoods. Nine Matsigenka communities are within the economic sphere of the Camisea Project, the largest of which is Kirigueti with 900 people. Relative newness is suggested by the names of three others: Nuevo Mundo, Nueva Luz, and Nueva Vida. Many people have come to these places from elsewhere on the river. Never before have Matsigenka lived in such large concentrations. Natural gas development has provided alternatives to a traditional way of life thanks to the substantial financial benefits that have accrued to native people. Royalties are shared among the Department of Cusco and political units within the region. In 2007, the District of Echarati, which has 18,000 inhabitants including those living near the Camisea fields, received 216 million soles (US$44,000,000) (Fontaine and Le Calvez 2010). A planned pipeline, the Gasoducto Sur Peruano, will move natural gas from Camisea to Quillabamba. From there the pipeline is programmed to reach Anta near Cusco; from there, gas will go to the coastal ports of Matarani and Ilo. Two giant thermoelectric plants are planned to consume 70 % of the gas; only 30 % of the fuel is expected to go to towns along the way.

The Camisea gas resource has benefited two cities far upstream. Quillabamba, capital of the Province of La Convención, now with a population close to 50,000, occupies a large flat tableland well above the Urubamba floodplain that was once the prime agricultural land of two haciendas. It hosts a radio station, hospital, several high schools, and a branch of a University of Cusco. Quillabamba is a wholesaling center for agricultural products of the Lower Urubamba. Since 1991 Quillabamba has received its electrical power from the hydroelectric complex below Machu Picchu, but now new plans are in the works for a huge dam on the Urubamba just north of Quillabamba. The 222-km highway linking Quillabamba to Cusco through the Málaga Pass has generated strong ties between the two cities. Echarati, capital of the huge district that encompasses Camisea, is now a town of 16,000 people; in 1964, it was a straggling village connected to the hacienda from which it got its name. The population trajectory of the Province of La Convención from 120,019 in 1981 to 163,938 in 1993 and 235,000 in 2005 reflects both substantial in-migration and natural increase.

9.8.4 Tribal Peoples of the Urubamba Today

Two groups, the Piro and the Matsigenka, have long associations with the Urubamba. New to the Urubamba are several hundred Ashaninka people who came as refugees from the violence wrought by Sendero Luminoso in the Ene-Tambo drainage to the northwest. Most Ashaninka are in the Rio Picha, a northward-flowing Urubamba tributary, and were brought there in small planes by Franciscan priests in the 1980s and 1990s. Members of two other tribes, the Amahuaca and the Yaminahua now live in Sepahua, having been pushed out of the Ucayali in the 1960s and 1970s by woodcutters.

9.8.4.1 The Piro

The Piro are now officially referred to as Yine, the name they use for themselves. They are four times as numerous as they were in the 1950s when their estimated population was between 400 and 500 (Alvarez 1960). The three largest of their ten settlements north of the Pongo are Bufeo Pozo, Sensa and Miaría. They also account for one-fourth of the population of Sepahua. Indigenous affairs officials consider Miaría to have a high level of socioeconomic development, reflecting a Piro/Yine openness to Western ideas of progress already evident in the nineteenth century. An entrepreneurial tribal psychology, apparent in the eighteenth century when they enslaved Matsigenka, continues to assert itself today. The Piro/Yine people have maintained their old affinity for canoes; for them, "to travel" is to move along the river (Gow 2012).

9.8.4.2 The Matsigenka

In 2012, the Matsigenka as a tribal entity numbered about 10,000 people, half of whom were under 15 years of age. This figure included several hundred who still chose to live in the traditional manner in the hills back from the main river. That figure contrasts with Antonio Raimondi's (1898) 1865 estimate of 1500–1600 Matsigenka over a distance of "100 leagues," which could not have included the people living away from the river. The 34 communities into which the Matsigenka are organized today include many people who have come from elsewhere on the river. Sixteen of these communities are between the Yanatile and the Pongo de Mainique; 18 others are below the Pongo. Their growing perception of themselves as a separate ethnic group with certain rights—notions attributable in part to outsider advocates—explains the formation in 1988 of an umbrella organization called, in Spanish, the Consejo Matisguenga del Rio Urubamba (COMARU).

Matsigenka life and livelihoods have changed substantially. Koribeni has become entirely different from the mission that I knew in the 1960s. Whereas only Matsigenka once occupied the zone, they now form an enclave surrounded by colonos of highland origin. Seven amplified family groups formed out of 163 families have legal title to designated parcels of land where both swidden farming and commercial agriculture are possible (Sanchez Vásquez 2010). About 60 Matsigenka families have their dwellings around the Dominican mission. Housing consists of traditional oval structures of palm wood and palm leaf thatch or, as is now favored, sheet metal roofs. Other dwellings show highland influence that goes back to the 1960s when some adobe constructions mimicked those found in Quillabamba. Koribeni has a potable water supply, septic tanks, electricity, a medical post and primary and secondary schools, which, together with its bridge across the Urubamba, almost guarantees that it will eventually grow into a city.

Livelihood can no longer depend on swidden agriculture and hunting. In a 15-km radius around the Koribeni mission, game animals are gone and fishing is poor. Decline of hunting and fishing is attributed to colonos whose use of firearms has wiped out larger game animals in the main valley. The widespread colono use of dynamite for fishing, a technique that indiscriminately kills big fish and small fry, has drastically reduced the fish catch in the Urubamba and, to a lesser degree, its tributaries. Older tribal people who remember the traditional way of life are prone to lament that "everything has gone bad." Hunters killed a tapir (*kemari*) fairly close to the mission when I was at Koribeni in 1968 and shared it with their neighbors. Today most children in that community, have only heard about, but never seen, that animal. An individualistic ethos has replaced sharing. Much food is no longer of local origin; canned fish from the coast feeds many. Lack of land and increasing population limit the options of the Matsigenka of Koribeni. Either young people must leave to seek better land farther downriver or more intense agricultural practices must be adopted to support more people. Although not at all part of any Matsigenka tradition, the idea of land use intensification was nevertheless accepted by the 1990s (Henrich 1997).

Most Matsigenka families now grow marketable perennials—coffee, cacao and achiote—crops that preclude land abandonment and periodic change of dwelling. In the 1960s, the Koribeni missionaries had urged Matsigenkas to cultivate coffee and to sell it in the wholesale markets in Quillabamba and Cusco. Coffee grown on Matsigenka and colono farms has become the most important commercial crop south of the Pongo. In Chirumbia today, the *comunidad nativa* had 150 ha in coffee. The reliance on coffee is risky as coffee rust (*Hemileia vastatrix*), a virulent fungus spread by wind and rain, has entered the valley. Of the 70,000 ha in coffee in La Convención in 2013, one third has been affected (La Republica 2014).

Although it is more lucrative than coffee, the Matsigenka produce little coca. The quality of coca grown below 500 m asl is inferior for mastication. Since 1978, all coca must be sold to the Empresa Nacional de la Coca (ENACO). Illegal cultivation risks seizure by government agents. Matsigenka farmers also grow subsistence crops, raise chickens, and guinea pigs. Some have gotten into cattle raising on permanent pastures. By 2010, the tropical Urubamba below Yanatile had more than 40 ranches.

Since the 1940s, Matsigenka also have cut timber for a living. That resource has now waned. So much cedro and mahogany, the two most sought after species, have been cut that these trees are now rare in the Urubamba. Whereas into the 1970s, wood cutters used simple tools, logging companies now use gasoline-powered machines (White 1978). These enterprises, which lease forest concessions from native communities in return for payment, have reached their tentacles into many Urubamba tributaries eliminating the cedro and mahogany formerly found there. The Alto Purus has one of the last stands of these woods in the Peruvian Amazon. In 2003–2004, large chunks of forested territory in the Lower Urubamba drainage were removed from concessions. The two most relevant were the Matsigenka communal reserve, 219,000 ha, west of the Urubamba River and the Matsigenka Megantoni Natural Sanctuary, 216,000 ha on the right side of the river adjoining the boundary of the Manu National Park. The Otishi National Park was formed partly out of the Urubamba drainage as was the Kugapakoire and Nahua Reserve in the Department of Ucayali. In 1999, the Matsigenka had defined this sanctuary as a protected area; in 2004 the Peruvian state acceded to that status. The inviolability of these lands set aside for nature protection cannot be assumed. The pipelines through these protected lands have leaked and broken several times. If new deposits of hydrocarbons are found, they would probably be exploited. If colonization pressures ratchet up, land invasions could decimate the trees and wildlife in these protected areas. In South America, many precedents exist of such lands losing protected status when development is at stake.

However, the future may be different. In the long-term, biotic diversity and the maintenance of a primary forest may have the highest economic value. The success of ecotourism in Manu National Park in the Department of Madre de Dios has provided a model of what the future could bring to the economy of the lower Urubamba by keeping these protected areas intact. The Matsigenka speak of an animal in the Sanctuary they called *segamai* (Shepard and Chicchón 2000). Their description of it calls to mind the extinct giant ground sloth. The number of visitors is likely to grow, especially to the Pongo de Mainique as a spinoff from Cusco-Machu Picchu tourism. Rumors of ancient stone ruins encourage those seeking the adventure of discovery. Rehabilitation of the Camino de Lámberri, now reclaimed by the jungle, as a hiking trail would neatly conjoin rubber boom history with ecotourism.

Within the last 50 years, the Lower Urubamba has undergone massive transformation. Population has increased fivefold and the Matsigenka and Piro/Yine have in large measure become part of the national economy and culture. Most of the 450,000 ha of land now cleared in the Urubamba below the Yanatile since the 1960s are degraded. About 110,000 ha of that amount have become totally eroded. Much of that can be attributed to colonos who cleared land unsuitable for cultivation and who then moved on when crops decreased. Many, still connected to highland communities, regard themselves as being impermanent, which affects their attitudes toward the land resource.

9.9 Conclusion

The geographical contact zone on the Urubamba where serranos and aboriginals once encountered one another has disappeared. The spatio-cultural division that held sway during the Inca period and survived for almost 400 years after the Spanish Conquest has been swept away. The entire Lower Urubamba is now enmeshed in the relentless advance of modernization and Hispanicization. Retrieval of the past would put the highland presence, both pre- and post-Conquest, in perspective. Archaeological investigation in the contact zone may reveal more about the time depth of the highland-lowland interchange and the early settlement locations of highlanders in the jungle, as well as of lowlanders in the valley above. From everything that is known now about the prehistory and history of the Lower Urubamba, the story is one of unidirectional impingement from high to low. Nothing in this river basin matched what happened on the Andean front 700 km to the south where over centuries the lowland Chiriguano people made frequent bellicose incursions into highland settlements.

The serrano invasions into this region after 1940 brought Western technology, notably the internal combustion engine and the vehicle road which opened access to the selva alta. Several thousand livelihoods have been created, but at substantial cost to the environment and tribal culture. Through that landscape change, traditional hunting, gathering, fishing and farming on swidden cycles have largely disappeared. Several large chunks of selva alta are in reserves, but by no means can they be seen as inviolable. Just as elsewhere in the Amazon Basin, cultural impoverishment accompanies landscape transformation. Most of the things that defined the Matsigenka and Piro as distinct ethnic entities are lost. The great paradox is that, politically, tribal identities have gained strength but culturally they are on the path to extinction and the environment has greatly deteriorated outside protected areas. Yet, demographically speaking, these peoples are more numerous than they ever were. The Lower Urubamba offers another example that collapsing space collapses culture.

References

Alvarez R (1960) Los Piros. Instituto de Estudios Tropicales Pio Aza, Lima

Bowman I (1916) The Andes of Southern Peru. Henry Holt & Co, New York

Buendia J (1874) Memoria presentada al supremo gobierno por el prefecto del Departamento del Cuzco. El Peruano 7 May, p 265

Camino A (1973) Algunos factores del cambio socio-economico en el Alto Urubamba. Estudios Andinos 3(3):119–138

Camino A (1978) Trueque, correrias e intercambios entre los Quechuas andinos y los Piro y Machiguenga de la montaña peruana. Amazonia Peruana 1(2):123–140

Camino A (1979) Sociocultural change in Upper Urubamba, Peru. In: Browman D, Schwartz R (eds) Peasants, primitives, and proletariats. Mouton Publishers, The Hague, pp 125–148

Cenitagoya V (1931) Los salvajes machiguengas. Misiones Dominicanas del Perú 7:218–226

Cieza de Leon P (1959) The Incas. In: von Hagen VW (ed) University of Oklahoma Press, Norman

Cook DN (ed) (1975) Tasa de la visita general de Francisco de Toledo. Universidad Nacional Mayor de San Marcos, Lima

Coverly-Price AV, Wood MM (1933) Professor J.W. Gregory's expedition to Peru, 1932: two papers read at the evening meeting of the Society on 1 May 1933. Geogr J 82:16–41

de Matienzo J (1967) Gobierno del Perú [1567]. In: Lohmann Villena G (ed) Institut Français d'Etudes Andines, Paris/Lima

de Ocampo BC (2013) Descripción de la Provincia de Vilcabamba [1611]. In: Baltasar de Ocampo Conejeros y la Provincia deVilcabamba. Ceques Editores, Cusco

de Sartigues E (1851) Voyáge dans les républiques de l'Amérique du Sud. Ecrit sous le pseudonyme de Lavandais E de. Revue des Deux Mondes 21(10):1019–1059

Egaña PA (ed) (1954) Monumenta peruana, 7 vols. Societas Iesu, Rome

Farabee WC (1922) Indian tribes of eastern Peru. Papers of the Peabody Museum of American Archaeology and Ethnology 10. Harvard University, Cambridge

Fernández Moro W (1952) Cincuenta años en la selva amazónica. Madrid

Ferrero A (1967) Los machiguengas: tribu selvática del sur-oriente peruano. Editorial OPE, Pamplona

Fontaine G, Le Calvez M (2010) Dos perspectivas indígenas en torno a la extracción de gas en la Amazonia andina: el proyecto Camisea (Perú) y la empresa comunitaria Amazonia Gas Ecuador. In: Gros C, Foyer J (eds) ¿Desarrollo con identidad? Gobernanza economica indigena: Siete ensayos de caso. FACSO-IFEA, Lima/Quito

Gade DW (1999) Nature and culture in the Andes. University of Wisconsin Press, Madison
Gade DW (2003) Tapir magic in the Andes and its shamanic origins. J Lat Am Lore 21(2):201–220
Garcilaso de la Vega EI (1960) Comentarios reales: Universidad Nacional del Cuzco, Cusco
Glave LM, Remy MI (1983) Estructura agraria y vida rural de una region andina: Ollantaytambo entre los siglos XVI-XIX. Centro de Estudios Rurales Andinos "Bartolome de las Casas," Cusco
González E (1833) Guía de forasteros del Departmento del Cuzco, Cusco
Gow P (2012) The Piro canoes: a preliminary ethnographic account. Journal de la Société des Américanistes 98(1):39–61
Guaman Poma de Ayala F (1980) La nueva coronica y buen gobierno. In: Murra J, Adorno R (eds) 3 vols. Siglo Veintiuno, Mexico City
Hardy O (1919) The Indians of the Department of Cuzco. Am Anthropol 21:1–27
Heim A (1957) Wunderland Peru: Naturerlebnisse. Verlag Hans Huber, Berne
Henrich F (1997) Market incorporation, agricultural change, and sustainability among the Machiguenga Indians of the Peruvian Amazon. Hum Ecol 25(2):319–351
Johnson A (2004) Families of the forest. University of California Press, Berkeley
Julien C (2001) Inca estates and the encomienda: Hernando Pizzarro's holdings in Cusco. Andean Past 6:229–275
Kerbey JA (1905) Paso del Pongo Mainique, carta de 31 de Agosto de 1897. Boletín de la Sociedad Geográfica de Lima 17:264–275
La Republica (2014) Lima
Leake BE (2012) The life and work of Professor J.W. Gregory FRS (1864–1932): geologist, writer and explorer. Geological Society of London, London
Lyon PJ (1981) An imaginary frontier: prehistoric highland-lowland interchange in the Southern Peruvian Andes. In: Francis PD, Kense FJ, Duke PG (eds) Networks of the past: regional interaction in archaeology. University of Calgary, Calgary, pp 2–16
Macera P (1968) Mapas coloniales de haciendas cuzqueñas. Seminario de Historia Rural Andina. Universidad Nacional Mayor San Marcos, Lima
Maurtua VM (ed) (1907) Juicio de límites entre el Perú y Bolivia. Tomo 7. Imprenta de M.A. Rozas, Buenos Aires
Menéndez Rua A (1948) Paso a la civilización. Padres Dominicos, Quillabamba
Miller J (1829) Memoirs of General Miller, in the service of the Republic of Peru, 2 vols, 2nd edn. Longman, London
Miller G (1836) Notice of a journey to the northward and also to the eastward of Cuzco and among the Chunchos Indians in July, 1835. J R Geogr Soc 6:174–186
Oricain P (1907) Compendio breve de discursos varios sobre diferentes materias y noticias geográficas (1790). In: Maurtua VM (ed) Juicio de límites entre el Perú y Bolivia, vol 5. Imprenta de Heinrich y Compañia, Barcelona, pp 319–377
Pio Aza J (1928) Apuntes para la historia del Madre de Dios. Libreria e Imprenta Gil, Lima
Quevedo C (1900) Exploración de los indios Antis del Cuzco, 1715. Revista de Archivos y Bibliotecas Nacionales 4:479–499
Raimondi A (1898) Itinerario de los viajes de Raimondi en el Perú: valle de Lares, Santa Ana y regreso por Mollepata y Limatambo [1865]. Boletin dela Sociedad Geográfica de Lima 8:1–61
Renard-Casewitz F-M (1972) Les Matsiguenga. Journal de la Sociétédes Américanistes 61:215–253
Renard-Casevitz FM, Saignes T, Taylor-Descola AC (1986) L'Inca, l'Espagnol et les sauvages. Editions Recherche sur les Civilisations, Paris
Rénique JL (1980) El centro científico del Cuzco (1897–1907). Historia, Lima 4(1):4–52
Sabaté L (1877) Viaje de los padres misioneros del Convento del Cusco a las tribus salvajes de los Campas, Piros, Cunibos, y Sipibos Tipografia. La Sociedad, Lima
Samanez y Ocampo JB (1885) Exploración de los ríos peruanos Apurímac, Eni, Tambo, Ucayali y Urubamba. Imp. de "El Pais." Lima

Sanchez Vásquez M (2010) Cambio social entre los machiguengas. El caso de las comunidades de Koribeni y Matoriato. Gazeta de Antropologia 26(1). http://www.hdl.handle.net/1048/6082

Santacruz Pachacuti Yanqui J de (1968) Relación de antiguedades deste Reyno del Perú. In: Estes Barba E (ed) Crónicas peruanas de interés indígena. Ediciones Atlas, Madrid

Schultes RE, Ruffauf R (1990) The healing forest: medicinal and toxic plants of the northwest Amazon. Dioscorides Press, Portland

Scipión Llona P (1903) Reseña histórico-geográfica de los ríos Paucartambo y Madre de Dios. Boletín de la Sociedad Geográfica de Lima 13:73–176

Shepard G Jr, Chicchón A (2000) Resource use and ecology of the Matsigenka of the eastern slopes of the Cordillera de Vilcabamba, Peru. In: Alonso LE (ed) Biological and Social Assessments of the Cordillera de Vilcabamba, Peru. Conservation International/Smithsonian Institution, Washington, pp 164–174

Siiräinen A, Pärssinen M (2001) The Amazonian interests of the Inca state (Tawantinsuyo). Baessler Archiv 49:45–78

Tamayo Herrera J (1992) Historia general del Qosqo: una historia regional desde el período lítico hasta el año 2000, 3 vols. Municipalidad del Qosqo, Cusco

Urton G (1981) At the crossroads of the earth and the sky: an Andean cosmology. University of Texas Press, Austin

Valdez y Palacios JM (1971) Viaje del Cuzco a Belén el Gran Pará por los rios Huillcamayo, Ucayali y Amazonas (trans: Pereira RM). Biblioteca Nacional, Lima

Villanueva Urteaga H, Macera P (eds) (1980) Cusco 1689: informes de los párrocos al Obispo Mollinedo: economia y societad en el sur andino. Centro de Estudios Rurales Andinos "Bartolome de las Casas," Cusco

White S (1978) Cedar and mahogany logging in eastern Peru. Geogr Rev 68(4):394–416

Chapter 10
Conclusion: The Spell Is Cast

Abstract The Urubamba Valley is a locus of larger ideas about Andean land, life, place, and memory. It is a vertical domain whose cascading environments acutely respond to and reflect the climatic implications of elevation above sea level (asl). The Urubamba attracts, drawing in people to settle and farm. Its magnetic quality is evident in the many visitors who come from afar seeking scenery, extinct civilization, folklore and/or adventure sports. The Urubamba is a container of secrets, harboring elements that only persistent inquiry reveals. Since the Inca period, the Urubamba's mid-section known as the Sacred Valley offers a case study of how the elite, whatever the culture, take possession of preferred territory. In this work the Urubamba is examined for changes on two very different scales: the effects of the Spanish Conquest that began 500 years ago and, more recently, the past half century in the life of the author. The Urubamba holds personal memories and offers a reflexive perspective on the transformative experience of doing fieldwork there. Lastly, the Urubamba Valley is discussed in terms of the meaning of place in understanding the world.

This multifaceted account of the Urubamba Valley combines survey information, personal experiences, landscape descriptions, commentaries on trends, analyses of past events, and detailed studies of specific elements to convey a diachronic semi-personalized view of a compelling region. This work offers reflections about the meaning of a given place as an ingathering of the scientific, phenomenological and reflexive dimensions of knowledge. Ideas recurring in the book fall into eight themes.

10.1 The Urubamba as a Vertical Domain

From end to end, the valley of the Urubamba is a grand kaleidoscope of cascading environments. Vegetation, land uses, crops and disease change, often subtly sometimes dramatically, along a gradient encompassing 4000 m of altitude. Altitude that differentiated climates divided the Urubamba into an upper and lower valley of divergent prehistories, historic cultures and resource endowments. Between 3700 and 2400 m asl a mesothermal upper valley zone has now and had in the past high agricultural productivity and population density. On the other hand, both the Incas

D.W. Gade, *Spell of the Urubamba*, DOI 10.1007/978-3-319-20849-7_10

and Spaniards avoided the selva except for its tropical fringe. Over the last half millennium, a fundamental agricultural verticality placed wheat, barley, flour maize and potatoes in the upper valley and coca, sugar cane alcohol, coffee and fruit below 1500 m asl, a division that gave rise to pendular migration and market exchange. In the late twentieth century, the distinctive separation of the highland peasantry of the upper valley and the swidden agriculturalist and hunter forest tribes of the lower valley has greatly diminished through migration of the former and acculturation of the latter. Verticality is not immutable: environmentally-related diseases have cures, people adapt or change their perceptions, human decisions change the distribution of crops, and global warming creates higher temperatures and lower water supplies. Nor does verticality override a certain peasant way of thinking; an Andean genre de vie is pervasive despite differences in environment. Colonos in the lower Urubamba, in spite of the tropical crops they grow and conditions they face, have thought patterns in common with peasants in the upper valley.

10.2 The Urubamba as a Magnetic Landscape

The Urubamba has an unusual capacity to draw in people. Its flattish valley corridor provides a path of least resistance in an otherwise fragmented highland landscape, to funnel people, goods and diseases. Transport infrastructures, including a piece of the Inca road system and, in the twentieth century, two railroads and then vehicular roads, have run through parts of the valley floor. Below 700 m asl, the river remained the means for human movement until the 1960s when a road started to penetrate the selva alta.

With livelihood assets unparalleled in the Central Andes, the Urubamba depression became an important zone of settlement. Food surpluses in the upper Urubamba sustained territorial expansions of the Incas elsewhere in the Andes. After the Conquest, the Toledan decree creating nucleated settlements brought large numbers of people who had lived on the slopes down to the valley floor. Spaniards gradually established haciendas over virtually all of the Urubamba Valley above 700 m asl. People have moved into the lower Urubamba in response to resource extraction booms based, at various times, on cinchona bark, rubber and lumber; now natural gas deposits are repeating that pattern.

The Urubamba especially after 1950 has been a magnet for travelers beguiled by Inca monuments and scenic mountain vistas. Machu Picchu, brought to world attention by a foreigner's unquenchable curiosity about ruins has become the goose that every day lays golden eggs. On that one Inca site, the lion's share of Cusco tourism still depends, though visitors come to the valley for other reasons as well. The transient flow, most of it non-Peruvian, is now the prime economic engine between 3000 and 1500 m asl. Among its positive consequences are motivation to restore the riverine environment, tidy up the cultural landscape, and improve the peasant standard of living.

10.3 The Urubamba as a Container of Mystery

Settled populations and visitor hordes have not prevented the Urubamba from possessing a cachet that hints at the mysterious. Well into the twentieth century, the rivers flowing into the Amazon had an aura of impalpability about them. In pure Eurocentric mode, ruins uncovered by foreigners were touted as "discoveries." In the Urubamba, a revelatory period starting in 1843 culminated in 1911 with Hiram Bingham's artful encounter with Machu Picchu. Subsequently Vilcabamba la Vieja in the jungle with its multiple enigmas contributed a sense of the hidden and cryptic. The pre-Conquest lack of a writing system guaranteed that there will always be more questions than answers about ancient Peru.

A secret dimension hovers over the case of the vilca, a plant which underwent violent suppression in the early colonial period. The Urubamba is also a stage for the spectacled bear, undoubtedly the Andes' most recondite megaspecies. Surprises will emerge from the Urubamba. One can expect to learn about nuclei of unassimilated Matsigenka living as their ancestors lived and Highland Indian communities above the valley floor still holding Inca vestiges. There are tantalizing rumors that the giant ground sloth is a not-yet-extinct species in an untrammeled zone of the Cordillera de Vilcabamba. The rumors may be based on Matsigenka myth, but mysteries of nature and culture draw many. More explorers and would-be discoverers have plunged into the Urubamba than perhaps any other valley in the Western Hemisphere.

10.4 The Urubamba as an Orbit of Productivity and Privilege

Everywhere in the world, wealth, power and status are used to take possession of certain places and not others. The Urubamba Valley mid-section, known as the Sacred Valley, was treated in that manner already in the Inca period. Over a period of 13 decades, a succession of Inca kings chose the heart of this valley for their royal domains and to memorialize their deification. Several rulers re-engineered the valley floor and sides to manifest their kingly prerogatives and enhance productivity and desirability. With the Conquest, the swine herder-turned-warrior Francisco Pizarro chose the Sacred Valley as his encomienda and later other Spaniards took over the floor of the depression for their haciendas. Flour maize, especially the giant white cultivar of it, became the focus at some point of Sacred Valley agriculture. While the land reform of the 1970s seemingly reversed the elitist tradition, the trend has been for those with money to buy their piece of this delectable stretch of the Urubamba. The land use dilemma which sets tourism and second home development against agriculture will come to a head.

10.5 The Urubamba as an Arena of Change

The two time dimensions featured in this book are the half millennium and the half century. The Spanish Conquest ushered in fundamental accretions of crops, tools, religion, and settlement patterns as well as European people themselves. Rather than being displaced, the native and the introduced have existed side by side for the last 500 years. Native foods, customs and practices have characterized some parts of the Urubamba more than others. Cross-adoptions were quite frequent; for example, Spanish hacendados cultivated maize and Indians adopted the oxen, plow, chicken and sheep.

A second time scale in this book, measured in decades rather than centuries, has a personal dimension. Between 1963 when I first went to Peru and 2013 when I last was there, changes have occurred that follow the modernization trajectory seen in so many places in the less-developed world. In 1963 farming technology was a mix of indigenous and Spanish colonial. Today neither of those have disappeared, but mechanized ideas about farming have entered the valley as have pesticides. Agriculture has become more commercial than it was 50 years ago. Advance of the money economy is seen in the greater use of packaged foods and packaged medicines. Extreme poverty in the valley has declined.

Class differences persist, but the abuses that defined the relationship between the haves and have-nots greatly lessened. Ironically, indigenous culture, though less conspicuous or vigorous than it was half a century ago, is now accepted with equanimity. In the early 1960s bilingualism was already widespread, but Quechua was still very much the home language of rural folk. Except in remote communities, Hispanicization has brought Spanish into a dominant position among people under 40. Intermarriage and acculturation, which occurred very slowly over four centuries, has now racially and culturally blended the European and the indigenous. The Urubamba of the twenty-first century is a hybrid reality.

Future change in the valley will come from several directions. Except for the valley's northern and southern extremities, inhabitants feel increasingly connected to Cusco. The city harbors many thousands of people whose relatives live in the Urubamba and who frequently go to the city. While no one would rightly call Cusco a modern city, it is no longer isolated from the innovations that filter into the country from outside the Andes. Valley people will accept innovations in their livelihoods, land uses and technology that 50 years ago were often met with resistance. Three specific developments are likely to accelerate change in Urubamba economy and society. Tourism, mainly between Pisac and Santa Teresa, will transform that zone into a service-oriented economy. Secondly, cheap sources of energy will transform household fuel use and encourage levels of manufacturing previously unknown. A pipeline slated to bring natural gas to the highlands from the Urubamba selva at Camisea and new dams planned for the main river and some of its tributaries to generate electrical power, will transform the Urubamba if all the proposed plans are carried out. Thirdly, export-driven agricultural specializations will change land uses beyond anything known there today.

10.6 The Urubamba as a Demonstration of a Goethe an Approach to Knowledge

Certain assumptions pervade this book: the inextricability of nature and culture; the indissociability of time and space; the blending of art and science; and the recognition that knowledge extends beyond information to experience. American cultural-historical geography is the salient scholarly tradition that has bundled together these and other assumptions. Its founder and chief protagonist, Carl O. Sauer (1889–1975), derived an important part of his intellectual inspiration from Johann Wolfgang von Goethe (1749–1832). Like Goethe, Sauer valued qualities more than quantities, the charm of local variation, holistic understanding, and an intuitive way of seeing (*Anschauung*). In turn, Sauer and Goethe have inspired me and I do not hesitate to acknowledge them as part of my intellectual inheritance.

The Urubamba was a perfect venue to train the eye to discern the subtle distinctions that occur with elevation and, at the same time, capture its wholeness. Without an a priori theoretical agenda, I have described the past and present of a range of phenomena that converge in the study area. Out of the factual base, larger ideas have emerged.

10.7 The Urubamba as a Purlieu of Memory

10.7.1 Reconstructing Image, Following Leads

The affective element in this book will be apparent to most readers. The Urubamba was the place that defined me as a seeker of knowledge. Youthful energy and enthusiasm enabled me to embark on a gigantic learning curve and, at the same time, to put to the test a passion for adventure in an exotic venue. The cues I took into the field, creating an image in the mind's eye and keeping my eyes open, became a *modus operandi* that allowed me to extract a range of facts and broader ideas. By focusing on a given mental picture, my questions had a target. The presence of columnar cacti as a component of the forested zone between 1500 and 700 m asl startled me into thinking about how different that area was from the tropical rainforest that began only 40 km or so down the valley. The happenstance sight of a small boy with mucocutaneous leishmaniasis on the street in Quillabamba haunted me so much I had to strive to understand that disease. Seeing potatoes carried on the backs of Indians on the trail that led from Chinchero on the plateau at 3700 m asl down to Urquillos at 2800 m asl near the valley floor became my mantra for thinking about the workings of ecological exchange. How many hundreds of years had farmers been doing that? When the Old World-New World agricultural contrast came to mind, I always went back to the simple scene of donkeys carrying loads of barley from the high fields to the threshing floor below. Native people had their own seed crops and beast of burden; why had Old World material cultural elements superseded the native among native people? Another image was a woman making rope by braiding grass that grew outside her door. Visual images are concrete facts from which larger meanings can be derived.

Another lesson I learned in the field was to look at everything. Reading Alexander von Humboldt it was apparent that in his travels through western South America at the end of the eighteenth century, he asked questions about each phenomenon that caught his eye. To make sense of the visual landscape, I stripped down its composite elements, human and non-human, that comprised the whole. That attention to components, each with its own history and its place in the whole, led me to multiple side projects, some of them discussed in chapter one. Each element of the landscape was created out of the past and once one becomes knowledgeable about a place, it is easy to link the different elements.

10.7.2 Research Beyond Information to Values

For many geographers and anthropologists, fieldwork offers the sought-for opportunity to expand knowledge about places and peoples. For them "being there" is a rite of passage in their scholarly trajectories. Formulating a project, pursuing the data for it in a setting outside one's culture and writing it up was a discipline that opened my thinking to wider intellectual benefits. In my own experience, diligent observation was important, but it was not enough. Imagination must be there to connect relevant aspects. Another realization is that a good deal of the knowledge about places is interpretational. Facts matter, but human minds work differently when evaluating facts. As Heraclitus (500 BC) is reputed to have said: "for those who go into the same river, different waters flow."

Only in retrospect could I appreciate that my research in the Urubamba was about more than information. That I accomplished what I had set out to do without much direction from others provided reassurance that I had crossed the threshold to intellectual autonomy. To become an autonomous learner on a self-directed path led to a sense of freedom to determine my research themes. Some take research choice for granted; others do not consider it in facultative terms at all. Research in the Urubamba also validated intellectual curiosity as a prerequisite. I came to appreciate the awesome power of curiosity as the source of the intellectual energy needed to pursue a project. Without an inquisitive spirit as its driving force, the research enterprise falters unless one substitutes for it the prospect of money or fame. From autonomy, freedom, and curiosity, the seeds of eternal growth are derived.

10.8 The Urubamba and the Imperative of Place in Understanding the World

Implicit in this book is the geographer's belief that the world is understandable and worth understanding. For any part of the earth to be comprehensible, one must discern spatial patterns and temporal change, Deriving the sense of what place means is also important. As the philosopher Edward Casey (2009:343) noted "place is there to be seen if only we have the vision to behold it." The challenge is to discern the diversity and the unity of a place at the same time. Drawing out the Urubamba as a place involved two kinds of imagination: one was the ability to bring together meaningful sense impressions; the other was to assemble the multiple fragments of places into space-time wholes. Stunning verticality, deeply engrained peasant culture, monumental prehistoric tradition and a vivid though painful history of cultural collision have together given the Urubamba much of its character. Joining the power of understanding with the power of interpretation brings the Urubamba—and any place—to life.

Reference

Casey ES (2009) Getting back into place, 2nd edn. Indiana University Press, Bloomington

Glossary (Q=Quechua; S=Spanish; E=English)

Adobe (S) Sundried brick made of a mixture of mud and straw.

Acculturation (E) Trend over time of indigenous people losing some or all of their indigenous ways.

Aluvión (S) Debris flow consisting of water, mud and rock propelled by gravity and covering everything in its path.

Antis (Q) Collective name used by the Incas and during the colonial period for people of forest tribes.

Antisuyo (Q) Name used by the Incas for the northeastern quarter of the Inca Empire, which included forested lowlands.

Apu (Q) Mountain spirits comparable to deities who control water flow.

Ayllu (Q) Basic social unit of Andean society bound together by kinship.

Cabildo (S) Town council.

Cacique (S) Indian chief.

Cal y canto (S) Refers to a Spanish colonial type of bridge supported by an arch made of stones held together by mortar with lime.

Cañari (Q) Ethnic group of Indians from Ecuador, some of whom the Incas moved to the Urubamba.

Capac ñan (Q) Name for the Inca highway system, including that part which ran through the upper Urubamba from Cusco to Titicaca.

Camelid (E) Member of the camel family: alpaca and llama (both domesticated) and vicuña and guanaco (both wild).

Camino de herradura (S) Trail or road used by mule trains.

Campesino (S) Peasant.

Ceja de la montaña (S) Literally "eyebrow of the mountain," it refers to a vegetation type found in the humid zone of the Urubamba between 2300 and 1800 m asl consisting of a dense elfin forest laden with epiphytes and tree ferns. Sometimes abbreviated to "ceja."

Cervecería (S) Brewery.

Chacra (Q) Cultivated field or small farm belonging to a smallholder.

D.W. Gade, *Spell of the Urubamba*, DOI 10.1007/978-3-319-20849-7

Chaquitaclla (Q) Tool (like a spade) used to prepare an agricultural plot and to harvest root crops.

Chaupiyunga (Q) Zone at elevations between ca. 1500 and 2400 m asl sharing some climatic characteristics of the yungas but not so hot.

Chicha (S) Maize beer.

Cholo (S) Person (woman: chola) of indigenous racial background but who has acculturated to Spanish ways to a greater or lesser degree.

Chuncho (Q/S) Highland term for a person belonging to a forest tribe.

Chuño (Q) Dehydrated potato.

Cocacamayoc (Q) Worker in a coca field during the Inca and colonial periods.

Cocal (S) Coca field.

Controlled appellation of origin (E) Label that indicates that an agricultural product is from a specific region.

Corregimiento (S) A colonial administrative unit corresponding to a province.

Creole (E) or Criollo (S) Person of Spanish blood born in the New World.

Cultivar (E) Vernacular variety of crop plant that is distinctive in character but is not so variable as a landrace.

Cuy (Q) Guinea pig.

Denuncio (S) A lease of jungle land to an individual from the government.

Encomienda (S) Royal grant of Indian labor to a Spanish settler.

Entheogen (E) Substance, usually from a plant, having psychoactive properties.

Flour maize (E) Corn with high starch content, characteristic of most maize grown in the upper Urubamba Valley.

Geltung (E) Psychological term referring to the human propensity to cultivate appearances and enhance one's reputation.

Germplasm (E) The genetic element of a crop plant used to breed new varieties.

Hacienda (S) Privately owned estate consisting of a dwelling and land.

Hanan (Q) A subdivision of an Inca town; (word literally means "upper").

Huaca (guaca; wak'a) (Q) Animistic spirits inhabiting a natural feature such as mountain peaks, caves, or rocks.

Huaquero (Q/S) Person who loots ancient sites.

Huayno (Q) Mountain music of the Central Andes played with a guitar and sung in high-pitched tones.

Hurin (Q) A subdivision of an Inca town; (word literally means "lower").

Killke (Q) Early Inca ceramic style.

Kuraka (curaca) (Q) Local hereditary nobility, used in both the Inca and colonial periods.

Indian (E) In the Central Andes, a name applied to a person who has indigenous cultural traits, though is not necessarily racially Amerindian.

Land race (E) A highly variable folk variety of a crop plant.

Llacta (Q) Town or settlement.

Llamero (Q/S) Indian who leads llama trains.

Lower Urubamba (E) The part of the valley below 2200 m asl, having a semi-tropical and tropical climate.

Macrothermal (E) Adjective referring to high temperatures, i.e. a tropical climate.

Masato (S/E) Manioc beer.

Mestizo (S) Person of mixed Spanish and indigenous ancestry.

Microthermal (E) Adjective referring to low temperatures; in the Urubamba corresponding to the cold zone above 3500 m asl.

Middle Horizon (E) In Andean culture history, the period 400 to 1000 AD. See Wari.

Mita (mi'ta) (Q) System of periodic forced labor.

Mitayo (S) Indian performing or paying his tribute (mita).

Mitmaqkuna (Q) People taken by the Incas from their place of origin and sent to do specific tasks, often in environments roughly similar to their place of origin.

Mitimaes (S) Hispanicized form of mitmaqkuna.

Onccoy (Q) Disease.

Originario (S) Adult male Indian who is officially a member of a community and who has access to community land.

Oroya (S) Basket, used to carry people or goods, pulled across a stream by a rope.

Peasant (E) Person who farms the land for both subsistence and trade or to sell to others.

Pishtaco (Q) In highland folklore, an evil foreigner who steals and kills children.

Piso del valle (S) Valley floor (floodplain).

Plaza (S) A square in a town a town or city; sometimes also refers to a market.

Puna (Q) The high-altitude zone, mostly grass-covered, of the Central Andes. In the Urubamba, corresponds to land above 3900 m asl.

Quebrada (S) Valley with steep sides; in the Urubamba, name often given to the main Urubamba Valley above 2400 m asl to refer to the depression.

Quechua (S/E) a) Lingua franca of the Inca Empire still spoken by ca. 6 million people in the Andes. Called runasimi by those who speak it; b) by extension, also used to refer to the people who speak it; and c) the temperate environmental zone where maize growing dominates.

Quepiri (Q) Person who carried coca on his back from its production zone to the highlands where the leaf was marketed.

Quipu (Q) Inca device made of knotted strings of various colors, tied together, used to encode information.

Reducción (S) New colonial settlement where Spanish authorities forced native people to live in a nucleated community.

Riego (S) Irrigation.

Secano (S) Refers to agriculture that depends solely on rainfall.

Serrano (S) Person from the highlands.

Shaman (E) Non-institutionalized religious practitioner who personally contacts the supernatural.

Snuff tray (E) A small flat tray used to hold vilca or other hallucinogenic snuffs.

Solimán (S) Substance used topically to heal wounds, of mineral or plant origin.

Tambo (Q/S) Way station (inn) along the Inca road system; the term was also used in the colonial period.

Topo (Q) Inca measure of various land dimensions.

Tribute (E) Tax on indigenous people during the colonial period.

Ucumari (Q) Spectacled bear.

Ukuku (Q) Masked dancer or guardian meant to evoke a spectacled bear.

Upper Urubamba (E) The valley above 2200 m asl, having in most of its area a temperate climate.

Usnu (Q) Platform, often artificially leveled, used for Inca ceremonies.

Verticality (E) The condition of several environmental zones being telescoped over short distances, a prime characteristic of Andean ecology. By extension, description of the use of multiple altitude belts and the social means to guarantee that households and communities have access to land in several of them.

Visita (S) Colonial census report of a particular community.

Yanacona (Q) In Inca usage, a servant of the Inca; in the colonial period, Indian who left his community to avoid the mita.

Yungas (Q/S) Hot country at elevations usually below 1600 m asl.

Wari (Huari) (Q) The name of a pre-Inca culture centered in Ayacucho, but carried to other places, including to part of the Urubamba. Flourished between 550 and 1000 AD.

Variations of Proper Names

[Quechua spellings follow varying protocols. In the book, I have chosen those spellings that seem most common or otherwise appropriate].

Amaybamba (valley); also: Lucumayo; Huayopta

Ccolpani (hacienda); also: Ccollpani, Colpani, Collpani

Chinchero (town); also: Chincheros

Choquequirao (site); also: Choqquequirao; Chokekiraw

Cusco (city); also: Cuzco; Qosqo

Echarati (town, district); also: Echarate

Huayopata (valley); also: Amaybamba; Lucumayo

Huiro (hacienda); also: Huyro

Lucumayo (valley); also: Amaybamba; Huayopata

Machu Picchu (site); also: Machupijchu

Pachacutec (Inca king); also: Pachacuti; Pachacutic

Pantiacolla (pass); also: Málaga

Qolla (people of the Altiplano); also: Colla

Quispicanchi (province); also: Quispicanchis

Salcca (river); also: Salca

Sillque (hacienda); also: Silque

Taki Onccoy (millenarist movement); also: Taki Unquy, Taqui Ongo, Taqui Onqoy

Tiwanaku (site); also: Tiahuanaco

Vilcanota (river); also: Vilcamayo

Vilcabamba (river); also: Vilcapampa; Wilcapampa,

Waqawillca (mountain); also: Wequewillca, Willcawiqui, Wikaywillka, Verónica

D.W. Gade, *Spell of the Urubamba*, DOI 10.1007/978-3-319-20849-7

Index

D.W. Gade, *Spell of the Urubamba*, DOI 10.1007/978-3-319-20849-7

CPSIA information can be obtained at www.ICGtesting.com
Printed in the USA
BVOW11*0242141015

422387BV00001B/2/P

9 783319 208480